I0034463

AWWA Water Operator Field Guide
Second Edition

Compiled by AWWA staff members:

William C. Lauer

Timothy J. McCandless

Dawn Flancher

American Water Works Association

Science and Technology

AWWA unites the drinking water community by developing and distributing authoritative scientific and technological knowledge. Through its members, AWWA develops industry standards for products and processes that advance public health and safety. AWWA also provides quality improvement programs for water and wastewater utilities.

Project Manager: Melissa Christensen, Senior Technical Editor
Produced by Glacier Publishing Services, Inc.

Disclaimer

Library of Congress Cataloging-in-Publication Data
Lauer, William C.
   AWWA water operator field guide / compiled by AWWA staff members
William C. Lauer, Dawn Flancher, Timothy J. McCandless. -- Second edition.
      pages cm
   Includes bibliographical references and index.
   ISBN 978-1-58321-904-1
   1. Water treatment plants--Handbooks, manuals, etc. 2. Water--Purification--Handbooks, manuals, etc. 3. Water--Distribution--Handbooks, manuals, etc. I. Lauer, Bill. II. McCandless, Timothy J. III. American Water
Works Association. IV. Title.
   TD434.S78 2012
   628.1'62--dc23

                          2012026716

ISBN: 9781583219041
           1-58321-904-8

eISBN: 9781613001998
           1-61300-199-1

**American Water Works
Association**

6666 West Quincy Avenue
Denver, CO 80235-3098
303.794.7711

Contents

Preface

This guide is a compilation of information, charts, graphs, formulas, and definitions that are used by water system operators in performing their daily duties. There is so much information, and contained in so many different sources, that finding it while in the field can be a problem. This guide compiles information mostly from AWWA manuals, books, and standards, but also from other generic information found in many publications.

The sections of this guide group the information based on how it would be used by the operator. The guide includes information for both water treatment and distribution system operators. Design engineers should also find this material helpful. Major sections include math, conversion factors, chemistry, safety, water quality, water treatment, distribution, wells, pumps, and pressure, flows, and meters. Perusing the guide will assist in finding handy information later.

This is the second edition of the field guide. tables have been updated to reflect information in the current AWWA standards and manuals. Many example calculations were converted to a more understandable format. Thank you to Tim McCandless, Bill Lauer, and Dawn Flancher for their efforts in this revision.

If you would like to suggest changes or additions to the guide, please submit them to AWWA at Publishing Group, 6666 W. Quincy Ave., Denver, CO 80235.

Basic Math

A number of calculations are used in the operation of small water and wastewater facilities. Some only need to be calculated once and recorded for future reference; others may need to be calculated more frequently. Operators need to be familiar with the formulas and basic calculations to carry out their duties properly. Note that the formulas in this section are basic and general; specific formulas for particular components of water systems can be found in the relevant sections of this guide.

SYSTÈME INTERNATIONAL UNITS

When performing calculations, water operators should pay particular attention not only to the numbers but also to the units involved. Where SI units and customary units are given, convert all units to one system, usually SI, *first*. Be sure to write the appropriate units with each number in the calculations for clarity. Inaccurate calculations and measurements can lead to incorrect reports and costly operational decisions. This section introduces the calculations that are the basic building blocks of the water/wastewater industry.

SI Prefixes

The SI is based on factors of ten, similar to the dollar. This allows the size of the unit of measurement to be increased or decreased while the base unit remains the same. The SI prefixes are

$$
\begin{aligned}
\text{mega, M} &= 1,000,000 \times \text{the base unit} \\
\text{kilo, k} &= 1,000 \times \text{the base unit} \\
\text{hecta, h} &= 100 \times \text{the base unit} \\
\text{deca, da} &= 10 \times \text{the base unit} \\
\text{deci, d} &= 0.1 \times \text{the base unit} \\
\text{centi, c} &= 0.01 \times \text{the base unit} \\
\text{milli, m} &= 0.001 \times \text{the base unit} \\
\text{micro, } \mu &= 0.000001 \times \text{the base unit}
\end{aligned}
$$

Base SI Units

Quantity	Unit	Abbreviation
length	meter	m
mass	kilogram	kg
time	second	sec
electric current	ampere	A
thermodynamic temperature	kelvin	K
amount of substance	mole	mol
luminous intensity	candela	cd

Supplementary SI Units

Quantity	Unit	Abbreviation
plane angle	radian	rad
solid angle	steradian	sr

Derived SI Units With Special Names

Quantity	Unit	Abbreviation	Equivalent-Units Abbreviation
frequency (of a periodic phenomenon)	hertz	Hz	sec^{-1}
force	newton	N	$kg \cdot m/sec^2$
pressure, stress	pascal	Pa	N/m^2
energy, work, quantity of heat	joule	J	$N \cdot m$
power, radiant flux	watt	W	J/sec
quantity of electricity, electric charge	coulomb	C	$A \cdot sec$
electric potential, potential difference, electromotive force	volt	V	W/A
electrical capacitance	farad	F	C/V
electrical resistance	ohm	Ω	V/A
electrical conductance	siemens	S	A/V
magnetic flux	weber	Wb	$V \cdot sec$
magnetic flux density	tesla	T	Wb/m^2
inductance	henry	H	Wb/A
luminous flux	lumen	lm	$cd \cdot Sr$
luminance	lux	lx	lm/m^2
activity (of a radionuclide)	becquerel	Bq	disintegrations/sec
absorbed ionizing radiation dose	gray	Gy	J/kg
ionizing radiation dose equivalent	sievert	Sv	J/kg

Some Common Derived SI Units

Quantity	Unit	Abbreviation
absorbed dose rate	grays per second	Gy/sec
acceleration	meters per second squared	m/sec^2
angular acceleration	radians per second squared	rad/sec^2
angular velocity	radians per second	rad/sec
area	square meter	m^2
concentration (amount of substance)	moles per cubic meter	mol/m^3
current density	amperes per square meter	A/m^2
density, mass	kilograms per cubic meter	kg/m^3
electric charge density	coulombs per cubic meter	C/m^3
electric field strength	volts per meter	V/m
electric flux density	coulombs per square meter	C/m^2
energy density	joules per cubic meter	J/m^3
entropy	joules per kelvin	J/K
exposure (X and gamma rays)	coulombs per kilogram	C/kg
heat capacity	joules per kelvin	J/K
heat flux density irradiance	watts per square meter	W/m^2
luminance	candelas per square meter	cd/m^2
magnetic field strength	amperes per meter	A/m
molar energy	joules per mole	J/mol
molar entropy	joules per mole per kelvin	J/(mol•K)
molar heat capacity	joules per mole per kelvin	J/(mol•K)
moment of force	newton-meter	N•m
permeability (magnetic)	henrys per meter	H/m
permittivity	farads per meter	F/m
power density	watts per square meter	W/m^2

Table continued on next page

Some Common Derived SI Units (continued)

Quantity	Unit	Abbreviation
radiance	watts per square meter per steradian	$W/(m^2{\cdot}sr)$
radiant intensity	watts per steradian	W/sr
specific energy	joules per kilogram	J/kg
specific entropy	joules per kilogram per kelvin	$J/(kg{\cdot}K)$
specific heat capacity	joules per kilogram per kelvin	$J/(kg{\cdot}K)$
specific volume	cubic meters per kilogram	m^3/kg
surface tension	newtons per meter	N/m
thermal conductivity	watts per meter per kelvin	$W/(m{\cdot}K)$
velocity	meters per second	m/sec
viscosity, absolute	pascal-second	$Pa{\cdot}sec$
viscosity, kinematic	square meters per second	m^2/sec
volume	cubic meter	m^3
wave number	per meter	m^{-1}

KEY FORMULAS FOR MATH

Area Formulas

Square

area = s × s
diagonal = 1.414 × s

Rectangle or Parallelogram

area = b × h
diagonal = square root $(b^2 + h^2)$

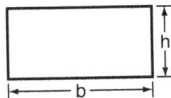

Trapezoid

$$\text{area} = \frac{(a+b)\,h}{2}$$

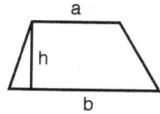

Any Triangle

$$\text{area} = \frac{b \times h}{2}$$

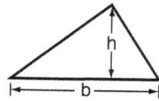

Right-Angle Triangle

$$a^2 + b^2 = c^2$$

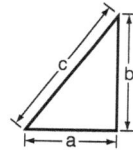

Circle

area = $\pi \times r^2$

circumference = $2 \times \pi \times r$

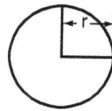

Sector of a Circle

$$\text{area} = \frac{\pi \times r \times r \times \alpha}{360}$$

length = $0.01745 \times r \times \alpha$

$$\text{angle} = \frac{1}{0.01745 \times r}$$

$$\text{radius} = \frac{1}{0.01745 \times \alpha}$$

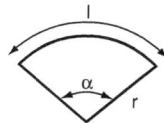

Ellipse

area = $\pi \times a \times b$

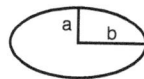

Volume Formulas

Rectangular Solid

volume = $h \times a \times b$
surface area = $(2 \times a \times b) + (2 \times b \times h) +$
$\qquad\qquad (2 \times a \times b)$

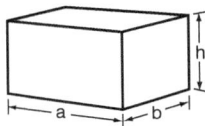

Cylinder

volume = $\pi \times r^2 \times h$
surface area = $2 \times \pi \times rh$
$\pi = 3.142$

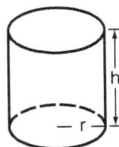

Elliptical Cylinder

volume = $\pi \times a \times b \times h$
area = $6.283 \times \dfrac{\sqrt{a^2 + b^2}}{2} \times h + 6.283 \times a \times b$

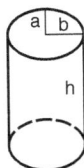

Sphere

volume = $\dfrac{4 \times \pi \times r^3}{3}$
surface area = $4 \times \pi \times r^2$

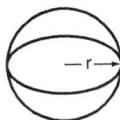

Cone

volume = $\dfrac{\pi \times r^2 \times h}{3}$
surface area = $\pi \times r \times \sqrt{r^2} \times (r + h) \times h$

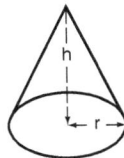

Pyramid

volume = $\dfrac{a \times b \times h}{3}$

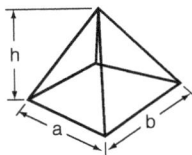

Other Formulas

$$\text{theoretical water horsepower} = \frac{\text{gal/min} \times \text{total head, ft}}{3,960}$$

$$= \frac{\text{gal/min} \times \text{lb/in.}}{1,715}$$

$$\text{brake horsepower} = \frac{\text{theoretical water horsepower}}{\text{pump efficiency}}$$

$$\text{detention time, min} = \frac{\text{volume of basin, gal}}{\text{flow rate, gpm}}$$

$$\text{filter backwash rate, gal/min/ft}^2 = \frac{\text{flow, gpm}}{\text{area of filter, ft}^2}$$

$$\text{surface overflow rate} = \frac{\text{flow, gpm}}{\text{area, ft}^2}$$

$$\text{weir overflow rate} = \frac{\text{flow, gpm}}{\text{weir length, ft}}$$

$$\text{pounds per mil gal} = \text{parts per million} \times 8.34$$

$$\text{parts per million} = \text{pounds per mil gal} \times 0.12$$

$$\text{parts per million} = \text{percent strength of solution} \times 10,000$$

$$\text{pounds per day} = \text{volume, mgd} \times \text{dosage, mg/L} \times 8.34 \text{ lb/gal}$$

$$\text{dosage, mg/L} = \frac{\text{feed, lb/day}}{\text{volume, mgd} \times 8.34 \text{ lb/gal}}$$

$$\text{rectangular basin volume, ft}^3 = \text{length, ft} \times \text{width, ft} \times \text{height, ft}$$

$$\text{rectangular basin volume, gal} = \text{length, ft} \times \text{width, ft} \times \text{height, ft} \times 7.48 \text{ gal/ft}^3$$

$$\text{right cylinder volume, ft}^3 = 0.785 \times \text{diameter}^2\text{, ft} \times \text{height or depth, ft}$$

$$\text{right cylinder volume, gal} = \frac{0.785 \times \text{diameter}^2\text{, ft} \times \text{height or depth, ft}}{\times 7.48 \text{ gal/ft}^3}$$

$$\text{gallons per capita per day, average water usage} = \frac{\text{volume, gpd}}{\text{population served/day}}$$

$$\text{supply, days (full to tank dry)} = \frac{\text{volume, gpd}}{\text{population served} \times \text{gpcd}}$$

$$\text{gallons per day of water consumption, (demand/day)} = \text{population} \times \text{gpcd}$$

Consumption Averages, per capita

winter = 170 gpcd
spring = 225 gpcd
summer = 325 gpcd

Units of Measure and Conversions

The ability to accurately and consistently measure such variables as flow and head, along with water quality indicators such as chemical and biological oxygen demand, total suspended solids, toxins, and pathogens is a key component of the successful operation of a water distribution system. Here are the most common units of measure and associated conversions typically used in the water industry.

acre An SI unit of area.

acre-foot (acre-ft) A unit of volume. One acre-foot is the equivalent amount or volume of water covering an area of 1 acre that is 1 foot deep.

ampere (A) An SI unit of that constant current that, if maintained in two straight parallel conductors of infinite length or negligible cross section and placed 1 meter apart in a vacuum, would produce a force equal to 2×10^{-7} newton per meter of length.

ampere-hour (A·hr) A unit of electric charge equal to 1 ampere flowing for 1 hour.

angstrom (Å) A unit of length equal to 10^{-10} meter.

atmosphere (atm) A unit of pressure equal to 14.7 pounds per square inch (101.3 kilopascals) at average sea level under standard conditions.

bar A unit of pressure defined as 100 kilopascals.

barrel (bbl) A unit of volume, frequently 42 gallons for petroleum or 55 gallons for water.

baud A measure of analog data transmission speed that describes the modulation rate of a wave, or the average frequency of the signal. One baud equals 1 signal unit per second. If an analog signal is viewed as an electromagnetic wave, one complete wavelength or cycle is equivalent to a signal unit. The term *baud* has often been used synonymously with *bits per second*. The baud rate may equal bits per second for some transmission techniques, but special modulation techniques frequently deliver a bits-per-second rate higher than the baud rate.

becquerel (Bq) An SI unit of the activity of a radionuclide decaying at the rate of one spontaneous nuclear transition per second.

billion electron volts (BeV) A unit of energy equivalent to 10^9 electron volts.

billion gallons per day (bgd) A unit for expressing the volumetric flow rate of water being pumped, distributed, or used.

binary digits (bits) per second (bps) A measure of the data transmission rate. A binary digit is the smallest unit of information or data, represented by a binary "1" or "0."

British thermal unit (Btu) A unit of energy. One British thermal unit was formerly defined as the quantity of heat required to raise the temperature of 1 pound of pure water 1° Fahrenheit; now defined as 1,055.06 joules.

bushel (bu) A unit of volume.

caliber (1) The diameter of a round body, especially the internal diameter of a hollow cylinder. (2) The diameter of a bullet or other projectile, or the diameter of a gun's bore. In US customary units, usually expressed in hundredths or thousandths of an inch and typically written as a decimal fraction (e.g., 0.32). In SI units, expressed in millimeters.

calorie (gram calorie) A unit of energy. One calorie is the amount of heat necessary to raise the temperature of 1 gram of pure water at 15° Celsius by 1° Celsius.

candela (cd) An SI unit of luminous intensity. One candela is the luminous intensity, in a given direction, of a source that emits monochromatic radiation of frequency 540×10^{12} hertz and that has a radiant intensity in that direction of $\frac{1}{683}$ watt per steradian.

candle A unit of light intensity. One candle is equal to 1 candela. Candelas are the preferred units.

candlepower A unit of light intensity. One candlepower is equal to 1 candela. Candelas are the preferred units.

centimeter (cm) A unit of length defined as one hundredth of a meter.

centipoise A unit of absolute viscosity equivalent to 10^{-2} poise. See also *poise*.

chloroplatinate (Co–Pt) unit (cpu) See *color unit*.

cobalt–platinum unit See *color unit*.

colony-forming unit (cfu) A unit of expression used in enumerating bacteria by plate-counting methods. A colony of bacteria develops from a single cell or a group of cells, either of which is a colony-forming unit.

color unit (cu) The unit used to report the color of water. Standard solutions of color are prepared from potassium chloroplatinate (K_2PtCl_6) and cobaltous chloride ($CoCl_2 \cdot 6H_2O$). Adding the following amounts in 1,000 milliliters of distilled water produces a solution with a color of 500 color units: 1.246 grams potassium chloroplatinate, 1.00 grams geobaltous chloride, and 100 milliliters concentrated hydrochloric acid (HCl).

coulomb (C) An SI unit of a quantity of electricity or electric charge. One coulomb is the quantity of electricity transported in 1 second by a current of 1 ampere, or about 6.25×10^{18} electrons. Coulombs are equivalent to ampere-seconds.

coulombs per kilogram (C/kg) A unit of exposure dose of ionizing radiation. See also *roentgen*.

cubic feet (ft^3) A unit of volume equivalent to a cube with a dimension of 1 foot on each side.

cubic feet per hour (ft^3/hr) A unit for indicating the rate of liquid flow past a given point.

cubic feet per minute (ft^3/min, CFM) A unit for indicating the rate of liquid flow past a given point.

cubic feet per second (ft^3/sec, cfs) A unit for indicating the rate of liquid flow past a given point.

cubic inch (in.^{3}) A unit of volume equivalent to a cube with a dimension of 1 inch on each side.

cubic meter (m^3) A unit of volume equivalent to a cube with a dimension of 1 meter on each side.

cubic yard (yd^3) A unit of volume equivalent to a cube with a dimension of 1 yard on each side.

curie (Ci) A unit of radioactivity. One curie equals 37 billion disintegrations per second, or approximately the radioactivity of 1 gram of radium.

cycles per second (cps) A unit for expressing the number of times something fluctuates, vibrates, or oscillates each second. These units have been replaced by hertz. One hertz equals 1 cycle per second.

dalton (D) A unit of weight. One dalton designates $\frac{1}{16}$ the weight of oxygen-16. One dalton is equivalent to 0.9997 atomic weight unit, or nominally 1 atomic weight unit.

darcy (da) The unit used to describe the permeability of a porous medium (e.g., the movement of fluids through underground formations studied by petroleum engineers, geologists or geophysicists, and groundwater specialists). A porous medium is said to have a permeability of 1 darcy if a fluid of 1-centipoise viscosity that completely fills the pore space of the medium will flow through it at a rate of 1 cubic centimeter per second per square centimeter of cross-sectional area under a pressure gradient of 1 atmosphere per centimeter of length. In SI units, 1 darcy = 9.87×10^{-13} square meters.

day A unit of time equal to 24 hours.

decibel (dB) A dimensionless ratio of two values expressed in the same units of measure. It is most often applied to a power ratio and defined as decibels = $10 \log_{10}$ (actual power level/ reference power level), or dB = $10 \log_{10} (W_2/W_1)$, where W is the power level in watts per square centimeter for sound. Power is proportional to the square of potential. In the case of sound, the potential is measured as a pressure, but the sound level is an energy level. Thus, dB = $10 \log_{10} (p_2/p_1)^2$ or dB = $20 \log_{10} (p_2/p_1)$, where p is the potential. The reference levels are not well standardized. For example, sound power is usually measured above 10^{-12} watts per square centimeter, but both 10^{-11} and 10^{-16} watts per square centimeter are used. Sound pressure is usually measured above 20 micropascals in air. The reference level is not important in most cases because one is usually concerned with the difference in levels, i.e., with a power ratio. A power ratio of 1.26 produces a difference of 1 decibel.

deciliter (dL) A unit of volume defined as one tenth of a liter. This unit is often used to express concentration in clinical chemistry. For example, a concentration of lead in blood would typically be reported in units of micrograms per deciliter.

degree (°) A measure of the phase angle in a periodic electrical wave. One degree is $\frac{1}{360}$ of the complete cycle of the periodic wave. Three hundred sixty degrees equals 2π radians.

degree Celsius (°C) A unit of temperature. The degree Celsius is exactly equal to the kelvin and is used in place of the kelvin for expressing Celsius temperature (symbol t) defined by the equation $t = T - T_0$, where T is the thermodynamic temperature in kelvin and $T_0 = 273.15$ kelvin by definition.

degree Fahrenheit (°F) A unit of temperature on a scale in which 32° marks the freezing point and 212° the boiling point of water at a barometric pressure of 14.7 pounds per square inch.

degree kelvin (K) See *kelvin*.

dram (dr) Small weight. Two different drams exist: the apothecary's dram (equivalent to 1/3.54 gram) and the avoirdupois dram (equivalent to 1/1.17 gram).

electron volt (eV) A unit of energy commonly used in the fields of nuclear and high-energy physics. One electron volt is the energy transferred to a charged particle with single charge when that particle falls through a potential of 1 volt. An electron volt is equal to 1.6×10^{-19} joule.

equivalents per liter (eq/L) An SI unit of an expression of concentration equivalent to normality. The normality of a solution (equivalent weights per liter) is a convenient way of expressing concentration in volumetric analyses.

farad (F) An SI unit of electrical capacitance. One farad is the capacitance of a capacitor between the plates of which a difference of potential of 1 volt appears to be present when the capacitor is charged by a quantity of electricity equal to 1 coulomb. Farads are equivalent to seconds to the fourth amperes squared per meter squared per kilogram.

fathom A unit of length equivalent to 6 feet, used primarily in marine measurements.

feet (ft) The plural form of length (the singular form is *foot*).

feet board measure (fbm) A unit of volume. One board foot is represented by a board measuring 1 foot by 1 foot by 1 inch thick (144 cubic inches). A board measuring 0.5 feet by 2 feet by 2 inches thick would equal 2 board feet.

feet per hour (ft/hr) A unit for expressing the rate of movement.

feet per minute (ft/min) A unit for expressing the rate of movement.

feet per second (ft/sec, fps) A unit for expressing the rate of movement.

feet per second squared (ft/sec²) A unit of acceleration (the rate of change of linear motion). For example, the acceleration caused by gravity is 32.2 ft/sec² at sea level.

feet squared per second (ft²/sec) A unit used in flux calculations.

fluid ounce (fl oz) A unit for expressing volume, equivalent to ¹⁄₁₂₈ of a gallon.

foot A unit of length, equivalent to 12 inches. See also *US customary system of units.*

foot of water (39.2° Fahrenheit) A unit for expressing pressure or elevation head.

foot per second per foot (ft/sec/ft; sec⁻¹) A unit for expressing velocity gradient.

foot-pound, torque A unit for expressing the energy used in imparting rotation, often associated with the power of engine-driven mechanisms.

foot-pound, work A unit of measure of the transference of energy when a force produces movement of an object.

formazin turbidity unit (ftu) A turbidity unit appropriate when a chemical solution of formazin is used as a standard to calibrate a turbidimeter. If a nephelometric turbidimeter is used, nephelometric turbidity units and formazin turbidity units are equivalent. See also *nephelometric turbidity unit.*

gallon (gal) A unit of volume, equivalent to 231 cubic inches. See also *Imperial gallon.*

gallons per capita per day (gpcd) A unit typically used to express the average number of gallons of water used by the average person each day in a water system. The calculation is made by dividing the total gallons of water used each day by the total number of people using the water system.

gallons per day (gpd) A unit for expressing the discharge or flow past a fixed point.

gallons per day per square foot (gpd/ft², gsfd) A unit of flux equal to the quantity of liquid in gallons per day through 1 square foot of area. It may also be expressed as a velocity in units of length per unit time. In pressure-driven membrane treatment processes, this unit is commonly used to describe the volumetric flow rate of permeate through a unit area of active membrane surface. In settling tanks, this rate is called the overflow rate.

gallons per flush (gal/flush) The number of gallons used with each flush of a toilet.

gallons per hour (gph) A unit for expressing the discharge or flow of a liquid past a fixed point.

gallons per minute (gpm) A unit for expressing the discharge or flow of a liquid past a fixed point.

gallons per minute per square foot (gpm/ft²) A unit for expressing flux, the discharge or flow of a liquid through a unit of area. In a filtration process, this unit is commonly used to describe the volumetric flow rate of filtrate through a unit of filter media surface area. It may also be expressed as a velocity in units of length per unit time.

gallons per second (gps) A unit for expressing the discharge or flow past a fixed point.

gallons per square foot (gal/ft²) A unit for expressing flux, the discharge or flow of a liquid through each unit of surface area of a granular filter during a filter run (between cleaning or backwashing).

gallons per square foot per day See *gallons per day per square foot*.

gallons per year (gpy) A unit for expressing the discharge or flow of a liquid past a fixed point.

gamma (γ) A symbol used to represent 1 microgram. Avoid using this symbol; the preferred symbol is µg.

gigabyte (GB) A unit of computer memory. One gigabyte equals 1 megabyte times 1 kilobyte, or 1,073,741,824 bytes (roughly 1 billion bytes).

gigaliter (GL) A unit of volume defined as 1 billion liters.

grad A unit of angular measure equal to $\frac{1}{400}$ of a circle.

grain (gr) A unit of weight.

grains per gallon (gpg) A unit sometimes used for reporting water analysis concentration results in the United States and Canada.

gram (g) A fractional unit of mass. One gram was originally defined as the weight of 1 cubic centimeter or 1 milliliter of water at 4° Celsius. Now it is $\frac{1}{1,000}$ of the mass of a certain block of platinum–iridium alloy known as the international prototype kilogram, preserved at Sèvres, France.

gram molecular weight The molecular weight of a compound in grams. For example, the gram molecular weight of CO_2 is 44.01 grams. See also *mole*.

gray (Gy) An SI unit of *absorbed* ionizing radiation dose. One gray, equal to 100 rad, is the absorbed dose when the energy per unit mass imparted to matter by ionizing radiation is 1 joule per kilogram. See also *rad*; *rem*; *sievert*.

hectare (ha) A unit of area equivalent to 10,000 square meters.

henry (H) An SI unit of electric inductance, equivalent to meters squared kilograms per second squared per ampere squared. One henry is the inductance of a closed circuit in which an electromotive force of 1 volt is produced when the electric current in the circuit varies uniformly at a rate of 1 ampere per second.

hertz (Hz) An SI unit of measure of the frequency of a periodic phenomenon in which the period is 1 second, equivalent to second^{-1}. Hertz units were formerly expressed as cycles per second.

horsepower (hp) A standard unit of power. See also *US customary system of units*.

horsepower-hour (hp·hr) A unit of energy or work.

hour (hr) An interval of time equal to ½₄ of a day.

Imperial gallon A unit of volume used in the United Kingdom, equivalent to the volume of 10 pounds of freshwater.

inch (in.) A unit of length.

inch of mercury (32° Fahrenheit) A unit of pressure or elevation head.

inch-pound (in.-lb) A unit of energy or torque.

inches per minute (in./min) A unit of velocity.

inches per second (in./sec) A unit of velocity.

International System of Units See *Système International*.

joule (J) An SI unit of the unit for energy, work, or quantity of heat, equivalent to meters squared kilograms per second squared. One joule is the work done when the point of application of a force of 1 newton is displaced a distance of 1 meter in the direction of the force (1 newton-meter).

kelvin (K) An SI unit of thermodynamic temperature. One kelvin is 1/273.16 of the thermodynamic temperature of the triple point of water. No degree sign (°) is used. Zero kelvin is absolute zero, the complete absence of heat.

kilobyte (kB) A unit of measurement for digital storage of data in various computer media, such as hard disks, random access memory, and compact discs. One kilobyte is 1,024 bytes.

kilograin A unit of weight equivalent to 1,000 grains.

kilogram (kg) An SI unit of mass. One kilogram is equal to the mass of a certain block of platinum–iridium alloy known as the international prototype kilogram (nicknamed Le Grand K), preserved at Sèvres, France. A new standard is expected early in the 21st century.

kilohertz (kHz) A unit of frequency equal to 1,000 hertz or 1,000 cycles per second.

kiloliter A unit of volume equal to 1,000 liters or 1 cubic meter.

kilopascal (kPa) A unit of pressure equal to 1,000 pascals.

kiloreactive volt-ampere (kvar) A unit of reactive power equal to 1,000 volt-ampere-reactive.

kilovolt (kV) A unit of electrical potential equal to 1,000 volts.

kilovolt-ampere (kVA) A unit of electrical power equal to 1,000 volt-amperes.

kilowatt (kW) A unit of electrical power equal to 1,000 watts.

kilowatt-hour (kW·hr) A unit of energy or work.

lambda (λ) A symbol used to represent 1 microliter. Avoid using this symbol; the preferred symbol is μL.

linear feet (lin ft) A unit of distance in feet along an object.

liter (L) A unit of volume. One liter of pure water weighs 1,000 grams at 4° Celsius at 1 atmosphere of pressure.

liters per day (L/day) A unit for expressing a volumetric flow rate past a given point.

liters per minute (L/min) A unit for expressing a volumetric flow rate past a given point.

lumen (lm) An SI unit of luminous flux equivalent to candela-steradian. One lumen is the luminous flux emitted in a solid angle of 1 steradian by a point source having a uniform intensity of 1 candela.

lux (lx) An SI unit of illuminance. One lux is the illuminance intensity given by a luminous flux of 1 lumen uniformly distributed over a surface of 1 square meter. One lux is equivalent to 1 candela-steradian per meter squared.

megabyte (MB) A unit of computer memory storage equivalent to 1,048,576 bytes.

megahertz (mHz) A unit of frequency equal to 1 million hertz, or 1 million cycles per second.

megaliter (ML) A unit of volume equal to 1 million liters.

megaohm (megohm) A unit of electrical resistance equal to 1 million ohms. This is the unit of measurement for testing the electrical resistance of water to determine its purity. The closer water comes to absolute purity, the greater its resistance to conducting an electric current. Absolutely pure water has a specific resistance of more than 18 million ohms across 1 centimeter at a temperature of 25° Celsius. See also *ohm*.

meter (m) An SI unit of length. One meter is the length of the path traveled by light in a vacuum during a time interval of 1/299,792,458 second.

meters per second per meter (m/sec/m; sec⁻¹) A unit for expressing velocity gradient.

metric system A system of units started in about 1900 based on three basic units: the meter for length, the kilogram for mass, and the second for time—the so-called MKS system. Decimal fractions and multiples of the basic units are used for larger and smaller quantities. The principal departure of the SI from the more familiar form of metric engineering units is the use of the newton as the unit of force instead of the kilogram-force. Likewise, the newton instead of kilogram-force is used in combination units including force; for example, pressure or stress (newton per square meter), energy (newton-meter = joule), and power (newton-meter per second = watt). See also *Système International*.

metric ton (t) A unit of weight equal to 1,000 kilograms.

mho A unit of electrical conductivity in US customary units equal to 1 siemens, which is an SI unit. See also *siemens*.

microgram (µg) A unit of mass equal to one millionth of a gram.

micrograms per liter (µg/L) A unit of concentration for dissolved substances based on their weights.

microhm A unit of electrical resistance equal to one millionth of an ohm.

micrometer (μm) A unit of length equal to one millionth of a meter.

micromho A unit of electrical conductivity equal to one millionth of an mho. See also *microsiemens.*

micromhos per centimeter (μmho/cm) A measure of the conductivity of a water sample, equivalent to microsiemens per centimeter. Absolutely pure water, from a mineral content standpoint, has a conductivity of 0.055 micromhos per centimeter at 25° Celsius.

micromolar (μ*M*) A concentration in which the molecular weight of a substance (in grams) divided by 10^6 (i.e., 1 μmol) is dissolved in enough solvent to make 1 liter of solution. See also *micromole; molar.*

micromole (μmol) A unit of weight for a chemical substance, equal to one millionth of a mole. See also *mole.*

micron (μ) A unit of length equal to 1 micrometer. Micrometers are the preferred units.

microsiemens (μS) A unit of conductivity equal to one millionth of a siemens. The microsiemens is the practical unit of measurement for conductivity and is used to approximate the total dissolved solids content of water. Water with 100 milligrams per liter of sodium chloride (NaCl) will have a specific resistance of 4,716 ohm-centimeters and a conductance of 212 microsiemens per centimeter. Absolutely pure water, from a mineral content standpoint, has a conductivity of 0.055 microsiemens per centimeter at 25° Celsius.

microwatt (μW) A unit of power equal to one millionth of a watt.

microwatt-seconds per square centimeter (μW-sec/cm²) A unit of measurement of irradiation intensity and retention or contact time in the operation of ultraviolet systems.

mil A unit of length equal to one thousandth of an inch.

mile (mi) A unit of length, equivalent to 5,280 feet.

miles per hour (mph) A unit of speed.

milliampere (mA) A unit of electrical current equal to one thousandth of an ampere.

milliequivalent (meq) A unit of weight equal to one thousandth the equivalent weight of a chemical.

milliequivalents per liter (meq/L) A unit of concentration for dissolved substances based on their equivalent weights.

milligram (mg) A unit of mass equal to one thousandth of a gram.

milligrams per liter (mg/L) The unit used in reporting the concentration of matter in water as determined by water analyses.

milliliter (mL) A unit of volume equal to one thousandth of a liter.

millimeter (mm) A unit of length equal to one thousandth of a meter.

millimicron (mμ) A unit of length equal to one thousandth of a micron. This unit is correctly called a nanometer.

millimolar (mM) A concentration in which the molecular weight of a substance (in grams) divided by 10^3 (i.e., 1 mmol) is dissolved in enough solvent to make 1 liter of solution. See also *millimole; molar.*

millimole (mmol) A unit of weight for a chemical substance, equal to one-thousandth of a mole. See also *mole.*

million electron volts (MeV) A unit of energy equal to 10^6 electron volts. This unit is commonly used in the fields of nuclear and high-energy physics. See also *electron volt.*

million gallons (mil gal, MG) A unit of volume equal to 10^6.

million gallons per day (mgd) A unit for expressing the flow rate past a given point.

mils per year (mpy) A unit for expressing the loss of metal resulting from corrosion. Assuming the corrosion process is uniformly distributed over the test surface, the corrosion rate of a metal coupon may be converted to a penetration rate (length per time) by dividing the unit area of metal loss by the metal density (mass per volume). The penetration rate, expressed as mils per year, describes the rate at which the metal surface is receding because of the corrosion-induced metal loss. See also *mil.*

minute (min) A unit of time equal to 60 seconds.

molar (M) A unit for expressing the molarity of a solution. A 1-molar solution consists of 1 gram molecular weight of a compound dissolved in enough water to make 1 liter of solution. A gram molecular weight is the molecular weight of a compound in grams. For example, the molecular weight of sulfuric acid (H_2SO_4) is 98. A 1-molar, or 1-mole-per-liter, solution of sulfuric acid would consist of 98 grams of H_2SO_4 dissolved in enough distilled water to make 1 liter of solution.

mole (mol) An SI unit of the amount of substance that contains as many elementary entities as atoms in 0.012 kilogram.

moles per liter (mol/L, M) A unit of concentration for a dissolved substance.

mrem An expression or measure of the extent of biological injury that would result from the absorption of a particular radionuclide at a given dosage over 1 year.

nanograms per liter (ng/L) A unit expressing the concentration of chemical constituents in solution as mass (nanograms) of solute per unit volume (liter) of water. One million nanograms per liter is equivalent to 1 milligram per liter.

nanometer (nm) A unit of length defined as 10^{-12} meter.

nephelometric turbidity unit (ntu) A unit for expressing the cloudiness (turbidity) of a sample as measured by a nephelometric turbidimeter. A turbidity of 1 nephelometric turbidity unit is equivalent to the turbidity created by a 1:4,000 dilution of a stock solution of 5.0 milliliters of a 1.000-gram hydrazine sulfate $((NH_2)_2 \cdot H_2SO_4)$ in 100 milliliters of distilled water solution plus 5.0 milliliters of a 10.00-gram hexamethylenetetramine $((CH_2)_6N_4)$ in 100 milliliters of distilled water solution that has stood for 24 hours at 25 ± 3° Celsius.

newton (N) An SI unit of force. One newton is equivalent to 1 kilogram-meter per second squared. It is that force that, when applied to a body having a mass of 1 kilogram, gives it an acceleration of 1 meter per second squared. The newton replaces the unit kilogram-force, which is the unit of force in the metric system.

ohm (Ω) An SI unit of electrical resistance, equivalent to meters squared kilograms per second cubed per ampere squared. One ohm is the electrical resistance between two points of a conductor when a constant difference of potential of 1 volt, applied between these two points, produces in this conductor a current of 1 ampere, with this conductor not being the source of any electromotive force.

one hundred cubic feet (ccf) A unit of volume.

ounce (oz) A unit of force, mass, and volume.

ounce-inch (ounce-in., ozf-in.) A unit of torque.

parts per billion (ppb) A unit of proportion, equal to 10^{-9}. This expression represents a measure of the concentration of a substance dissolved in water on a weight-per-weight basis or the concentration of a substance in air on a weight-per-volume basis. One liter of water at 4° Celsius has a mass equal to 1.000 kilogram (specific gravity equal to 1.000, or 1 billion micrograms). Thus, when 1 microgram of a substance is dissolved in 1 liter of water with a specific gravity of 1.000 (1 microgram per liter), this would be one part of substance per billion parts of water on a weight-per-weight basis. This terminology is now obsolete, and the term *micrograms per liter (μg/L)* should be used for concentrations in water.

parts per million (ppm) A unit of proportion, equal to 10^{-6}. This terminology is now obsolete, and the term *milligrams per liter (mg/L)* should be used for concentrations in water. See also *parts per billion*.

parts per thousand (ppt) A unit of proportion, equal to 10^{-3}. This terminology is now obsolete, and the term *grams per liter (g/L)* should be used for concentrations in water. See also *parts per billion*.

parts per trillion (ppt) A unit of proportion, equal to 10^{-12}. This terminology is now obsolete, and the term *nanograms per liter (ng/L)* should be used for concentrations in water. See also *parts per billion*.

pascal (Pa) An SI unit of pressure or stress equivalent to newtons per meter per second squared. One pascal is the pressure or stress of 1 newton per square meter.

pascal-second (Pa·sec) A unit of absolute viscosity equivalent to kilogram per second per meter cubed. The viscosity of pure water at 20° Celsius is 0.0010087 pascal-second.

pi (π) The ratio of the circumference of a circle to the diameter of that circle, approximately equal to 3.14159, or about $^{22}/_{7}$.

picocurie (pCi) A unit of radioactivity. One picocurie represents a quantity of radioactive material with an activity equal to one millionth of one millionth of a curie, i.e., 10^{-12} curie.

picocuries per liter (pCi/L) A radioactivity concentration unit.

picogram (pg) A unit of mass equal to 10^{-12} gram or 10^{-15} kilogram.

picosecond (ps) A unit of time equal to one trillionth (10^{-12}) of a second.

plaque-forming unit (pfu) A unit expressing the number of infectious virus particles. One plaque-forming unit is equivalent to one virus particle.

platinum–cobalt (Pt–Co) color unit (PCU) See *color unit.*

poise A unit of absolute viscosity, equivalent to 1 gram mass per centimeter per second.

pound (lb) A unit used to represent either a mass or a force. This can be a confusing unit because two terms actually exist, *pound mass* (lbm) and *pound force* (lbf). One pound force is the force with which a 1 pound mass is attracted to the earth. In equation form, pounds force =

$$(\text{pounds mass}) \left(\frac{\text{local acceleration resulting from gravity}}{\text{standard acceleration resulting from gravity}} \right)$$

One pound mass, on the other hand, is the mass that will accelerate at 32.2 feet per second squared when a 1-pound force is applied to it. As an example of the effect of the local acceleration resulting from gravity, at 10,000 feet (3,300 meters) above sea level, where the acceleration resulting from gravity is 32.17 feet per second squared (979.6 centimeters per second squared) instead of the sea level value of 32.2 feet per second squared (980.6 centimeters per second squared), the force of gravity on a 1-pound mass would be 0.999 pounds force. On the surface of the earth at sea level, pound mass and pound force are numerically the same because the acceleration resulting from gravity is applied to an object, although they are quite different physical quantities. This may lead to confusion.

pound force (lbf) See *pound.*

pound mass (lbm) See *pound.*

pounds per day (lb/day) A unit for expressing the rate at which a chemical is added to a water treatment process.

pounds per square foot (lb/ft²) A unit of pressure.

pounds per square inch (psi) A unit of pressure.

pounds per square inch absolute (psia) A unit of pressure reflecting the sum of gauge pressure and atmospheric pressure.

pounds per square inch gauge (psig) A unit of pressure reflecting the pressure measured with respect to that of the atmosphere. The gauge is adjusted to read zero at the surrounding atmospheric pressure.

rad (radiation absorbed dose) A unit of adsorbed dose of ionizing radiation. Exposure of soft tissue or similar material to 1 roentgen results in the absorption of about 100 ergs (10^{-5} joules) of energy per gram, which is 1 rad. See also *gray*; *rem*; *sievert*.

radian (rad) An SI unit of measure of a plane angle between two radii of a circle that cut off on the circumference an arc equal in length to the radius. This unit is also used to measure the phase angle in a periodic electrical wave. Note that 2π radians is equivalent to $360°$.

radians per second (rad/sec) A unit of angular frequency.

rem (roentgen equivalent man [person]) A unit of *equivalent* dose of ionizing radiation, developed by the International Commission on Radiation Units and Measurements in 1962 to reflect the finding that the biological effects of ionizing radiation were dependent on the nature of the radiation as well as other factors. For X- and gamma radiation, the weighting factor is 1; thus, 1 rad equals 1 rem. For alpha radiation, however, 1 rad equals 20 rem. See also *gray*; *rad*; *sievert*.

revolutions per minute (rpm) A unit for expressing the frequency of rotation, or the number of times a fixed point revolves around its axis in 1 minute.

revolutions per second (rps) A unit for expressing the frequency of rotation, or the number of times a fixed point revolves around its axis in 1 second.

roentgen (r) The quantity of electrical charge produced by X- or gamma radiation. One roentgen of exposure will produce about

2 billion ion pairs per cubic centimeter of air. It was first introduced at the Radiological Congress held in Stockholm as the special unit for expressing exposure to ionizing radiation. It is now obsolete. See also *gray*; *rad*; *rem*; *sievert*.

second (sec) An SI unit of the duration of 9,192,631,770 periods of radiation corresponding to the transition between the two hyperfine levels of the ground state of the cesium-133 atom.

second feet A unit of flow equivalent to cubic feet per second.

second-foot day A unit of volume. One second-foot day is the discharge during a 24-hour period when the rate of flow is 1 second foot (i.e., 1 cubic foot per second). In ordinary hydraulic computations, 1 cubic foot per second flowing for 1 day is commonly taken as 2 acre-feet. The US Geological Survey now uses the term *cfs day* (cubic feet per second day) in its published reports.

section A unit of area in public land surveying. One section is a land area of 1 square mile.

SI See *Système International*.

siemens (S) An SI unit of the derived unit for electrical conductance, equivalent to seconds cubed amperes squared per meter squared per kilogram. One siemens is the electrical conductance of a conductor in which a current of 1 ampere is produced by an electric potential difference of 1 volt.

sievert (Sv) An SI unit of *equivalent* ionizing radiation dose. One sievert is the dose equivalent when the adsorbed dose of ionizing radiation multiplied by the dimensionless factors Q (quality factors) and N (product of any other multiplying factors) is 1 joule per kilogram. One sievert is equal to 100 rem. See also *gray*; *rad*; *rem*.

slug The base unit of mass. A slug is a mass that will accelerate at 1 foot per second squared when 1 pound force is applied.

square foot (ft²) A unit of area equivalent to that of a square, 1 foot on each side.

square inch (in.²) A unit of area equivalent to that of a square, 1 inch on each side.

square meter (m²) A unit of area equivalent to that of a square, 1 meter on each side.

square mile (mi²) A unit of area equivalent to that of a square, 1 mile on each side.

standard cubic feet per minute (SCFM) A unit for expressing the flow rate of air. This unit represents cubic feet of air per minute at standard conditions of temperature, pressure, and humidity (32° Fahrenheit, 14.7 pounds per square inch absolute, and 50% relative humidity).

steradian (sr) An SI unit of measure of a solid angle which, having its vertex in the center of a sphere, cuts off an area on the surface of the sphere equal to that of a square with sides of length equal to the radius of the sphere.

Système International (SI) The International System of Units of measure as defined by the periodic meeting of the General Conference on Weights and Measures. This system is sometimes called the international metric system or Le Système International d'Unités. The SI is a rationalized selection of units from the metric system with seven base units for which names, symbols, and precise definitions have been established. Many derived units are defined in terms of the base units, with symbols assigned to each and, in some cases, given names, e.g., the newton (N). The great advantage of SI is its establishment of one and only one unit for each physical quantity—the meter for length, the kilogram (not the gram) for mass, the second for time, and so on. From these elemental units, units for all other mechanical quantities are derived. Another advantage is the ease with which unit conversions can be made, as few conversion factors need to be invoked.

tesla (T) An SI unit of magnetic flux density, equivalent to kilograms per second squared per ampere. One tesla is the magnetic flux density given by a magnetic flux of 1 weber per square meter.

ton A unit of force and mass defined as 2,000 pounds.

tonne (t) A unit of mass defined as 1,000 kilograms. A tonne is sometimes called a metric ton.

torr A unit of pressure. One torr is equal to 1 centimeter of mercury at 0° Celsius.

true color unit (tcu) A unit of color measurement based on the platinum–cobalt color unit. This unit is applied to water samples in which the turbidity has been removed. One true color unit equals 1 color unit. See also *color unit*.

turbidity unit See *nephelometric turbidity unit*.

US customary system of units A system of units based on the yard and the pound, commonly used in the United States and defined in "Unit of Weights and Measures (United States Customary and Metric): Definitions and Tables of Equivalents," *National Bureau of Standards Miscellaneous Publication MP 233*, Dec. 20, 1960. Most of the units have a historical origin from the United Kingdom; e.g., the length of a king's foot for the length of 1 foot, the area a team of horses could plow in a day—without getting tired—for an acre, the load a typical horse could lift in a minute for horsepower, and so forth. No organized method of multiples and fractions is involved. See also *Système International*.

volt (V) An SI unit of electrical potential, potential difference, and electromotive force, equivalent to meters squared kilograms per second cubed per ampere. One volt is the difference of electric potential between two points of a conductor, carrying a constant current of 1 ampere, when the power dissipated between these points is equal to 1 watt.

volt-ampere (VA) A unit used for expressing apparent power and complex power.

volt-ampere-reactive (VAR) A unit used for expressing reactive power.

watt (W) An SI unit of power and radiant flux, equivalent to meters squared kilograms per second cubed. One watt is the power that gives rise to the production of energy at the rate of 1 joule per second. Watts represent a measure of active power and instantaneous power.

weber (Wb) An SI unit of magnetic flux, equivalent to meters squared kilograms per second squared per ampere. One weber is the magnetic flux that, linking a circuit of one turn, produces in the circuit an electromotive force of 1 volt as the magnetic flux is reduced to zero at a uniform rate in 1 second.

yard (yd) A unit of length equal to 3 feet.

CONVERSION OF US CUSTOMARY UNITS _____

Linear Measurement

fathoms	× 6	= feet (ft)
feet (ft)	× 12	= inches (in.)
inches (in.)	× 0.0833	= feet (ft)
miles (mi)	× 5,280	= feet (ft)
yards (yd)	× 3	= feet (ft)
yards (yd)	× 36	= inches (in.)

Circular Measurement

| degrees (angle) | × 60 | = minutes (angle) |
| degrees (angle) | × 0.01745 | = radians |

Area Measurement

acres	× 43,560	= square feet (ft^2)
square feet (ft^2)	× 144	= square inches ($in.^2$)
square inches ($in.^2$)	× 0.00695	= square feet (ft^2)
square miles (mi^2)	× 640	= acres
square miles (mi^2)	× 27,880,000	= square feet (ft^2)
square miles (mi^2)	× 3,098,000	= square yards (yd^2)
square yards (yd^2)	× 9	= square feet (ft^2)

Volume Measurement

acre-feet (acre-ft)	× 43,560	= cubic feet (ft^3)
acre-feet (acre-ft)	× 325,851	= gallons (gal)
barrels (bbl)	× 42	= gallons (gal)
board foot (fbm)		= 144 square inches × 1 inch
cubic feet (ft^3)	× 1,728	= cubic inches ($in.^3$)
cubic feet (ft^3)	× 7.48052	= gallons (gal)
cubic feet (ft^3)	× 29.92	= quarts (qt)
cubic feet (ft^3)	× 59.84	= pints (pt)
cubic feet (ft^3)	× 0.000023	= acre feet (acre-ft)
cubic inches ($in.^3$)	× 0.00433	= gallons (gal)
cubic inches ($in.^3$)	× 0.00058	= cubic feet (ft^3)
drops	× 60	= teaspoons (tsp)
gallons (gal)	× 0.1337	= cubic feet (ft^3)
gallons (gal)	× 231	= cubic inches ($in.^3$)
gallons (gal)	× 0.0238	= barrels (bbl)
gallons (gal)	× 4	= quarts (qt)
gallons (gal)	× 8	= pints (pt)
gallons, US	× 0.83267	= gallons, Imperial
gallons (gal)	× 0.00000308	= acre-feet (acre-ft)

gallons (gal)	$\times 128$	= ounces (oz)
gallons (gal)	$\times 0.0238$	= barrels (42 gal) (bbl)
gallons, Imperial	$\times 1.20095$	= gallons, US
pints (pt)	$\times 2$	= quarts (qt)
quarts (qt)	$\times 4$	= gallons (gal)
quarts (qt)	$\times 57.75$	= cubic inches (in.^{3})

Pressure Measurement

atmospheres	$\times 29.92$	= inches of mercury
atmospheres	$\times 33.90$	= feet of water
atmospheres	$\times 14.70$	= pounds per square inch (lb/in.^{2})
feet of water	$\times 0.8826$	= inches of mercury
feet of water	$\times 0.02950$	= atmospheres
feet of water	$\times 0.4335$	= pounds per square inch (lb/in.^{2})
feet of water	$\times 62.43$	= pounds per square foot (lb/ft^2)
feet of water	$\times 0.8876$	= inches of mercury
inches of mercury	$\times 1.133$	= feet of water
inches of mercury	$\times 0.03342$	= atmospheres
inches of mercury	$\times 0.4912$	= pounds per square inch (lb/in.^{2})
inches of water	$\times 0.002458$	= atmospheres
inches of water	$\times 0.07355$	= inches of mercury
inches of water	$\times 0.03613$	= pounds per square inch (lb/in.^{2})
pounds/square in. (lb/in.^{2})	$\times 0.01602$	= feet of water
pounds/square foot (lb/ft^2)	$\times 6,954$	= pounds per square inch (lb/in.^{2})
pounds/square in. (lb/in.^{2})	$\times 2.307$	= feet of water
pounds/square inch (lb/in.^{2})	$\times 2.036$	= inches of mercury
pounds/square inch (lb/in.^{2})	$\times 27.70$	= inches of water
feet suction lift of water	$\times 0.882$	= inches of mercury

Weight Measurement

cubic feet of ice	$\times 57.2$	= pounds (lb)
cubic feet of water (50°F)	$\times 62.4$	= pounds of water
cubic inches of water	$\times 0.036$	= pounds of water
gallons water (50°F)	$\times 8.3453$	= pounds of water
milligrams/liter (mg/L)	$\times 0.0584$	= grains per gallon (US) (gpg)
milligrams/liter (mg/L)	$\times 0.07016$	= grains per gallon (Imp)
milligrams/liter (mg/L)	$\times 8.345$	= pounds per million gallons (lb/mil gal)
ounces (oz)	$\times 437.5$	= grains (gr)
parts per million (ppm)	$\times$	= milligrams per liter (mg/L) (for normal water applications)
grains per gallon (gpg)	$\times 17.118$	= parts per million (ppm)

grains per gallon (gpg)	× 142.86	= pounds per million gallons (lb/mil gal)
percent solution	× 10,000	= milligrams per liter (mg/L)
pounds (lb)	× 16	= ounces (oz)
pounds (lb)	× 7,000	= grains (gr)
pounds (lb)	× 0.0004114	= tons (short)
pounds/cubic inch (lb/in.^{3})	× 1,728	= pounds per cubic foot (lb/ft^3)
pounds of water	× 0.0166032	= cubic feet (ft^3)
pounds of water	× 2,768	= cubic inches (in.^{3})
pounds of water	× 0.1198	= gallons (gal)
tons (short)	× 2,000	= pounds (lb)
tons (short)	× 0.89287	= tons (long)
tons (long)	× 2,240	= pounds (lb)
cubic feet air (@ 60°F and 29.92 in. mercury)	× 0.0763	= pounds (lb)

Flow Measurement

barrels per hour (bbl/hr)	× 0.70	= gallons per minute (gpm)
acre-feet/minute	× 325.851	= gallons per minute (gpm)
acre-feet/minute	× 726	= cubic feet per second (ft^3/sec)
cubic feet/minute (ft^3/min)	× 0.1247	= gallons per second (gps)
cubic feet/minute (ft^3/min)	× 62.43	= pounds of water per minute
cubic feet/second (ft^3/sec)	× 448.831	= gallons per minute (gpm)
cubic feet/second (ft^3/sec)	× 0.646317	= million gallons per day (mgd)
cubic feet/second (ft^3/sec)	× 1.984	= acre-feet per day (acre-ft/day)
gallons/minute (gpm)	× 1,440	= gallons per day (gpd)
gallons/minute (gpm)	× 0.00144	= million gallons per day (mgd)
gallons/minute (gpm)	× 0.00223	= cubic feet per second (ft^3/sec)
gallons/minute (gpm)	× 0.1337	= cubic feet per minute (ft^3/min)
gallons/minute (gpm)	× 8.0208	= cubic feet per hour (ft^3/hr)
gallons/minute (gpm	× 0.00442	= acre-feet per day (acre-ft/day)
gallons/minute (gpm)	× 1.43	= barrels (42 gal) per day (bbl/day)
gallons water/minute	× 6.0086	= tons of water per 24 hours
million gallons/day (mgd)	× 1.54723	= cubic feet per second (ft^3/sec)
million gallons/day (mgd)	× 92.82	= cubic feet per minute (ft^3/min)
million gallons/day (mgd)	× 694.4	= gallons per minute (gpm)
million gallons/day (mgd)	× 3.07	= acre-feet per day (acre-ft/day)
pounds of water/minute	× 26.700	= cubic feet per second (ft^3/sec)
miner's inch		= flow through an orifice of 1 in.^{2} under a head of 4 to 6 in.
miner's inches (9 gpm)	× 8.98	= gallons per minute (gpm)

miner's inches (9 gpm)	× 1.2	= cubic feet per minute (ft³/min)
miner's inches (11.25 gpm)	× 11.22	= gallons per minute (gpm)
miner's inches (11.25 gpm)	× 1.5	= cubic feet per minute (ft³/min)

Work Measurement

British thermal units (Btu)	× 777.5	= foot-pounds (ft-lb)
British thermal units (Btu)	× 39,270	= horsepower-hours (hp·hr)
British thermal units (Btu)	× 29,280	= kilowatt-hours (kW·hr)
foot-pounds (ft-lb)	× 1,286	= British thermal units (Btu)
foot-pounds (ft-lb)	× 50,500,000	= horsepower-hours (hp·hr)
foot-pounds (ft-lb)	× 37,660,000	= kilowatt-hours (kW·hr)
horsepower-hours (hp·hr)	× 2,547	= British thermal units (Btu)
horsepower-hours (hp·hr)	× 0.7457	= kilowatt-hours (kW·hr)
kilowatt-hours (kW·hr)	× 3,415	= British thermal units (Btu)
kilowatt-hours (kW·hr)	× 1.241	= horsepower-hours (hp·hr)

Power Measurement

boiler horsepower	× 33,480	= British thermal units per hour (Btu/hr)
boiler horsepower	× 9.8	= kilowatts (kW)
British thermal units/second (Btu/sec)	× 1.0551	= kilowatts (kW)
British thermal units/minute (Btu/min)	× 12.96	= foot-pounds per second (ft-lb/sec)
British thermal units/minute (Btu/min)	× 0.02356	= horsepower (hp)
British thermal units/minute (Btu/min)	× 0.01757	= kilowatts (kW)
British thermal units/hour (Btu/hr)	× 0.293	= watts (W)
British thermal units/hour (Btu/hr)	× 12.96	= foot-pounds per minute (ft-lb/min)
British thermal units/hour (Btu/hr)	× 0.00039	= horsepower (hp)
foot-pounds per second (ft-lb/sec)	× 771.7	= British thermal units per minute (Btu/min)
foot-pounds per second (ft-lb/sec)	× 1,818	= horsepower (hp)
foot-pounds per second (ft-lb/sec)	× 1,356	= kilowatts (kW)
foot-pounds per minute (ft-lb/min)	× 303,000	= horsepower (hp)

foot-pounds per minute (ft-lb/min)	× 226,000	= kilowatts (kW)
horsepower (hp)	× 42.44	= British thermal units per minute (Btu/min)
horsepower (hp)	× 33,000	= foot-pounds per minute (ft-lb/min)
horsepower (hp)	× 550	= foot-pounds per second (ft-lb/sec)
horsepower (hp)	× 1,980,000	= foot-pounds per hour (ft-lb/hr)
horsepower (hp)	× 0.7457	= kilowatts (kW)
horsepower (hp)	× 745.7	= watts (W)
kilowatts (kW)	× 0.9478	= British thermal units per second (Btu/sec)
kilowatts (kW)	× 56.92	= British thermal units per minute (Btu/min)
kilowatts (kW)	× 3,413	= British thermal units per hour (Btu/hr)
kilowatts (kW)	× 44,250	= foot-pounds per minute (ft-lb/min)
kilowatts (kW)	× 737.6	= foot-pounds per second (ft-lb/sec)
kilowatts (kW)	× 1.341	= horsepower (hp)
tons of refrig. (US)	× 288,000	= British thermal units per 24 hours
watts (W)	× 0.05692	= British thermal units per minute (Btu/min)
watts (W)	× 0.7376	= foot-pounds (force) per second (ft-lb/sec)
watts (W)	× 44.26	= foot-pounds per minute (ft-lb/min)
watts (W)	× 1,341	= horsepower (hp)

Velocity Measurement

feet/minute (ft/min)	× 0.01667	= feet per second (ft/sec)
feet/minute (ft/min)	× 0.01136	= miles per hour (mph)
feet/second (ft/sec)	× 0.6818	= miles per hour (mph)
miles/hour (mph)	× 88	= feet per minute (ft/min)
miles/hour (mph)	× 1.467	= feet per second (ft/sec)

Miscellaneous

grade: 1 percent (or 0.01)		= 1 foot per 100 feet

METRIC CONVERSIONS

Linear Measurement

inch (in.)	× 25.4	= millimeters (mm)
inch (in.)	× 2.54	= centimeters (cm)
foot (ft)	× 304.8	= millimeters (mm)
foot (ft)	× 30.48	= centimeters (cm)
foot (ft)	× 0.3048	= meters (m)
yard (yd)	× 0.9144	= meters (m)
mile (mi)	× 1,609.3	= meters (m)
mile (mi)	× 1.6093	= kilometers (km)
millimeter (mm)	× 0.03937	= inches (in.)
centimeter (cm)	× 0.3937	= inches (in.)
meter (m)	× 39.3701	= inches (in.)
meter (m)	× 3.2808	= feet (ft)
meter (m)	× 1.0936	= yards (yd)
kilometer (km)	× 0.6214	= miles (mi)

Area Measurement

square meter (m²)	× 10,000	= square centimeters (cm²)
hectare (ha)	× 10,000	= square meters (m²)
square inch (in.²)	× 6.4516	= square centimeters (cm²)
square foot (ft²)	× 0.092903	= square meters (m²)
square yard (yd²)	× 0.8361	= square meters (m²)
acre	× 0.004047	= square kilometers (km²)
acre	× 0.4047	= hectares (ha)
square mile (mi²)	× 2.59	= square kilometers (km²)
square centimeter (cm²)	× 0.16	= square inches (in.²)
square meters (m²)	× 10.7639	= square feet (ft²)
square meters (m²)	× 1.1960	= square yards (yd²)
hectare (ha)	× 2.471	= acres
square kilometer (km²)	× 247.1054	= acres
square kilometer (km²)	× 0.3861	= square miles (mi²)

Volume Measurement

cubic inch (in.³)	× 16.3871	= cubic centimeters (cm³)
cubic foot (ft³)	× 28,317	= cubic centimeters (cm³)
cubic foot (ft³)	× 0.028317	= cubic meters (m³)
cubic foot (ft³)	× 28.317	= liters (L)
cubic yard (yd³)	× 0.7646	= cubic meters (m³)

acre foot (acre-ft)	× 1233.48	=	cubic meters (m³)
ounce (US fluid) (oz)	× 0.029573	=	liters (L)
quart (liquid) (qt)	× 946.9	=	milliliters (mL)
quart (liquid) (qt)	× 0.9463	=	liters (L)
gallon (gal)	× 3.7854	=	liters (L)
gallon (gal)	× 0.0037854	=	cubic meters (m³)
peck (pk)	× 0.881	=	decaliters (dL)
bushel (bu)	× 0.3524	=	hectoliters (hL)
cubic centimeters (cm³)	× 0.061	=	cubic inches (in.³)
cubic meter (m³)	× 35.3183	=	cubic feet (ft³)
cubic meter (m³)	× 1.3079	=	cubic yards (yd³)
cubic meter (m³)	× 264.2	=	gallons (gal)
cubic meter (m³)	× 0.000811	=	acre-feet (acre-ft)
liter (L)	× 1.0567	=	quart (liquid) (qt)
liter (L)	× 0.264	=	gallons (gal)
liter (L)	× 0.0353	=	cubic feet (ft³)
decaliter (dL)	× 2.6417	=	gallons (gal)
decaliter (dL)	× 1.135	=	pecks (pk)
hectoliter (hL)	× 3.531	=	cubic feet (ft³)
hectoliter (hL)	× 2.84	=	bushels (bu)
hectoliter (hL)	× 0.131	=	cubic yards (yd³)
hectoliter (hL)	× 26.42	=	gallons (gal)

Pressure Measurement

pound/square inch (psi)	× 6.8948	=	kilopascals (kPa)
pound/square inch (psi)	× 0.00689	=	pascals (Pa)
pound/square inch (psi)	× 0.070307	=	kilograms/square centimeter (kg/cm²)
pound/square foot (lb/ft²)	× 47.8803	=	pascals (Pa)
pound/square foot (lb/ft²)	× 0.000488	=	kilograms/square centimeter (kg/cm²)
pound/square foot (lb/ft²)	× 4.8824	=	kilograms/square meter (kg/m²)
inches of mercury	× 3,376.8	=	pascals (Pa)
inches of water	× 248.84	=	pascals (Pa)
bar	× 100,000	=	newtons per square meter
pascals (Pa)	× 1	=	newtons per square meter
pascals (Pa)	× 0.000145	=	pounds/square inch (psi)
kilopascals (kPa)	× 0.145	=	pounds/square inch (psi)
pascals (Pa)	× 0.000296	=	inches of mercury (at 60°F)

kilogram/square centimeter (kg/cm^2)	$\times$ 14.22	= pounds/square inch (psi)
kilogram/square centimeter (kg/cm^2)	$\times$ 28.959	= inches of mercury (at 60°F)
kilogram/square meter (kg/m^2)	$\times$ 0.2048	= pounds per square foot (lb/ft^2)
centimeters of mercury	$\times$ 0.4461	= feet of water

Weight Measurement

ounce (oz)	$\times$ 28.3495	= grams (g)
pound (lb)	$\times$ 0.045359	= grams (g)
pound (lb)	$\times$ 0.4536	= kilograms (kg)
ton (short)	$\times$ 0.9072	= megagrams (metric ton)
pounds/cubic foot (lb/ft^3)	$\times$ 16.02	= grams per liter (g/L)
pounds/million gallons (lb/mil gal)	$\times$ 0.1198	= grams per cubic meter (g/m^3)
gram (g)	$\times$ 15.4324	= grains (gr)
gram (g)	$\times$ 0.0353	= ounces (oz)
gram (g)	$\times$ 0.0022	= pounds (lb)
kilograms (kg)	$\times$ 2.2046	= pounds (lb)
kilograms (kg)	$\times$ 0.0011	= tons (short)
megagram (metric ton)	$\times$ 1.1023	= tons (short)
grams/liter (g/L)	$\times$ 0.0624	= pounds per cubic foot (lb/ft^3)
grams/cubic meter (g/m^3)	$\times$ 8.3454	= pounds/million gallons (lb/mil gal)

Flow Rates

gallons/second (gps)	$\times$ 3.785	= liters per second (L/sec)
gallons/minute (gpm)	$\times$ 0.00006308	= cubic meters per second (m^3/sec)
gallons/minute (gpm)	$\times$ 0.06308	= liters per second (L/sec)
gallons/hour (gph)	$\times$ 0.003785	= cubic meters per hour (m^3/hr)
gallons/day (gpd)	$\times$ 0.000003785	= million liters per day (ML/day)
gallons/day (gpd)	$\times$ 0.003785	= cubic meters per day (m^3/day)
cubic feet/second (ft^3/sec)	$\times$ 0.028317	= cubic meters per second (m^3/sec)
cubic feet/second (ft^3/sec)	$\times$ 1,699	= liters per minute (L/min)
cubic feet/minute (ft^3/min)	$\times$ 472	= cubic centimeters/second (cm^3/sec)
cubic feet/minute (ft^3/min)	$\times$ 0.472	= liters per second (L/sec)
cubic feet/minute (ft^3/min)	$\times$ 1.6990	= cubic meters per hour (m^3/hr)

million gallons/day (mgd)	$\times$ 43.8126	= liters per second (L/sec)
million gallons/day (mgd)	$\times$ 0.003785	= cubic meters per day (m^3/day)
million gallons/day (mgd)	$\times$ 0.043813	= cubic meters per second (m^3/sec)
gallons/square foot (gal/ft^2)	$\times$ 40.74	= liters per square meter (L/m^2)
gallons/acre/day (gal/acre/day)	$\times$ 0.0094	= cubic meters/hectare/day (m^3/ha/day)
gallons/square foot/day (gal/ft^2/day)	$\times$ 0.0407	= cubic meters/square meter/day (m^3/m^2/day)
gallons/square foot/day (gal/ft^2/day)	$\times$ 0.0283	= liters/square meter/day (L/m^2/day)
gallons/square foot/minute (gal/ft^2/min)	$\times$ 2.444	= cubic meters/square meter/hour (m^3/m^2/hr) = m/hr
gallons/square foot/minute (gal/ft^2/min)	$\times$ 0.679	= liters/square meter/second (L/m^2/sec)
gallons/square foot/minute (gal/ft^2/min)	$\times$ 40.7458	= liters/square meter/minute (L/m^2/min)
gallons/capita/day (gpcd)	$\times$ 3.785	= liters/day/capita (L/d per capita)
liters/second (L/sec)	$\times$ 22,824.5	= gallons per day (gpd)
liters/second (L/sec)	$\times$ 0.0228	= million gallons per day (mgd)
liters/second (L/sec)	$\times$ 15.8508	= gallons per minute (gpm)
liters/second (L/sec)	$\times$ 2.119	= cubic feet per minute (ft^3/min)
liters/minute (L/min)	$\times$ 0.0005886	= cubic feet per second (ft^3/sec)
cubic centimeters/second (cm^3/sec)	$\times$ 0.0021	= cubic feet per minute (ft^3/min)
cubic meters/second (m^3/sec)	$\times$ 35.3147	= cubic feet per second (ft^3/sec)
cubic meters/second (m^3/sec)	$\times$ 22.8245	= million gallons per day (mgd)
cubic meters/second (m^3/sec)	$\times$ 15,850.3	= gallons per minute (gpm)
cubic meters/hour (m^3/hr)	$\times$ 0.5886	= cubic feet per minute (ft^3/min)
cubic meters/hour (m^3/hr)	$\times$ 4.403	= gallons per minute (gpm)
cubic meters/day (m^3/day)	$\times$ 264.1720	= gallons per day (gpd)
cubic meters/day (m^3/day)	$\times$ 0.00026417	= million gallons per day (mgd)
cubic meters/hectare/day (m^3/ha/day)	$\times$ 106.9064	= gallons per acre per day (gal/acre/day)
cubic meters/square meter/day (m^3/m^2/day)	$\times$ 24.5424	= gallons/square foot/day (gal/ft^2/day)
liters/square meter/minute (L/m^2/min)	$\times$ 0.0245	= gallons/square foot/minute (gal/ft^2/min)
liters/square meter/minute (L/m^2/min)	$\times$ 35.3420	= gallons/square foot/day (gal/ft^2/day)

Work, Heat, and Energy

British thermal units (Btu)	× 1.0551	=	kilojoules (kJ)
British thermal units (Btu)	× 0.2520	=	kilogram-calories (kg-cal)
foot-pound (force) (ft-lb)	× 1.3558	=	joules (J)
horsepower-hour (hp·hr)	× 2.6845	=	megajoules (MJ)
watt-second (W-sec)	× 1.000	=	joules (J)
watt-hour (W·hr)	× 3.600	=	kilojoules (kJ)
kilowatt-hour (kW·hr)	× 3,600	=	kilojoules (kJ)
kilowatt-hour (kW·hr)	× 3,600,000	=	joules (J)
British thermal units per pound (Btu/lb)	× 0.5555	=	kilogram-calories per kilogram (kg-cal/kg)
British thermal units per cubic foot (Btu/ft³)	× 8.8987	=	kilogram-calories/cubic meter (kg-cal/m³)
kilojoule (kJ)	× 0.9478	=	British thermal units (Btu)
kilojoule (kJ)	× 0.00027778	=	kilowatt-hours (kW·hr)
kilojoule (kJ)	× 0.2778	=	watt-hours (W·hr)
joule (J)	× 0.7376	=	foot-pounds (ft-lb)
joule (J)	× 1.0000	=	watt-seconds (W-sec)
joule (J)	× 0.2399	=	calories (cal)
megajoule (MJ)	× 0.3725	=	horsepower-hour (hp·hr)
kilogram-calories (kg-cal)	× 3.9685	=	British thermal units (Btu)
kilogram-calories per kilogram (kg-cal/kg)	× 1.8000	=	British thermal units per pound (Btu/lb)
kilogram-calories per liter (kg-cal/L)	× 112.37	=	British thermal units per cubic foot (Btu/ft³)
kilogram-calories/cubic meter (kg-cal/m³)	× 0.1124	=	British thermal units per cubic foot (Btu/ft³)

Velocity, Acceleration, and Force

feet per minute (ft/min)	× 18.2880	=	meters per hour (m/hr)
feet per hour (ft/hr)	× 0.3048	=	meters per hour (m/hr)
miles per hour (mph)	× 44.7	=	centimeters per second (cm/sec)
miles per hour (mph)	× 26.82	=	meters per minute (m/min)
miles per hour (mph)	× 1.609	=	kilometers per hour (km/hr)
feet/second/second (ft/sec²)	× 0.3048	=	meters/second/second (m/sec²)
inches/second/second (in./sec²)	× 0.0254	=	meters/second/second (m/sec²)
pound-force (lbf)	× 4.44482	=	newtons (N)
centimeters/second (cm/sec)	× 0.0224	=	miles per hour (mph)

meters/second (m/sec)	× 3.2808	= feet per second (ft/sec)
meters/minute (m/min)	× 0.0373	= miles per hour (mph)
meters per hour (m/hr)	× 0.0547	= feet per minute (ft/min)
meters per hour (m/hr)	× 3.2808	= feet per hour (ft/hr)
kilometers/second (km/sec)	× 2.2369	= miles per hour (mph)
kilometers/hour (km/hr)	× 0.0103	= miles per hour (mph)
meters/second/second (m/sec^2)	× 3.2808	= feet/second/second (ft/sec^2)
meters/second/second (m/sec^2)	× 39.3701	= inches/second/second (in./sec^2)
newtons (N)	× 0.2248	= pounds force (lbf)

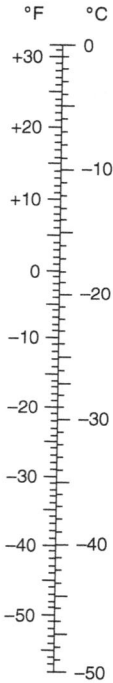

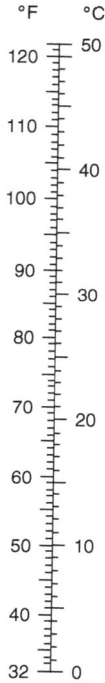

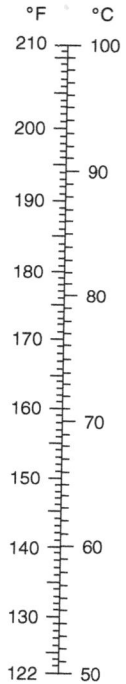

°F	°C		°F	°C		°F	°C
+30	0		120	50		210	100
+20			110			200	
	−10			40			90
+10			100			190	
0			90			180	
	−20			30			80
−10			80			170	
−20	−30		70	20		160	70
−30			60			150	
−40	−40		50	10		140	60
−50			40			130	
	−50		32	0		122	50

$$0.555 \, (°F - 32) = \text{degrees Celsius (°C)}$$
$$(1.8 \times °C) + 32 = \text{degrees Fahrenheit (°F)}$$
$$°C + 273.15 = \text{kelvin (K)}$$

boiling point* = 212°F
= 100°C
= 373 K

freezing point* = 32°F
= 0°C
= 273 K

*At 14.696 psia, 101.325 kPa.

Celsius/Fahrenheit Comparison Graph

Decimal Equivalents of Fractions

Fraction	Decimal	Fraction	Decimal
1/64	0.01563	33/64	0.51563
1/32	0.03125	17/32	0.53125
3/64	0.04688	35/64	0.54688
1/16	0.06250	9/16	0.56250
5/64	0.07813	37/64	0.57813
3/32	0.09375	19/32	0.59375
7/64	0.10938	39/64	0.60938
1/8	0.12500	5/8	0.62500
9/64	0.14063	41/64	0.64063
5/32	0.15625	21/32	0.65625
11/64	0.17188	43/64	0.67188
3/16	0.18750	11/16	0.68750
13/64	0.20313	45/64	0.70313
7/32	0.21875	23/32	0.71875
15/64	0.23438	47/64	0.73438
1/4	0.25000	3/4	0.75000
17/64	0.26563	49/64	0.76563
9/32	0.28125	25/32	0.78125
19/64	0.29688	51/64	0.79688
10/32	0.31250	13/16	0.81250
21/64	0.32813	53/64	0.82813
11/32	0.34375	27/32	0.84375
23/64	0.35938	55/64	0.85938
3/8	0.37500	7/8	0.87500
25/64	0.39063	57/64	0.89063
13/32	0.40625	29/32	0.90625
27/64	0.42188	59/64	0.92188
7/16	0.43750	15/16	0.93750
29/64	0.45313	61/64	0.95313
15/32	0.46875	31/32	0.96875
31/64	0.48438	63/64	0.98438
1/2	0.50000		

Units of Measure and Conversions

WATER EQUIVALENTS AND DATA

- 1 US gallon of water weighs 8.345 pounds.
- 1 cubic foot of water equals 7.48 gallons.
- 1 foot head of water develops 0.433 pounds per square inch.
- Pounds per hour times 0.12 equals gallons per hour.
- Grains per gallon times 0.143 equals pounds per 1,000 gallons.
- Parts per million divided by 120 equals pounds per 1,000 gallons.
- 1 grain per gallon equals 17.1 parts per million.
- Estimated flow in gallons per minute equals pipe diameter in inches squared times 20.
- 1 boiler horsepower based on 10 square feet of heating surface requires 4 gallons per hour of feedwater.
- 1 pound of coal will produce 7 to 10 pounds of steam.
- 1 gallon of oil will produce 70 to 120 pounds of steam.
- 1,000 cubic feet of natural gas will produce 600 pounds of steam.
- Saturated salt brine for zeolite regeneration contains 2.48 pounds of salt per gallon or 18.5 pounds per cubic foot.
- Refrigeration tonnage is gallons per minute of cooling water times increased temperature divided by 24.
- Cooling tower makeup is estimated at 1½ gallons per hour per ton of refrigeration.
- 1 ton of refrigeration is 288,000 Btu.

WATER CONVERSIONS

Water is composed of two gases, hydrogen and oxygen, in the ratio of two volumes of the former to one of the latter. It is never found pure in nature because of the readiness with which it absorbs impurities from the air and soil.

- One foot of water column at 39.1°F = 62.425 pounds on the square foot.

- One foot of water column at 39.1°F = 0.4335 pound on the square inch.

- One foot of water column at 39.1°F = 0.0295 atmospheric pressure.

- One foot of water column at 39.1°F = 0.8826 inch mercury column at 32°F.

- One foot of water column at 39.1°F = 773.3 feet of air column at 32°F and atmospheric pressure.

- One pound pressure per square foot = 0.01602 foot water column at 39.1°F.

- One pound pressure per square foot = 2.307 feet water column at 39.1°F.

- One atmospheric pressure = 29.92 inches mercury column = 33.9 feet water column.

- One inch of mercury column at 32°F = 1.133 feet water column.

- One foot of air column at 32°F and 1 atmospheric pressure = 0.001293 foot water column.

Chemistry

*The science of chemistry deals with the structure,
composition, and changes in composition of
matter, as well as with the laws that govern these
changes. To understand and work successfully
with the chemical phases of water treatment
such as coagulation, sedimentation, softening,
disinfection, and chemical removal of
various undesirable substances,
a water operator needs to know
some basic chemistry concepts.*

Periodic Table of Elements

1	2	3	4	5	6	7	8	9	10	11	12	13	14	15	16	17	18
1 H 1.0079																	2 He 4.0026
3 Li 6.941	4 Be 9.0122											5 B 10.811	6 C 12.011	7 N 14.007	8 O 15.999	9 F 18.998	10 Ne 20.180
11 Na 22.990	12 Mg 24.305											13 Al 26.982	14 Si 28.086	15 P 30.974	16 S 32.065	17 Cl 35.453	18 Ar 39.948
19 K 39.098	20 Ca 40.078	21 Sc 44.956	22 Ti 47.867	23 V 50.942	24 Cr 51.996	25 Mn 54.938	26 Fe 55.845	27 Co 58.933	28 Ni 58.693	29 Cu 63.546	30 Zn 65.409	31 Ga 69.723	32 Ge 72.64	33 As 74.922	34 Se 78.96	35 Br 79.904	36 Kr 83.798
37 Rb 85.468	38 Sr 87.62	39 Y 88.906	40 Zr 91.224	41 Nb 92.906	42 Mo 95.94	43 Tc (98)	44 Ru 101.07	45 Rh 102.91	46 Pd 106.42	47 Ag 107.87	48 Cd 112.41	49 In 114.82	50 Sn 118.71	51 Sb 121.76	52 Te 127.60	53 I 126.90	54 Xe 131.29
55 Cs 132.91	56 Ba 137.33	57–71 *	72 Hf 178.49	73 Ta 180.95	74 W 183.84	75 Re 186.21	76 Os 190.23	77 Ir 192.22	78 Pt 195.08	79 Au 196.97	80 Hg 200.59	81 Tl 204.38	82 Pb 207.2	83 Bi 208.98	84 Po (209)	85 At (210)	86 Rn (222)
87 Fr (223)	88 Ra (226)	89–103 #	104 Rf (261)	105 Db (262)	106 Sg (266)	107 Bh (264)	108 Hs (277)	109 Mt (268)	110 Ds (281)	111 Uuu (272)	112 Uub (285)	113 Uut (284)	114 Uuq (289)	115 Uup (288)	116 Uuh (289)	117 Uus	118 Uuo (293)

*Lanthanide series

57 La 138.91	58 Ce 140.12	59 Pr 140.91	60 Nd 144.24	61 Pm (145)	62 Sm 150.36	63 Eu 151.96	64 Gd 157.25	65 Tb 158.93	66 Dy 162.50	67 Ho 164.93	68 Er 167.26	69 Tm 168.93	70 Yb 173.04	71 Lu 174.97

#Actinide series

89 Ac (227)	90 Th 232.04	91 Pa 231.04	92 U 238.03	93 Np (237)	94 Pu (244)	95 Am (243)	96 Cm (247)	97 Bk (247)	98 Cf (251)	99 Es (252)	100 Fm (257)	101 Md (258)	102 No (259)	103 Lr (262)

NOTE: For elements with no stable nuclides, the mass of the longest-lived isotope is in parentheses.

List of Elements

Name	Symbol	Atomic Number	Atomic Weight
Actinium	Ac	89	227*
Aluminum	Al	13	26.98
Americium	Am	95	243*
Antimony	Sb	51	121.75
Argon	Ar	18	39.95
Arsenic	As	33	74.92
Astatine	At	85	210*
Barium	Ba	56	137.34
Berkelium	Bk	97	247*
Beryllium	Be	4	9.01
Bismuth	Bi	83	208.98
Boron	B	5	10.81
Bromine	Br	35	79.90
Cadmium	Cd	48	112.41
Calcium	Ca	20	40.08
Californium	Cf	98	251*
Carbon	C	6	12.01
Cerium	Ce	58	140.12
Cesium	Cs	55	132.91
Chlorine	Cl	17	35.45
Chromium	Cr	24	52.00
Cobalt	Co	27	58.93
Copper	Cu	29	63.55
Curium	Cm	96	247*
Dubnium	Db	105	262*
Dysprosium	Dy	66	162.50
Einsteinium	Es	99	252*
Erbium	Er	68	167.26
Europium	Eu	63	151.96
Fermium	Fm	100	257*
Fluorine	F	9	19.00

Table continued on next page

List of Elements (continued)

Name	Symbol	Atomic Number	Atomic Weight
Francium	Fr	87	223*
Gadolinium	Gd	64	157.25
Gallium	Ga	31	69.72
Germanium	Ge	32	72.64
Gold	Au	79	196.97
Hafnium	Hf	72	178.49
Hassium	Hs	108	265*
Helium	He	2	4.00
Holmium	Ho	67	164.93
Hydrogen	H	1	1.01
Indium	In	49	114.82
Iodine	I	53	126.90
Iridium	Ir	77	192.22
Iron	Fe	26	55.85
Krypton	Kr	36	83.80
Lanthanum	La	57	138.91
Lawrencium	Lr	103	262*
Lead	Pb	82	207.2
Lithium	Li	3	6.94
Lutetium	Lu	71	174.97
Magnesium	Mg	12	24.31
Manganese	Mn	25	54.94
Meitnerium	Mt	109	265*
Mendelevium	Md	101	258*
Mercury	Hg	80	200.59
Molybdenum	Mo	42	95.94
Neodymium	Nd	60	144.24
Neon	Ne	10	20.18
Neptunium	Np	93	237.05†
Nickel	Ni	28	58.69

Table continued on next page

List of Elements (continued)

Name	Symbol	Atomic Number	Atomic Weight
Niobium	Nb	41	92.91
Nitrogen	N	7	14.01
Nobelium	No	102	259*
Osmium	Os	76	190.23
Oxygen	O	8	16.00
Palladium	Pd	46	106.42
Phosphorus	P	15	30.97
Platinum	Pt	78	195.08
Plutonium	Pu	94	244*
Polonium	Po	84	209*
Potassium	K	19	39.10
Praseodymium	Pr	59	140.91
Promethium	Pm	61	145*
Protactinium	Pa	91	231.04†
Radium	Ra	88	226.03†
Radon	Rn	86	222*
Rhenium	Re	75	186.21
Rhodium	Rh	45	102.91
Rubidium	Rb	37	85.47
Ruthenium	Ru	44	101.07
Rutherfordium	Rf	104	261*
Samarium	Sm	62	150.36
Scandium	Sc	21	44.96
Seaborgium	Sg	106	263*
Selenium	Se	34	78.96
Silicon	Si	14	28.09
Silver	Ag	47	107.87
Sodium	Na	11	22.99
Strontium	Sr	38	87.62
Sulfur	S	16	32.06

Table continued on next page

List of Elements (continued)

Name	Symbol	Atomic Number	Atomic Weight
Tantalum	Ta	73	180.95
Technetium	Tc	43	98.91[†]
Tellurium	Te	52	127.60
Terbium	Tb	65	158.93
Thallium	Tl	81	204.38
Thorium	Th	90	232.04[†]
Thulium	Tm	69	168.93
Tin	Sn	50	118.69
Titanium	Ti	22	47.90
Tungsten	W	74	183.85
Ununbium	Uub	112	27*
Ununnillium	Uun	110	269*
Ununhexium	Uuh	116	289*
Ununoctium	Uuo	118	293*
Ununquadium	Uuq	114	285*
Unununium	Uuu	111	27*
Uranium	U	92	238.03
Vanadium	V	23	50.94
Xenon	Xe	54	131.29
Ytterbium	Yb	70	173.04
Yttrium	Y	39	88.91
Zinc	Zn	30	65.38
Zirconium	Zr	40	91.22

* Mass number of most stable or best-known isotope.
† Mass of most commonly available, long-lived isotope.

Compounds Common in Water Treatment

Chemical Name	Common Name	Chemical Formula
Aluminum hydroxide	Alum floc	$Al(OH)_3$
Aluminum sulfate	Filter alum	$Al_2(SO_4)_3 \cdot 14H_2O$
Ammonia	Ammonia	NH_3 (ammonia gas)
Calcium bicarbonate	—	$Ca(HCO_3)_2$
Calcium carbonate	Limestone	$CaCO_3$
Calcium chloride	—	$CaCl_2$
Calcium hydroxide	Hydrated lime (slaked lime)	$Ca(OH)_2$
Calcium hypochlorite	HTH	$Ca(OCl)_2$
Calcium oxide	Unslaked lime (quicklime)	CaO
Calcium sulfate	—	$CaSO_4$
Carbon	Activated carbon	C
Carbon dioxide	—	CO_2
Carbonic acid	—	H_2CO_3
Chlorine	—	Cl_2
Chlorine dioxide	—	ClO_2
Copper sulfate	Blue vitriol	$CuSO_4 \cdot 5H_2O$
Dichloramine	—	$NHCl_2$
Ferric chloride	—	$FeCl_3 \cdot 6H_2O$
Ferric hydroxide	Ferric hydroxide floc	$Fe(OH)_3$
Ferric sulfate	—	$Fe_2(SO_4)_3 \cdot 3H_2O$
Ferrous bicarbonate	—	$Fe(HCO_3)_2$
Ferrous hydroxide	—	$Fe(OH)_2$
Fluosilicic acid (hydrofluosilicic acid)	—	H_2SiF_6
Hydrochloric acid	Muriatic acid	HCl
Hydrofluosilicic acid (fluosilicic acid)	—	H_2SiF_6
Hydrogen sulfide	—	H_2S
Hypochlorous acid	—	$HOCl$

Table continued on next page

Chemistry

Compounds Common in Water Treatment (continued)

Chemical Name	Common Name	Chemical Formula
Magnesium bicarbonate	—	$Mg(HCO_3)_2$
Magnesium carbonate	—	$MgCO_3$
Magnesium chloride	—	$MgCl_2$
Magnesium hydroxide	—	$Mg(OH)_2$
Manganese dioxide	—	MnO_2
Manganous bicarbonate	—	$Mn(HCO_3)_2$
Manganous sulfate	—	$MnSO_4$
Monochloramine	—	NH_2Cl
Potassium bicarbonate	—	$KHCO_3$
Potassium permanganate	—	$KMnO_4$
Sodium bicarbonate	Soda	$NaHCO_3$
Sodium carbonate	Soda ash	Na_2CO_3
Nitrogen trichloride (trichloramine)	—	NCl_3
Sodium chloride	Salt	$NaCl$
Sodium chlorite	—	$NaClO_2$
Sodium fluoride	—	NaF
Sodium fluosilicate (sodium silicofluoride)	—	Na_2SiF_6
Sodium hydroxide	Lye	$NaOH$
Sodium hypochlorite	Bleach	$NaOCl$
Sodium phosphate	—	$Na_3PO_4 \cdot 12H_2O$
Sodium silicofluoride (sodium fluosilicate)	—	Na_2SiF_6
Sodium bisulfite	—	$NaHSO_3$
Sodium sulfate	—	Na_2SO_4
Sodium sulfite	—	Na_2SO_3
Sodium thiosulfate	—	$Na_2S_2O_3 \cdot 5H_2O$
Sulfur dioxide	—	SO_2
Sulfuric acid	Oil of vitriol	H_2SO_4
Trichloramine (nitrogen trichloride)	—	NCl_3

KEY FORMULAS FOR CHEMISTRY _____

$$\text{mg total suspended solids/L} = \frac{(A - B) \times 1{,}000}{\text{sample volume, mL}}$$

A = weight of filter + dry residue sample, mg

B = weight of filter, mg

A is the total weight wet.

B is the filter weight dry.

Dilutions

concentration 1 × volume 1 = concentration 2 × volume 2

$$\text{concentration 1} = \frac{\text{concentration 2} \times \text{volume 2}}{\text{volume 1}}$$

$$\text{volume 1} = \frac{\text{concentration 2} \times \text{volume 2}}{\text{concentration 1}}$$

concentration = mg/L

volume = L

CONDUCTIVITY AND DISSOLVED SOLIDS _____

Electrical conductivity is the ability of a solution to conduct an electric current and it can be used as an indirect measure of the total dissolved solids (TDS) in a water sample.

The unit of measure commonly used is siemens per centimeter (S/cm). The conductivity of water is usually expressed as microsiemens per centimeter (μS/cm) which is 10^{-6} S/cm. The relationship between conductivity and dissolved solids is approximately:

2 μS/cm = 1 ppm (which is the same as 1 mg/L)

The conductivity of water from various sources is

Absolutely pure water	= 0.055 μS/cm
Distilled water	= 0.5 μS/cm
Mountain water	= 1.0 μS/cm
Most drinking water sources	= 500 to 800 μS/cm
Seawater	= 56 mS/cm
Maximum for potable water	= 1,055 μS/cm

Some common conductivity conversion factors are

mS/cm	×	1,000	=	μS/cm
μS/cm	×	0.001	=	mS/cm
μS/cm	×	1	=	μmhos/cm
μS/cm	×	0.5	=	mg/L of TDS
mS/cm	×	0.5	=	g/L of TDS
mg/L TDS	×	0.001	=	g/L of TDS
mg/L TDS	×	0.05842	=	gpg TDS

Densities of Various Substances

Substance	Density	
	lb/ft³	lb/gal
Solids		
Activated carbon*†	8–28 (avg. 12)	
Lime*†	20–50	
Dry alum*†	60–75	
Aluminum (at 20°C)	168.5	
Steel (at 20°C)	486.7	
Copper (at 20°C)	555.4	
Liquids		
Propane (–44.5°C)	36.5	4.88
Gasoline†	43.7	5.84
Water (4°C)	62.4	8.34
Fluorosilicic acid (30%, –8.1°C)	77.8–79.2	10.4–10.6
Liquid alum (36°Bé, 15.6°C)	83.0	11.09
Liquid chlorine (–33.6°C)	97.3	13.01
Sulfuric acid (18°C)	114.2	15.3
Gases		
Methane (0°C, 14.7 psia)	0.0344	
Air (20°C, 14.7 psia)	0.075	
Oxygen (0°C, 14.7 psia)	0.089	
Hydrogen sulfide†	0.089	
Carbon dioxide†	0.115	
Chlorine gas (0°C, 14.7 psia)	0.1870	

* Bulk density of substance.
† Temperature and/or pressure not given.

Recommendations for Sampling and Preservation of Samples According to Measurement*

Measurement	Volume Required, mL	Container†	Preservative	Maximum Holding Time‡
Physical Properties				
Color	500	P, G	Refrigerate	48 hours
Conductance	100	P, G	Cool, 4°C	28 days
Hardness§	100	P, G	HNO_3 to pH <2, H_2SO_4 to pH <2	6 months
Odor	500	G only	As soon as possible refrigerate	6 hours
pH§	50	P, G	Determined onsite	Immediately
Residue, filterable	100	P, G	Cool, 4°C	48 hours
Temperature	1,000	P, G	Determined onsite	Immediately
Turbidity	100	P, G	Cool, 4°C	48 hours
Metals (Fe, Mn)				
Dissolved	200	P, G	Filter onsite, HNO_3 to pH <2	6 months
Suspended	200		Filter onsite	6 months
Total	100	P, G	HNO_3 to pH <2	6 months
Inorganics, Nonmetallics				
Acidity	100	P, G (B)	Refrigerate	14 days
Alkalinity	200	P, G	Refrigerate	14 days
Bromide	100	P, G	None required	28 days

Table continued on next page

Chemistry

57

Recommendations for Sampling and Preservation of Samples According to Measurement* (continued)

Measurement	Volume Required, mL	Container†	Preservative	Maximum Holding Time‡
Chloride	50	P, G	None required	28 days
Chlorine	200	P, G	Determined onsite	Immediately
Cyanide	500	P, G	Cool, 4°C; NaOH to pH >12	14 days
Fluoride	300	P, G	None required	28 days
Iodide	100	P, G	Cool, 4°C	24 hours
Nitrogen:				
Ammonia	400	P, G	Cool, 4°C; H_2SO_4 to pH <2	28 days
Kjeldahl, total	500	P, G	Cool, 4°C; H_2SO_4 to pH <2	28 days
Nitrate	100	P, G	Cool, 4°C	48 hours
Nitrate plus nitrite	100	P, G	Cool, 4°C; H_2SO_4 to pH <2	28 days
Nitrite	50	P, G	Cool, 4°C	48 hours
Dissolved Oxygen:				
Probe	300	G only	Determined onsite	Immediately
Winkler	300	G only	Fix onsite	8 hours
Phosphorus:				
Hydrolyzable	50	P, G	Cool, 4°C; H_2SO_4 to pH <2	24 hours
Orthophosphate, dissolved	50	P, G	Filter onsite; Cool, 4°C	24 hours

Table continued on next page

Recommendations for Sampling and Preservation of Samples According to Measurement* (continued)

Measurement	Volume Required, mL	Container†	Preservative	Maximum Holding Time‡
Total	50	P, G	Cool, 4°C; H_2SO_4 to pH <2	28 days
Total, dissolved	50	P, G	Filter onsite; Cool, 4°C; H_2SO_4 to pH <2	24 hours
Silica	50	P only	Cool, 4°C	28 days
Sulfate	50	P, G	Cool, 4°C	28 days
Sulfide	500	P, G	Cool, 4°C; 2 mL zinc acetate	7 days
Sulfite	50	P, G	Determined onsite	Immediately

NOTE: Whenever a sample is collected for a bacteriological test (coliforms), a sterile plastic or glass bottle must be used. If the sample contains any chlorine residual, sufficient sodium thiosulfate should be added to neutralize all of the chlorine residual. Usually two drops (0.1 mL) of 10% sodium thiosulfate for every 100 mL of sample is sufficient, unless you are disinfecting mains or storage tanks.

* "Guidelines Establishing Test Procedures for the Analysis of Pollutants Under the Clean Water Act" by USEPA, *Federal Register*, Part VIII, Vol. 49, No. 209, Friday, Oct. 26, 1984; 40 CFR Part 136.

† Polyethylene (P) or Glass (G). For metals, polyethylene with a polypropylene cap (no liner) is preferred.

‡ Holding times listed above are recommended for properly preserved samples based on currently available data. For some sample types, extension of these times may be possible; for other types, these times may be too long. Where shipping regulations prevent the use of the proper preservation technique or the holding time is exceeded, such as the case of a 24-hour composite, the final reported data for these samples should indicate the specific variance.

§ Hardness and pH are usually considered chemical properties of water rather than physical properties.

Safety

Water operators are exposed to a number of occupational hazards. In fact, water and wastewater treatment ranks high on the national listings of industrial occupations where on-the-job injuries can occur. Whether regulated by OSHA or dictated by common sense and plant policy, safe working practices are an important part of the water operator's job.

OSHA SAFETY REGULATIONS

Confined Space Entry

Beginning in April 1993, the Occupational Safety and Health Administration (OSHA) implemented and started enforcing comprehensive regulations governing confined spaces. Most states and municipalities have adopted these standards, even if OSHA does not regulate them directly.

Virtually all access entrances now come under OSHA standard 29 CFR 1910.146, Permit-Required Confined Spaces. These standards formally implement requirements and clarify previous recommendations and suggestions made by industry representatives.

Emergency Rescue

As of April 15, 1993, a mechanical device for rescue became required for all vertical-type, permit-required confined spaces deeper than 5 ft [1910.146(k)(3)(ii)]. A safety line and human muscles are no longer acceptable means of rescue for most confined spaces with the potential for vertical rescue. Systems that were used in the past, including "boat winches," should no longer be used. Today, "human-rated" alternatives are available that satisfy the OSHA requirements. This means that the manufacturer has designed the system specifically for lifting people rather than materials.

Nonemergency Ingress/Egress

Means for safe entry and exit by authorized personnel are just as important, per 1910.146(d)(4), as rescue systems. Most tripod/winch systems used for nonemergency work positioning and support applications (such as lowering a worker into an access space that does not contain a ladder) are defined as "single-point adjustable suspension scaffolds." Tripods and davit-arms are examples. Both general industry standards (OSHA 1910) and construction industry standards (OSHA 1926) stipulate specific requirements that must be satisfied when a tripod/manually operated winch system is the primary means used to suspend or support workers. Excerpts from the standards follow.

Utility owners and operators are also now clearly responsible for contractor or subcontractor activities in and around confined spaces. Contractors should be trained in following proper procedures and using the right equipment.

I. a. OSHA 1910.28(i)(1) Single-point adjustable suspension scaffolds. The scaffolding [tripod, davit-arm], including power units or manually operated winches, shall be a type tested and listed by a nationally recognized testing laboratory.

   b. OSHA 1926.451(k)(1) Single-point adjustable suspension scaffolds. The scaffolding [tripod, davit-arm], including power units or manually operated winches shall be a type tested and listed by Underwriters Laboratories or Factory Mutual Engineering Corporation.

Confined Space Entry Procedure

Job: Manhole Inspection and Cleaning   Employee: _____   Date: _____

Dept: _____   Foreman: _____   Review Date: _____

Municipality: _____

Required and/or Recommended PPE: Coveralls, rubber gloves, safety boots, safety glasses, hard hat, immunizations.

Sequence of Basic Job Steps	Potential Accidents or Hazards	Safe Job Procedure
1. Secure the work site to ensure traffic and public safety.	Injury or damage to equipment by contact with vehicles. Injury to public, either pedestrians or vehicle occupants.	Follow traffic control plan.
2. Check manhole for hazardous gases before removing access cover.	Ignition of gases that may be present and toxic vapors.	Follow procedures for confined space entry.
3. Remove access cover.	Injury to back or foot; slips and falls.	Always use proper access cover lifting tools.

Table continued on next page

Confined Space Entry Procedure (continued)

Sequence of Basic Job Steps	Potential Accidents or Hazards	Safe Job Procedure
4. Before entering confined space, use flashlight or mirror to visually check condition of manhole and ladder rungs. Ensure that testing of hazardous gases is continuous and ventilation is in use where entry is required.	Falls, hazardous gases, and infection.	Test the atmosphere of confined space for oxygen deficiency, explosive or toxic gas (confined space entry plan). Provide adequate lighting with explosion-proof fixture. Always wear hard hat. Wear rubber gloves. Wherever possible, carry out the job in such a manner so that entry of personnel into the manhole is not necessary. Ensure gas mask (self-contained breathing apparatus plan) and other safety equipment are operational and available. Ensure that life support and rescue equipment is available.
5. Perform routine flushing operation, removing debris and sediment as necessary.	Hazardous gases may be released from disturbed sediments. Surcharging of collection system. Slips, falls, and infection.	Wear hard hat at all times (personal protection equipment plan). Where entry into manhole is necessary, provide full body harness and lifeline and approved equipment for removing debris.
6. Replace access cover.	Injury to back or foot; slips and falls.	Use proper tools to clean the ring to allow the cover to fit snugly. Replace the cover and ensure that it fits properly.

65

TRENCH SHORING CONDITIONS*

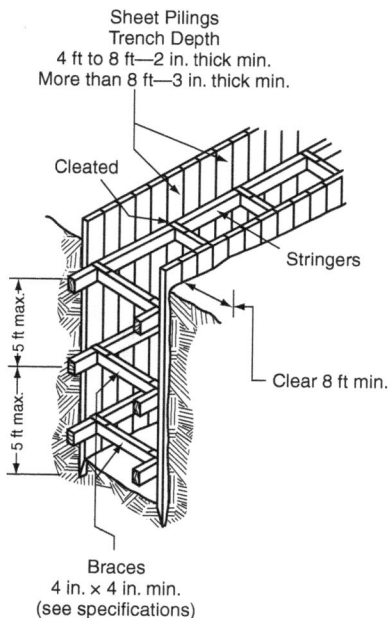

Sheet Pilings
Trench Depth
4 ft to 8 ft—2 in. thick min.
More than 8 ft—3 in. thick min.

Cleated

Stringers

5 ft max.

5 ft max.

Clear 8 ft min.

Braces
4 in. × 4 in. min.
(see specifications)

Sheet piling or equivalent solid sheeting is required for trenches 4 ft or more deep.

Longitudinal-stringer dimensions depend on the strut braces, the stringer spacing, and the depth of stringer below the ground surface.

Greater loads are encountered as the depth increases, so more or stronger stringers and struts are required near the trench bottom.

Running Material

* *This section adapted from Office of Water Programs, California State University, Sacramento Foundation, in* Small Water System Operation and Maintenance. *For additional information, visit <www. owp.csus.edu> or call 916-278-6142.*

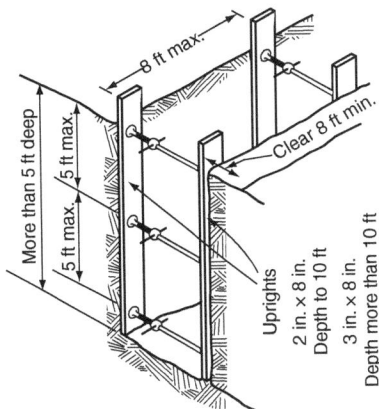

Trenches 5 ft or more deep and more than 8 ft long must be braced at intervals of 8 ft or less.

Hard Compact Ground (5 ft or more in depth)

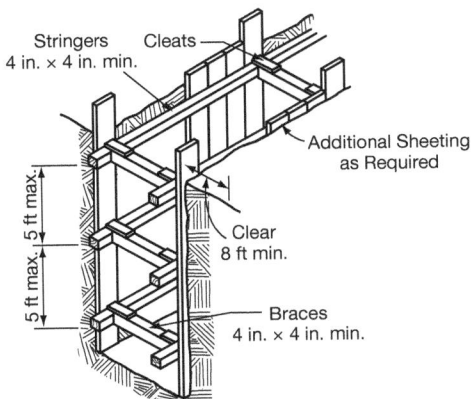

Sheeting must be provided and must be sufficient to hold the material in place.

Longitudinal-stringer dimensions depend on the strut and stringer spacing and on the degree of instability encountered.

Saturated, Filled, or Unstable Ground (additional sheeting as required)

67

ROADWAY, TRAFFIC, AND VEHICLE SAFETY[*] _____

Recommended Barricade Placement for Working in a Roadway

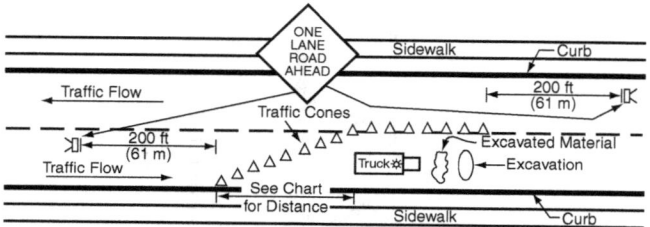

NOTE: If traffic is heavy or construction work causes interference in the open lane, one or more flaggers should be used.

Speed Limit, mph (km/hr)	Lane Width,						Minimum Number of Cones Required
	10 ft (3 m)		11 ft (3.4 m)		12 ft (3.7 m)		
	Taper Length,						
	ft	(m)	ft	(m)	ft	(m)	
20 (32)	70	(21)	75	(23)	80	(24)	5
25 (40)	105	(32)	115	(35)	125	(38)	6
30 (48)	150	(46)	165	(50)	180	(55)	7
35 (56)	205	(62)	225	(69)	245	(75)	8
40 (64)	270	(82)	295	(90)	320	(98)	9
45 (72)	450	(137)	495	(151)	540	(165)	13
50 (81)	500	(152)	550	(168)	600	(183)	13
55 (89)	550	(168)	605	(184)	660	(201)	13

*This section adapted from Office of Water Programs, California State University, Sacramento Foundation, in Small Water System Operation and Maintenance. For additional information, visit <www.owp.csus.edu> or call 916-278-6142.

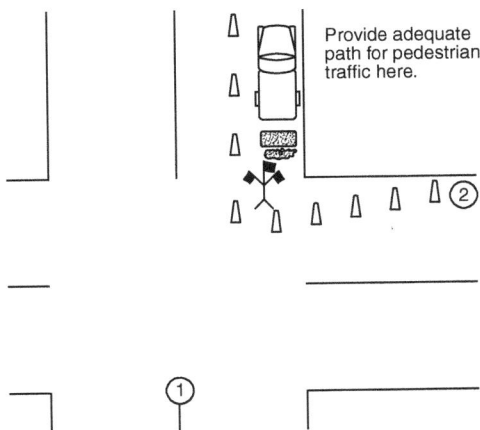

Provide adequate path for pedestrian traffic here.

Placement near intersection. Some locations may require high-level warnings at points 1 and 2.

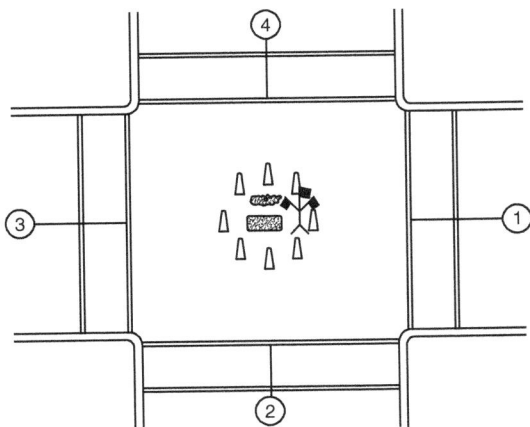

Placement at major traffic signal–controlled intersection where congestion is extreme. Some locations may permit warnings at points 1, 2, 3, and 4.

Placement of Traffic Cones and Signs

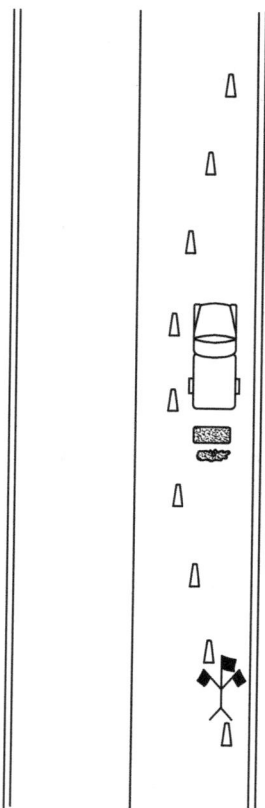

Placement for normal service, leak, or construction. See table on page 68 for distances.

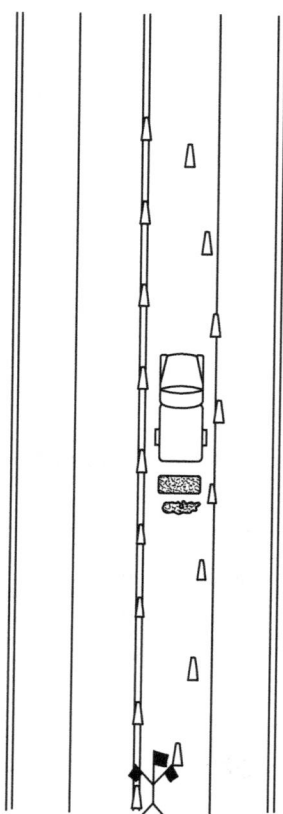

Placement for multilane highway. Place high-level warning in same lane as obstruction. See table on page 68 for distances.

Placement of Traffic Cones and Signs (continued)

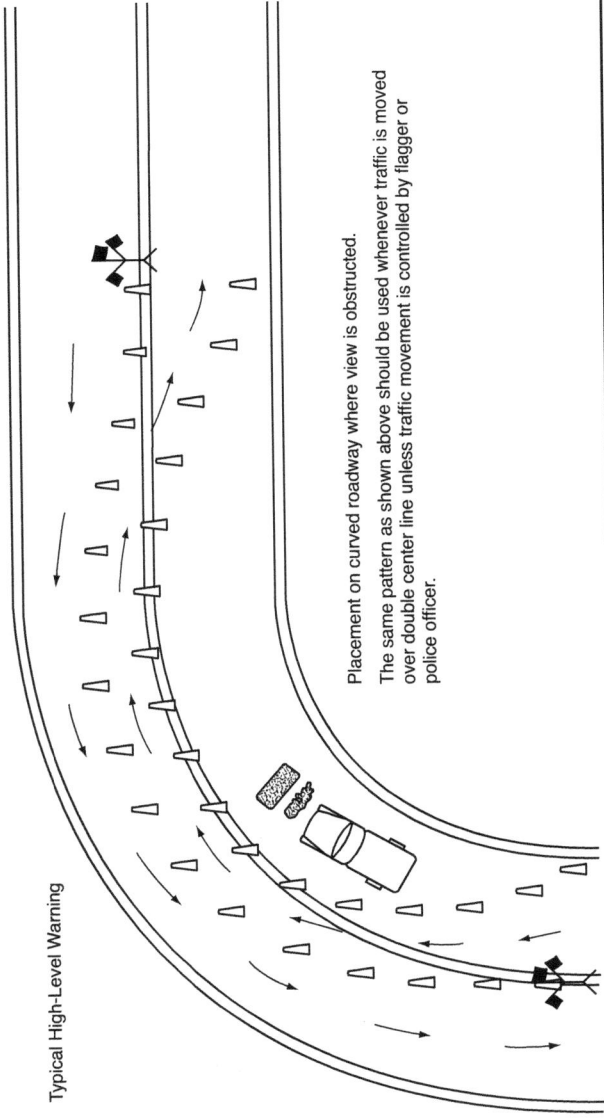

Typical High-Level Warning

Placement on curved roadway where view is obstructed.

The same pattern as shown above should be used whenever traffic is moved over double center line unless traffic movement is controlled by flagger or police officer.

Placement of Traffic Cones and Signs (continued)

71

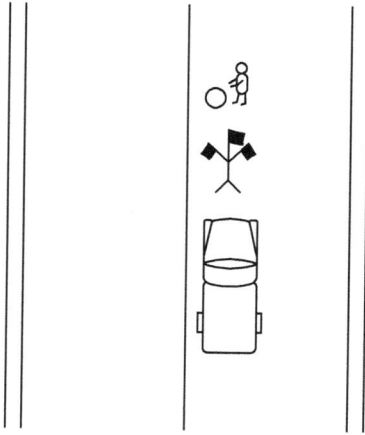

Placement for gate operation or other jobs of short duration.
Employee must wear high-visibility vest or jacket.

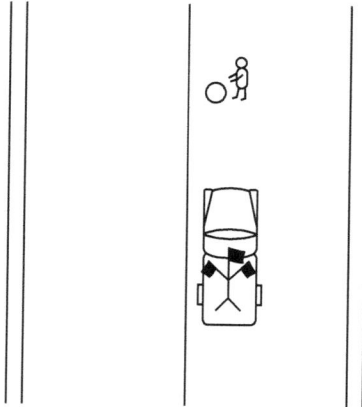

Alternate placement for operation described at left.
High-level warning is mounted on rear of vehicle that is
parked in advance of work location. Employee must wear
high-visibility vest or jacket.

Placement of Traffic Cones and Signs (continued)

Road Work Ahead

150 ft min.

Work Area

Work Space

High-Level Warning Device

Single Lane Ahead

Road Work Ahead

100 ft min.

150 ft min.

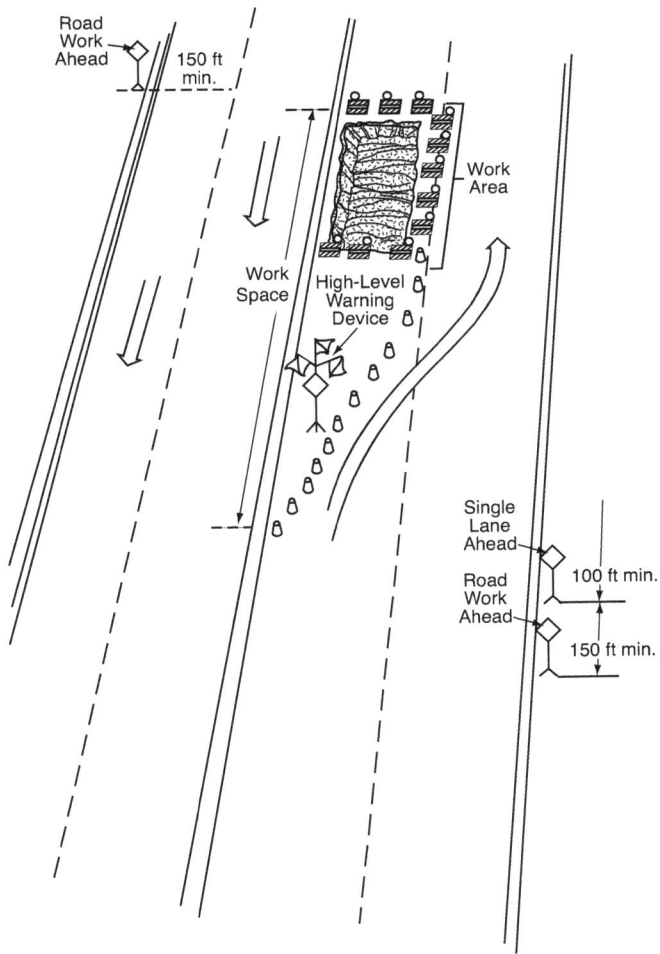

Closing of Left Lane

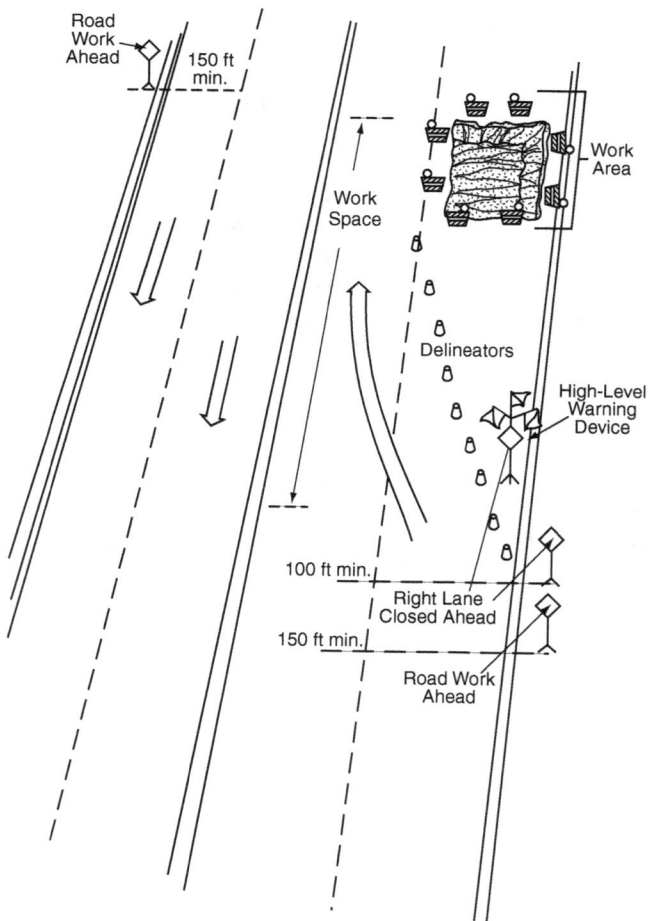

Road Work Ahead

150 ft min.

Work Space

Work Area

Delineators

High-Level Warning Device

100 ft min.

Right Lane Closed Ahead

150 ft min.

Road Work Ahead

Closing of Right Lane

1. Truck and spoil bank placed ahead of excavation for employee protection.

2. Cone pattern arranged with gentle curves—traffic adjusts smoothly.

3. Pipe blocked to prevent rolling into street. Barricades warn pedestrians.

4. Material is neatly stacked.

5. High-level warning or barricades of solid material to give audible warning of vehicles entering work area.

6. Pedestrian bridge over excavation.

7. Left side of truck protected by cone pattern; work area entirely outlined.

8. Tools out of way of pedestrians; tools not in use replaced in truck.

9. Pickup parked in work area or on street away from work area.

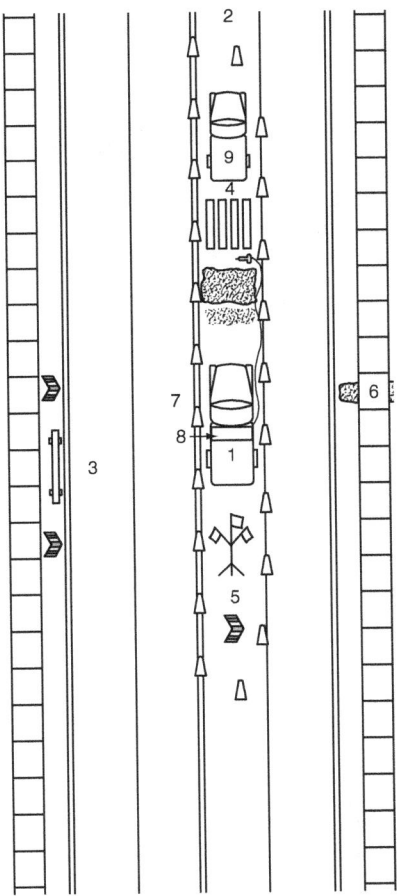

Good Practices in Work Area Protection

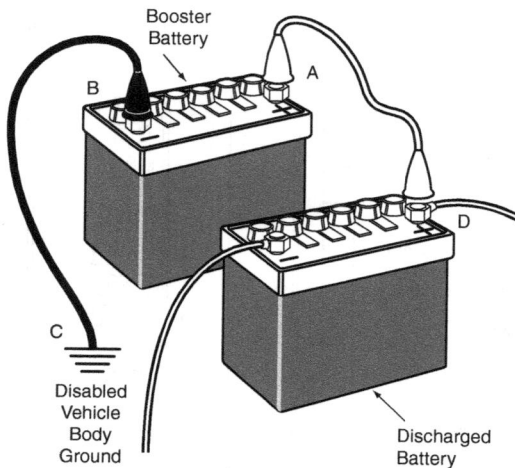

Booster
Battery

B

A

C

D

Disabled
Vehicle
Body
Ground

Discharged
Battery

Proper Booster Cable Hookup

To boost the battery of a disabled vehicle from that of another vehicle, follow this procedure.

For maximum eye safety, wear protective goggles around vehicle batteries to keep flying battery fragments and chemicals out of the eyes. Should battery acid get into the eyes, immediately flush them with water continuously for 15 minutes, then see a doctor.

First, extinguish all cigarettes and flames. A spark can ignite hydrogen gas from the battery fluid. Next, take off the battery caps, if removable, and add distilled water if it is needed. Check for ice in the battery fluid. Never jump-start a frozen battery! Replace the caps.

Next, park the vehicle with the "live" battery close enough so the cables will reach between the batteries of the two vehicles. The vehicles can be parked close, but do not allow them to touch. If they touch, this can create a dangerous situation. Now set each vehicle's parking brake. Be sure that an automatic transmission is set in park; put a manual-shift transmission in neutral. Make sure your headlights, heater, and all other electrical accessories are off (you don't want to sap electricity away from the discharged [dead] battery while you're trying to start the vehicle). If the two batteries have vent caps, remove them. Then lay a cloth over the open holes. This will reduce the risk of explosion (relieves pressure within the battery).

Attach one end of the jumper cable to the positive terminal of the booster battery (A) and the other end to the positive terminal of the discharged battery (D). The positive terminal is identified by a + sign, a red color, or a "P" on the battery in each vehicle. Each of the two booster cables has an alligator clip at each end. To attach, simply squeeze the clip, place it over the terminal, and let it shut. Now attach one end of the remaining booster cable to the negative terminal of the booster battery (B). The negative terminal is marked with a – sign, a black color, or the letter "N." Attach the other end of the cable to a metal part on the engine of the disabled vehicle (C). Many mechanics simply attach it to the negative post of the battery, but this is not recommended because a resulting arc could ignite hydrogen gas present at the battery surface and cause an explosion. Be sure that the cables do not interfere with the fan blades or belts. The engine in the booster vehicle should be running, although it is not an absolute necessity.

Get in the disabled vehicle and start the engine. After it starts, remove the booster cables. Removal is the exact reverse of attachment. Remove the black cable attached to the previously disabled vehicle, then remove it from the negative terminal of the booster battery. Next, remove the remaining cable from the positive terminal of the dead battery and then from the booster vehicle. Replace the vent caps and you're done. Have the battery and/or charging system of the vehicle checked by a mechanic to correct any problems.

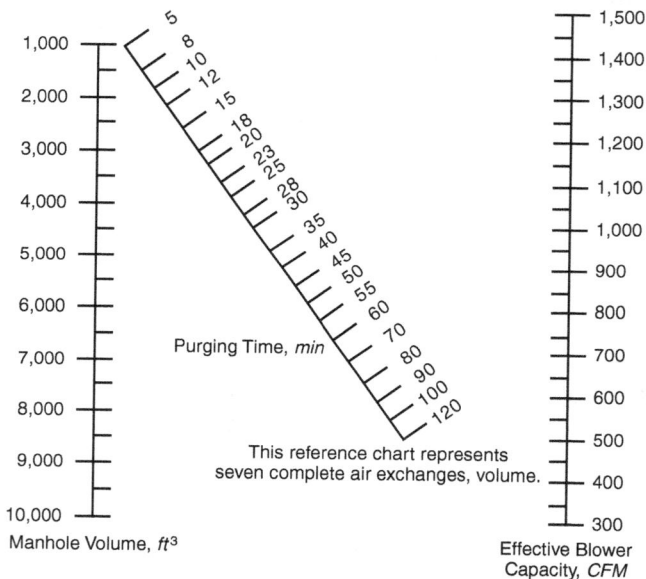

Manhole Volume, ft^3

Purging Time, *min*

This reference chart represents
seven complete air exchanges, volume.

Effective Blower
Capacity, *CFM*

Use of alignment chart:
1. Place straightedge on manhole volume (left scale).
2. Place either end of straightedge on blower capacity (right scale).
3. Read required purging time, in minutes, on diagonal scale.
4. If two blowers are used, add the two capacities, then proceed as above.
5. When basic gases are encountered, increase purging time by 50%.
6. Effective blower capacity is measured with one or two 90° bends in
   standard 15-ft blower hose.

Ventilation Nomograph

Hazardous Location Information

A hazardous location is an area where the possibility of explosion and fire is created by the presence of flammable gases, vapors, dusts, fibers, or flyings. (Fibers and flyings are not likely to be suspended in the air but can collect around machinery or on lighting fixtures where heat, a spark, or hot metal can ignite them.)

Class I

[NEC-500.5 (B)]

Those areas in which flammable gases or vapors may be present in the air in sufficient quantities to be explosive or ignitable.

Class II

[NEC-500.5 (C)]

Those areas made hazardous by the presence of combustible dust.

Class III

[NEC-500.5 (D)]

Those areas in which there are easily ignitable fibers or flyings present due to type of material being handled, stored, or processed.

Division I

(NEC-800-5, 6, 7)

In the normal situation, hazard would be expected to be present in everyday production operations or during frequent repair and maintenance activity.

Division II

(NEC-500-5, 6, 7)

In the abnormal situation, material is expected to be confined within closed containers or closed systems and will be present only through accidental rupture, breakage, or unusual faulty operation.

Groups

(NEC-500-3)

The gases and vapors of class locations are broken into four groups by the code: A, B, C, and D. These materials are grouped according to the ignition temperature of the substance, its explosion pressure, and other flammable characteristics.

The dust locations of Class II are designated E, F, and G. These groups are classified according to the ignition temperature and the conduction of the hazardous substance.

NOTE: For detailed group descriptions, refer to NEC-500-3.

Table continued on next page

Safety

79

Hazardous Location Information (continued)

Typical Class I Locations

- Petroleum refineries, and gasoline storage and dispensing areas.
- Industrial firms that use flammable liquids in dip tanks for parts cleaning or other operations.
- Petrochemical companies that manufacture chemicals from gas and oil.
- Dry-cleaning plants where vapors from cleaning fluids can be present.
- Companies that have spraying areas where they coat product with paint or plastics.
- Aircraft hangars and fuel servicing areas.
- Utility gas plants, and operations involving storage and handling of liquefied petroleum gas or natural gas.

Typical Class II Locations

- Grain elevators, flour and feed mills.
- Plants that manufacture, use, or store magnesium or aluminum powders.
- Plants that have chemical or metallurgical processes; producers of plastics, medicines, and fireworks.
- Producers of starch or candies.
- Spice-grinding plants, sugar plants, and cocoa plants.
- Coal preparation plants and other carbon-handling or processing areas.

Typical Class III Locations

- Textile mills, cotton gins, cotton seed mills, and flax processing plants.
- Any plant that shapes, pulverizes, or cuts wood and creates sawdust or flyings.

Source: Explosion Proof Blowers: 9503 and 9515-01 NEC. (Warning: Explosion-proof blowers must be used with statically conductive ducting.)

Types of Fires and Fire Extinguishers

Combustible Material	Class of Fire and Extinguisher Marking	Extinguish With
Paper, wood, cloth	A (ordinary combustibles)	Water, soda-acid, and dry chemical rated A, B, C
Oil, tar, gasoline, paint	B (flammable liquids)	Foam, carbon dioxide, liquid gas (Halon™), and dry chemical rated B, C, or A, B, C
Electric motors, power cords, wiring, and transformer boxes	C (electrical equipment)	Carbon dioxide, liquid gas (Halon™), and dry chemical rated B, C, or A, B, C
Sodium, zinc phosphorus, magnesium, potassium, and titanium, especially as dust or turnings	D (special metals)	Only special dry-powder extinguishers marked for this purpose

Hazard Classification

Class 1   Explosives

Class 2   Gas

Class 3   Flammable liquid

Class 4   Flammable solids (potential spontaneous combustion, or emission of flammable gases when in contact with water)

Class 5   Oxidizing substances and organic peroxides

Class 6   Toxic (poisonous) and infectious substances

Class 7   Radioactive material

Class 8   Corrosives

Class 9   Miscellaneous dangerous goods

Common Dangerous Gases Encountered in Water Supply Systems and

Name of Gas	Chemical Formulae	Specific Gravity of Vapor Density* (Air = 1)	Explosive Range (% by Volume in Air)	
			Lower Limit	Upper Limit
Carbon dioxide	CO_2	1.53	Not flammable.	Not flammable.
Carbon monoxide	CO	0.97	12.5	74.2
Chlorine	Cl_2	2.5	Not flammable. Not explosive.	Not flammable. Not explosive.
Ethane	C_2H_4	1.05	3.1	15.0
Gasoline vapor	C_5H_{12} to C_9H_{20}	3.0 to 4.0	1.3	7.0
Hydrogen	H_2	0.07	4.0	74.2

at Water Treatment Plants

Common Properties (percentages below are percent in air by volume)	Physiological Effects (percentages below are percent in air by volume)	Most Common Sources in Sewers	Simplest and Cheapest Safe Method of Testing[†]
Colorless, odorless, nonflammable. Not generally present in dangerous amounts unless there is already an oxygen deficiency.	10% cannot be endured for more than a few minutes. Acts on nerves of respiration.	Issues from carbona-ceous strata. Sewer gas.	Oxygen deficiency indicator.
Colorless, odorless, nonirritating, tasteless. Flammable. Explosive.	Hemoglobin of blood has strong affinity for gas causing oxygen starvation. 0.2% to 0.25% causes unconsciousness in 30 minutes.	Manufactured fuel gas.	CO ampoules.
Greenish yellow gas, or amber color liquid under pressure. Highly irritating and penetrating odor. Highly corrosive in presence of moisture.	Respiratory irritant, irritating to eyes and mucous membranes. 30 ppm causes coughing. 40–60 ppm dangerous in 30 minutes. 1,000 ppm likely to be fatal in a few breaths.	Leaking pipe connections. Overdosage.	Chlorine detector. Odor, strong. Ammonia on swab gives off white fumes.
Colorless, tasteless, odorless, nonpoison-ous. Flammable. Explosive.	See hydrogen.	Natural gas.	Combustible gas indicator.
Colorless. Odor noticeable in 0.03%. Flammable. Explosive.	Anesthetic effects when inhaled. 2.43% rapidly fatal. 1.1% to 2.2% dangerous for even short exposure.	Leaking storage tanks, discharges from garages, and commercial or home dry-cleaning operations.	1. Combustible gas indicator. 2. Oxygen deficiency indicator for concentrations >30%.
Colorless, odorless, tasteless, nonpoi-sonous. Flammable. Explosive. Propagates flame rapidly; very dangerous.	Acts mechanically to deprive tissues of oxygen. Does not support life. A simple asphyxiant.	Manufactured fuel gas.	Combustible gas indicator.

Table continued on next page

83

Common Dangerous Gases Encountered in Water Supply Systems and

Name of Gas	Chemical Formulae	Specific Gravity of Vapor Density* (Air = 1)	Explosive Range (% by volume in air)	
			Lower Limit	Upper Limit
Hydrogen sulfide	H_2S	1.19	4.3	46.0
Methane	CH_4	0.55	5.0	15.0
Nitrogen	N_2	0.97	Not flammable.	Not flammable.
Oxygen (In air)	O_2	1.11	Not flammable.	Not flammable.

*Gases with a specific gravity less than 1.0 are lighter than air; those with a specific gravity more than 1.0 are heavier than air.

† The first method given is the preferable testing procedure.

‡ Never enter a 12% atmosphere. Use detection meters with alarm warning devices rather than meters.

at Water Treatment Plants (continued)

Common Properties (percentages below are percent in air by volume)	Physiological Effects (percentages below are percent in air by volume)	Most Common Sources in Sewers	Simplest and Cheapest Safe Method of Testing[†]
Rotten egg odor in small concentrations but sense of smell rapidly impaired. Odor not evident at high concentrations. Colorless. Flammable. Explosive. Poisonous.	Death in a few minutes at 0.2%. Paralyzes respiratory center.	Petroleum fumes, from blasting, sewer gas.	1. H_2S analyzer. 2. H_2S ampoules.
Colorless, tasteless, odorless, nonpoisonous. Flammable. Explosive.	See hydrogen.	Natural gas, marsh gas, manufacturing fuel gas, sewer gas.	1. Combustible gas indicator. 2. Oxygen deficiency indicator.
Colorless, tasteless, odorless. Nonflammable. Nonpoisonous. Principal constituent of air (about 79%).	See hydrogen.	Issues from some rock strata. Sewer gas.	Oxygen deficiency indicator.
Colorless, odorless, tasteless, nonpoisonous gas. Supports combustion.	Normal air contains 20.93% of O_2. Humans tolerate down to 12%.[‡] Below 5% to 7%, likely to be fatal.	Oxygen depletion from poor ventilation and absorption or chemical consumption of available O_2.	Oxygen deficiency indicator.

HEALTH EFFECTS OF TOXIN EXPOSURE

Potential Effects of Hydrogen Sulfide Exposure

ppm	Effects and Symptoms	Time
1,000 or more	Unconsciousness, death	Minutes
500–700	Unconsciousness, death	½–1 hour
200–300	Marked eye and respiratory irritations	1 hour
50–100	Mild eye and respiratory irritations	1 hour
10	Permissible exposure level	8 hours

Although the foul odor (rotten eggs) of hydrogen sulfide is easily detected at low concentrations, it is an unreliable warning because the gas rapidly desensitizes the sense of smell, leading to a false sense of security. In high concentrations of hydrogen sulfide, a worker may collapse with little or no warning.

Potential Effects of Carbon Monoxide Exposure

ppm	Effects and Symptoms	Time
4,000	Fatal	<1 hour
2,000–2,500	Unconsciousness	30 minutes
1,000–2,000	Slight heart palpitation	30 minutes
1,000–2,000	Tendency to stagger	1½ hours
1,000–2,000	Confusion, headache, nausea	2 hours
600	Headache, discomfort	1 hour
400	Headache, discomfort	2 hours
200	Slight headache, discomfort	3 hours
50	Permissible exposure limit	8 hours

Carbon monoxide is an odorless, colorless gas that may build up in a confined space. In high concentrations of carbon monoxide a worker may collapse with little or no warning.

Effects of Chlorine Gas Exposure

ppm	Effects and Symptoms
1	Slight symptoms after several hours' exposure
3	Detectable odor
4	60-minute inhalation without serious effects
5	Noxiousness
15	Throat irritation
30	Coughing
40	Dangerous from ½–1 hour
1,000	Death after a few deep breaths

Water Quality

Because water quality analysis and monitoring is an important part of the operation of every public water system, water operators should understand how drinking water regulations are administered and why compliance is essential. Regulations on drinking water quality set the treatment goals for the water supply industry, ensuring that safe and aesthetically pleasing drinking water is consistently supplied to the public.

STATE PRIMACY

The intent of the Safe Drinking Water Act (SDWA) is for each state to accept primary enforcement responsibility (primacy) for the operation of the state's drinking water program. Under the provisions of the delegation, the state must establish requirements for public water systems that are at least as stringent as those set by USEPA. The primacy agency in each state has been designated by the state governor. In some states the primacy agency is the state health department, and in others it is the state environmental protection agency, department of natural resources, or pollution control agency.

PUBLIC WATER SYSTEMS

The basic definition of a public water system in the SDWA is, in essence, a system that supplies piped water for human consumption and that has at least 15 service connections or serves 25 or more persons for 60 or more days of the year. Examples of water systems that would not fall under the federal definition are private homes, groups of fewer than 15 homes using the same well, and summer camps that operate for fewer than 60 days per year. These systems are, however, generally under some degree of supervision by a local, area, or state health department.

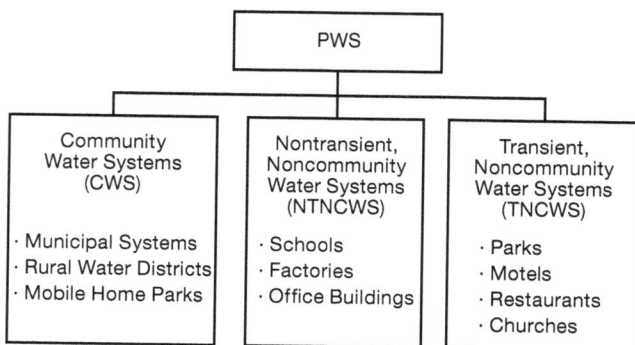

```
                         ┌─────────────┐
                         │     PWS     │
                         └──────┬──────┘
        ┌───────────────────────┼───────────────────────┐
┌───────────────┐      ┌────────────────┐      ┌────────────────┐
│  Community    │      │ Nontransient,  │      │  Transient,    │
│ Water Systems │      │ Noncommunity   │      │ Noncommunity   │
│    (CWS)      │      │ Water Systems  │      │ Water Systems  │
│               │      │   (NTNCWS)     │      │    (TNCWS)     │
│               │      │                │      │                │
│ · Municipal   │      │ · Schools      │      │ · Parks        │
│   Systems     │      │ · Factories    │      │ · Motels       │
│ · Rural Water │      │ · Office       │      │ · Restaurants  │
│   Districts   │      │   Buildings    │      │ · Churches     │
│ · Mobile Home │      │                │      │                │
│   Parks       │      │                │      │                │
└───────────────┘      └────────────────┘      └────────────────┘
```

Classification of Public Water Systems

List of Contaminants and Their Maximum Contaminant Levels

Contaminant	MCLG,* mg/L†	MCL or TT,* mg/L†	Potential Health Effects From Ingestion of Water	Sources of Contaminant in Drinking Water
Microorganisms				
Cryptosporidium	0	TT‡	Gastrointestinal illness (e.g., diarrhea, vomiting, cramps)	Human and animal fecal waste
Giardia lamblia	0	TT	Gastrointestinal illness (e.g., diarrhea, vomiting, cramps)	Human and animal fecal waste
Heterotrophic plate count (HPC)	NA	TT	HPC has no health effects; it is an analytic method used to measure the variety of bacteria that are common in water. The lower the concentration of bacteria in drinking water, the better maintained the water system.	HPC measures a range of bacteria that are naturally present in the environment
Legionella	0	TT	Legionnaires' disease, a type of pneumonia	Found naturally in water; multiplies in heating systems
Total coliforms (including fecal coliform and *Escherichia coli*)	0	5.0%§	Not a health threat in itself; it is used to indicate whether other potentially harmful bacteria may be present.**	Coliforms are naturally present in the environment, as well as feces; fecal coliforms and *E. coli* come only from human and animal fecal waste.
Turbidity	NA	TT	Turbidity is a measure of the cloudiness of water. It is used to indicate water quality and filtration effectiveness (e.g., whether disease-causing organisms are present). Higher turbidity levels are often associated with higher levels of disease-causing microorganisms such as viruses, parasites, and some bacteria. These organisms can cause symptoms such as nausea, cramps, diarrhea, and headaches.	Soil runoff

Table continued on next page

List of Contaminants and Their Maximum Contaminant Levels (continued)

Contaminant	MCLG,* mg/L†	MCL or TT,* mg/L†	Potential Health Effects From Ingestion of Water	Sources of Contaminant in Drinking Water
Viruses (enteric)	0	TT	Gastrointestinal illness (e.g., diarrhea, vomiting, cramps)	Human and animal fecal waste
Disinfection by-products				
Bromate	0	0.010	Increased risk of cancer	By-product of drinking water disinfection
Chlorite	0.8	1.0	Anemia; infants & young children, nervous system effects	By-product of drinking water disinfection
Haloacetic acids (HAA5)	NA††	0.060	Increased risk of cancer	By-product of drinking water disinfection
Total trihalomethanes	None‡‡ NA	0.080 0.080	Liver, kidney, or central nervous system problems; increased risk of cancer	By-product of drinking water disinfection
Disinfectants				
Chloramines (as Cl_2)	MRDLG = 4	MRDL = 4.0	Eye/nose irritation; stomach discomfort; anemia	Water additive used to control microbes
Chlorine (as Cl_2)	MRDLG = 4	MRDL = 4.0	Eye/nose irritation; stomach discomfort	Water additive used to control microbes
Chlorine dioxide (as ClO_2)	MRDLG = 0.8	MRDL = 0.8	Anemia; infants & young children, nervous system effects	Water additive used to control microbes
Inorganic Chemicals				
Antimony	0.006	0.006	Increase in blood cholesterol; decrease in blood sugar	Discharge from petroleum refineries; fire retardants; ceramics; electronics; solder
Arsenic	0	0.010 as of 1/23/06	Skin damage or problems with circulatory systems; possible increased risk of cancer	Erosion of natural deposits; runoff from orchards; runoff from glass and electronics production wastes

Table continued on next page

List of Contaminants and Their Maximum Contaminant Levels (continued)

Contaminant	MCLG,* mg/L†	MCL or TT,* mg/L†	Potential Health Effects From Ingestion of Water	Sources of Contaminant in Drinking Water
Asbestos (fiber >10 μm)	7 million fibers per liter (MFL)	7 MFL	Increased risk of benign intestinal polyps	Decay of asbestos cement in water mains; erosion of natural deposits
Barium	2	2	Increase in blood pressure	Discharge of drilling wastes; discharge from metal refineries; erosion of natural deposits
Beryllium	0.004	0.004	Intestinal lesions	Discharge from metal refineries and coal-burning factories; discharge from electrical, aerospace, and defense industries
Cadmium	0.005	0.005	Kidney damage	Corrosion of galvanized pipes; erosion of natural deposits; discharge from metal refineries; runoff from waste batteries and paints
Chromium (total)	0.1	0.1	Allergic dermatitis	Discharge from steel and pulp mills; erosion of natural deposits
Copper	1.3	TT§§; action level = 1.3	Short-term exposure—gastrointestinal distress. Long-term exposure—liver or kidney damage. People with Wilson's disease should consult their personal physician if the amount of copper in their water exceeds the action level.	Corrosion of household plumbing systems; erosion of natural deposits
Cyanide (as free cyanide)	0.2	0.2	Nerve damage or thyroid problems	Discharge from steel/metal factories; discharge from plastics and fertilizer factories

Table continued on next page

Water Quality

List of Contaminants and Their Maximum Contaminant Levels (continued)

Contaminant	MCLG,* mg/L†	MCL or TT,* mg/L†	Potential Health Effects From Ingestion of Water	Sources of Contaminant in Drinking Water
Fluoride	4.0	4.0	Bone disease (pain and tenderness of the bones); children may get mottled teeth	Water additive that promotes strong teeth; erosion of natural deposits; discharge from fertilizer and aluminum factories
Lead	0	TT; action level = 0.015	Infants and children—delays in physical or mental development; children could show slight deficits in attention span and learning abilities. Adults—kidney problems; high blood pressure.	Corrosion of household plumbing systems; erosion of natural deposits
Mercury (inorganic)	0.002	0.002	Kidney damage	Erosion of natural deposits; discharge from refineries and factories; runoff from landfills and croplands
Nitrate (measured as nitrogen)	10	10	Infants below the age of 6 months who drink water containing nitrate in excess of the MCL could become seriously ill and, if untreated, may die. Symptoms include shortness of breath and blue-baby syndrome.	Runoff from fertilizer use; leaching from septic tanks, sewage; erosion of natural deposits
Nitrite (measured as nitrogen)	1	1	Infants below the age of 6 months who drink water containing nitrite in excess of the MCL could become seriously ill and, if untreated, may die. Symptoms include shortness of breath and blue-baby syndrome.	Runoff from fertilizer use; leaching from septic tanks, sewage; erosion of natural deposits
Selenium	0.05	0.05	Hair or fingernail loss; numbness in fingers or toes; circulatory problems	Discharge from petroleum refineries; erosion of natural deposits; discharge from mines

Table continued on next page

List of Contaminants and Their Maximum Contaminant Levels (continued)

Contaminant	MCLG,* mg/Lt	MCL or TT,* mg/Lt	Potential Health Effects From Ingestion of Water	Sources of Contaminant in Drinking Water
Thallium	0.0005	0.002	Hair loss; changes in blood; kidney, intestine, or liver problems	Leaching from ore-processing sites; discharge from electronics, glass, and drug factories
Organic Chemicals				
Acrylamide	0	TT***	Nervous system or blood problems; increased risk of cancer	Added to water during sewage/wastewater treatment
Alachlor	0	0.002	Eye, liver, kidney, or spleen problems; anemia; increased risk of cancer	Runoff from herbicide used on row crops
Atrazine	0.003	0.003	Cardiovascular system or reproductive problems	Runoff from herbicide used on row crops
Benzene	0	0.005	Anemia; decrease in blood platelets; increased risk of cancer	Discharge from factories; leaching from gas storage tanks and landfills
Benzo(a)pyrene (poly-aromatic hydrocarbons)	0	0.0002	Reproductive difficulties; increased risk of cancer	Leaching from linings of water storage tanks and distribution lines
Carbofuran	0.04	0.04	Problems with blood, nervous system, or reproductive system	Leaching of soil fumigant used on rice and alfalfa
Carbon tetrachloride	0	0.005	Liver problems; increased risk of cancer	Discharge from chemical plants and other industrial activities
Chlordane	0	0.002	Liver or nervous system problems; increased risk of cancer	Residue of banned termiticide
Chlorobenzene	0.1	0.1	Liver or kidney problems	Discharge from chemical and agricultural chemical factories

Table continued on next page

Water Quality

95

List of Contaminants and Their Maximum Contaminant Levels (continued)

Contaminant	MCLG,* mg/L†	MCL or TT,* mg/L†	Potential Health Effects From Ingestion of Water	Sources of Contaminant in Drinking Water
2,4-D	0.07	0.07	Kidney, liver, or adrenal gland problems	Runoff from herbicide used on row crops
Dalapon	0.2	0.2	Minor kidney changes	Runoff from herbicide used on rights of way
1,2-Dibromo-3-chloropropane	0	0.0002	Reproductive difficulties; increased risk of cancer	Runoff/leaching from soil fumigant used on soybeans, cotton, pineapples, and orchards
o-Dichlorobenzene	0.6	0.6	Liver, kidney, or circulatory system problems	Discharge from industrial chemical factories
p-Dichlorobenzene	0.075	0.075	Anemia; liver, kidney, or spleen damage; changes in blood	Discharge from industrial chemical factories
1,2-Dichloroethane	0	0.005	Increased risk of cancer	Discharge from industrial chemical factories
1,1-Dichloroethylene	0.007	0.007	Liver problems	Discharge from industrial chemical factories
cis-1,2-Dichloroethylene	0.07	0.07	Liver problems	Discharge from industrial chemical factories
trans-1,2-Dichloroethylene	0.1	0.1	Liver problems	Discharge from industrial chemical factories
Dichloromethane	0	0.005	Liver problems; increased risk of cancer	Discharge from drug and chemical factories
1,2-Dichloropropane	0	0.005	Increased risk of cancer	Discharge from industrial chemical factories
Di(2-ethylhexyl) adipate	0.4	0.4	Weight loss, liver problems, or possible reproductive difficulties	Discharge from chemical factories
Di(2-ethylhexyl) phthalate	0	0.006	Reproductive difficulties; liver problems; increased risk of cancer	Discharge from rubber and chemical factories

Table continued on next page

List of Contaminants and Their Maximum Contaminant Levels (continued)

Contaminant	MCLG,* mg/L†	MCL or TT,* mg/L†	Potential Health Effects From Ingestion of Water	Sources of Contaminant in Drinking Water
Dinoseb	0.007	0.007	Reproductive difficulties	Runoff from herbicide used on soybeans and vegetables
Dioxin (2,3,7,8-TCDD)	0	0.00000003	Reproductive difficulties; increased risk of cancer	Emissions from waste incineration and other combustion; discharge from chemical factories
Diquat	0.02	0.02	Cataracts	Runoff from herbicide use
Endothall	0.1	0.1	Stomach and intestinal problems	Runoff from herbicide use
Endrin	0.002	0.002	Liver problems	Residue of banned insecticide
Epichlorohydrin	0	TT	Increased cancer risk; over a long period of time, stomach problems	Discharge from industrial chemical factories; an impurity of some water treatment chemicals
Ethylbenzene	0.7	0.7	Liver or kidney problems	Discharge from petroleum refineries
Ethylene dibromide	0	0.00005	Problems with liver, stomach, reproductive system, or kidneys; increased risk of cancer	Discharge from petroleum refineries
Glyphosate	0.7	0.7	Kidney problems; reproductive difficulties	Runoff from herbicide use
Heptachlor	0	0.0004	Liver damage; increased risk of cancer	Residue of banned termiticide
Heptachlor epoxide	0	0.0002	Liver damage; increased risk of cancer	Breakdown of heptachlor
Hexachlorobenzene	0	0.001	Liver or kidney problems; reproductive difficulties; increased risk of cancer	Discharge from metal refineries and agricultural chemical factories

Table continued on next page

Water Quality

List of Contaminants and Their Maximum Contaminant Levels (continued)

Contaminant	MCLG,* mg/L†	MCL or TT,* mg/L†	Potential Health Effects From Ingestion of Water	Sources of Contaminant in Drinking Water
Hexachlorocyclopentadiene	0.05	0.05	Kidney or stomach problems	Discharge from chemical factories
Lindane	0.0002	0.0002	Liver or kidney problems	Runoff/leaching from insecticide used on cattle, lumber, gardens
Methoxychlor	0.04	0.04	Reproductive difficulties	Runoff/leaching from insecticide used on fruits, vegetables, alfalfa, livestock
Oxamyl (Vydate)	0.2	0.2	Slight nervous system effects	Runoff/leaching from insecticide used on apples, potatoes, and tomatoes
Pentachlorophenol	0	0.001	Liver or kidney problems; increased cancer risk	Discharge from wood-preserving factories
Picloram	0.5	0.5	Liver problems	Herbicide runoff
Polychlorinated biphenyls	0	0.0005	Skin changes; thymus gland problems; immune deficiencies; reproductive or nervous system difficulties; increased risk of cancer	Runoff from landfills; discharge of waste chemicals
Simazine	0.004	0.004	Problems with blood	Herbicide runoff
Styrene	0.1	0.1	Liver, kidney, or circulatory system problems	Discharge from rubber and plastics factories; leaching from landfills
Tetrachloroethylene	0	0.005	Liver problems; increased risk of cancer	Discharge from factories and dry cleaners
Toluene	1	1	Nervous system, kidney, or liver problems	Discharge from petroleum factories
Toxaphene	0	0.003	Kidney, liver, or thyroid problems; increased risk of cancer	Runoff/leaching from insecticide used on cotton and cattle

Table continued on next page

List of Contaminants and Their Maximum Contaminant Levels (continued)

Contaminant	MCLG,* mg/L†	MCL or TT,* mg/L†	Potential Health Effects From Ingestion of Water	Sources of Contaminant in Drinking Water
2,4,5-TP (Silvex)	0.05	0.05	Liver problems	Residue of banned herbicide
1,2,4-Trichlorobenzene	0.07	0.07	Changes in adrenal glands	Discharge from textile finishing factories
1,1,1-Trichloroethane	0.2	0.2	Liver, nervous system, or circulatory problems	Discharge from metal degreasing sites and other factories
1,1,2-Trichloroethane	0.003	0.005	Liver, kidney, or immune system problems	Discharge from industrial chemical factories
Trichloroethylene	0	0.005	Liver problems; increased risk of cancer	Discharge from metal degreasing sites and other factories
Vinyl chloride	0	0.002	Increased risk of cancer	Leaching from PVC pipes; discharge from plastics factories
Xylenes (total)	10	10	Nervous system damage	Discharge from petroleum factories; discharge from chemical factories
Radionuclides				
Alpha particles	none — 0	15 pCi/L	Increased risk of cancer	Erosion of natural deposits of certain minerals that are radioactive and may emit a form of radiation known as alpha radiation
Beta particles and photon emitters	none — 0	4 mrem/yr	Increased risk of cancer	Decay of natural and synthetic deposits of certain minerals that are radioactive and may emit forms of radiation known as photons and beta radiation

Table continued on next page

Water Quality

List of Contaminants and Their Maximum Contaminant Levels (continued)

Contaminant	MCLG,* mg/L†	MCL or TT,* mg/L†	Potential Health Effects From Ingestion of Water	Sources of Contaminant in Drinking Water
Radium 226 and radium 228 (combined)	None 0	5 pCi/L	Increased risk of cancer	Erosion of natural deposits
Uranium	0	30 µg/L as of 12/8/03	Increased risk of cancer, kidney toxicity	Erosion of natural deposits

* Maximum contaminant level (MCL)—The highest level of a contaminant that is allowed in drinking water. MCLs are set as close to MCLGs as feasible using the best available treatment technology and taking cost into consideration. MCLs are enforceable standards.

Maximum contaminant level goal (MCLG)—The level of a contaminant in drinking water below which there is no known or expected risk to health. MCLGs allow for a margin of safety and are nonenforceable public health goals.

Maximum residual disinfectant level (MRDL)—The highest level of a disinfectant allowed in drinking water. There is convincing evidence that addition of a disinfectant is necessary for control of microbial contaminants.

Maximum residual disinfectant level goal (MRDLG)—The level of a drinking water disinfectant below which there is no known or expected risk to health. MRDLGs do not reflect the benefits of the use of disinfectants to control microbial contaminants.

Treatment technique (TT)—A required process intended to reduce the level of a contaminant in drinking water.

† Units are in milligrams per liter (mg/L) unless otherwise noted. Milligrams per liter is equivalent to parts per million.

‡ USEPA's Surface Water Treatment Rules require systems using surface water or groundwater under the direct influence of surface water to (1) disinfect their water, and (2) filter their water or meet criteria for avoiding filtration so that the following contaminants are controlled at the following levels:

- *Cryptosporidium* (as of 1/1/02 for systems serving >10,000 and 1/14/05 for systems serving <10,000): 99% removal.
- *Giardia lamblia*: 99.9% removal/inactivation.
- Viruses: 99.99% removal/inactivation.
- *Legionella*: No limit, but USEPA believes that if *Giardia* and viruses are removed/inactivated, *Legionella* will also be controlled.

Table continued on next page

List of Contaminants and Their Maximum Contaminant Levels (continued)

- Turbidity: At no time can turbidity (cloudiness of water) go above 5 ntu; systems that filter must ensure that the turbidity goes no higher than 1 ntu (0.5 ntu for conventional or direct filtration) in at least 95% of the daily samples in any month. As of 1/1/02, turbidity may never exceed 1 ntu and must not exceed 0.3 ntu in 95% of daily samples in any month.

- HPC: No more than 500 bacterial colonies per milliliter.

- Long-Term 1 Enhanced Surface Water Treatment Rule (effective date: 1/14/05); surface water systems or GWUDI systems serving fewer than 10,000 people must comply with the applicable Long-Term 1 Enhanced Surface Water Treatment Rule provisions (e.g., turbidity standards, individual filter monitoring, *Cryptosporidium* removal requirements, updated watershed control requirements for unfiltered systems).

- Filter Backwash Recycling: The Filter Backwash Recycling Rule requires systems that recycle to return specific recycle flows through all processes of the system's existing conventional or direct filtration system or at an alternate location approved by the state.

§ More than 5.0% samples total coliform-positive in 1 month. (For water systems that collect fewer than 40 routine samples per month, no more than one sample can be total coliform-positive per month.) Every sample that has total coliform must be analyzed for either fecal coliforms or *E. coli*; if two consecutive total coliform-positive samples are found and one is also positive for *E. coli* fecal coliforms, the system has an acute MCL violation.

** Fecal coliform and *E. coli* are bacteria whose presence indicates that the water may be contaminated with human or animal wastes. Disease-causing microbes (pathogens) in these wastes can cause diarrhea, cramps, nausea, headaches, or other symptoms. These pathogens may pose a special health risk for infants, young children, and people with severely compromised immune systems.

†† Although there is no collective MCLG for this contaminant group, there are individual MCLGs for some of the individual contaminants:

- Trihalomethanes: bromodichloromethane (0); bromoform (0); dibromochloromethane (0.06 mg/L). Chloroform is regulated with this group but has no MCLG.
- Haloacetic acids: dichloroacetic acid (0); trichloroacetic acid (0.3 mg/L). Monochloroacetic acid, bromoacetic acid, and dibromoacetic acid are regulated with this group but have no MCLGs.

‡‡ MCLGs were not established before the 1986 amendments to the Safe Drinking Water Act. Therefore, there is no MCLG for this contaminant.

§§ Lead and copper are regulated by a treatment technique that requires systems to control the corrosiveness of their water. If more than 10% of tap water samples exceed the action level, water systems must take additional steps. The action level for copper is 1.3 mg/L, and lead is 0.015 mg/L.

*** Each water system must certify, in writing, to the state (using third-party or manufacturer's certification) that when acrylamide and epichlorohydrin are used in drinking water systems, the combination (or product) of dose and monomer level does not exceed the levels specified, as follows:

- Acrylamide = 0.05% dosed at 1 mg/L (or equivalent).
- Epichlorohydrin = 0.01% dosed at 20 mg/L (or equivalent).

National Secondary Drinking Water Regulations

Contaminant	Secondary Standard
Aluminum	0.05 to 0.2 mg/L
Chloride	250 mg/L
Color	15 cu
Copper	1.0 mg/L
Corrosivity	noncorrosive
Fluoride	2.0 mg/L
Foaming agents	0.5 mg/L
Iron	0.3 mg/L
Manganese	0.05 mg/L
Odor	3 threshold odor number
pH	6.5–8.5
Silver	0.10 mg/L
Sulfate	250 mg/L
Total dissolved solids	500 mg/L
Zinc	5 mg/L

Summary of Notification Requirements

Category of Violation	Mandatory Health Effects Information Required (all public water supplies)	Notice to New Billing Units (community water supplies only)
Tier 1		
Maximum contaminant level	Yes	Yes
Treatment technique	Yes	Yes
Variance of exemption schedule violation	Yes	Yes
Tier 2		
Monitoring*	No	No
Testing procedures	No	No
Variance of exemption issued	Yes	No

* Continuous reportt required if posting is used; quarterly report required if hand delivery is used.

Record-Keeping Requirements

Type of Record	Time Period
Bacteriological and turbidity analyses	5 years
Chemical analyses	10 years
Actions taken to correct violations	3 years
Sanitary survey reports	10 years
Exemptions	5 years following expiration

Guidelines on Water Turnover Rates

- Daily turnover goal equals 50% of storage facility volume; minimum desired turnover equals 30% of storage facility volume.
- Complete turnover recommended every 72 hours.
- Required daily turnover of 20%; recommended daily turnover of 25%.
- Maximum 5–7-day turnover.
- 50% reduction of water depth during a 24-hour cycle.
- Maximum 1–3-day turnover.

Summary of Special Sampling and Handling Requirements*

Determination	Container†	Minimum Sample Size, mL	Sample Type‡	Preservation§	Maximum Storage Recommended	Regulatory**
Acidity	P, G(B)	100	g	Refrigerate	24 hours	14 days
Alkalinity	P, G	200	g	Refrigerate	24 hours	14 days
Biochemical oxygen demand	P, G	1,000	g, c	Refrigerate	6 hours	48 hours
Boron	P (PTFE) or quartz	1,000	g, c	HNO_3 to pH <2	28 days	6 months
Bromide	P, G	100	g, c	None required	28 days	28 days
Carbon, organic, total	G(B)	100	g, c	Analyze immediately or refrigerate and add HCl, H_3PO_4, or H_2SO_4 to pH <2	7 days	28 days
Carbon dioxide	P, G	100	g	Analyze immediately	0.25 hour	N.S.††
Chemical oxygen demand	P, G	100	g, c	Analyze as soon as possible, or add H_2SO_4 to pH <2; refrigerate	7 days	28 days
Chloride	P, G	50	g, c	None required	N.S.††	28 days
Chlorine, total, residual	P, G	500	g	Analyze immediately	0.25 hour	0.25 hour
Chlorine dioxide	P, G	500	g	Analyze immediately	0.25 hour	N.S.††

Table continued on next page

104

Summary of Special Sampling and Handling Requirements* (continued)

Determination	Container†	Minimum Sample Size, mL	Sample Type‡	Preservation§	Maximum Storage Recommended	Regulatory**
Chlorophyll	P, G	500	g	Unfiltered, dark, 4°C	24–48 hours	
				Filtered, dark, −20°C (do not store in frost-free freezer)	28 days	
Color	P, G	500	g, c	Refrigerate	48 hours	48 hours
Specific conductance	P, G	500	g, c	Refrigerate	28 days	28 days
Cyanide						
Total	P, G	1,000	g, c	Add NaOH to pH >12, refrigerate in dark	24 hours	14 days; 24 hours if sulfide present
Amenable to chlorination	P, G	1,000	g, c	Add 0.6 g ascorbic acid if chlorine is present and refrigerate	stat‡‡	14 days; 24 hours if sulfide present
Fluoride	P	100	g, c	None required	28 days	28 days
Hardness	P, G	100	g, c	Add HNO_3 or H_2SO_4 to pH <2	6 months	6 months
Iodine	P, G	500	g	Analyze immediately	0.25 hour	N.S.††
Metals, general	P(A), G(A)	1,000	g, c	For dissolved metals filter immediately, add HNO_3 to pH <2	6 months	6 months
Chromium VI	P(A), G(A)	1,000	g	Refrigerate	24 hours	24 hours

Table continued on next page

Water Quality

105

Summary of Special Sampling and Handling Requirements* (continued)

Determination	Container†	Minimum Sample Size, mL	Sample Type‡	Preservation§	Maximum Storage Recommended	Regulatory**
Copper by colorimetry*			g, c			
Mercury	P(A), G(A)	1,000	g, c	Add HNO_3 to pH <2, 4°C, refrigerate	28 days	28 days
Nitrogen						
Ammonia	P, G	500	g, c	Analyze as soon as possible or add H_2SO_4 to pH <2, refrigerate	7 days	28 days
Nitrate	P, G	100	g, c	Analyze as soon as possible; refrigerate	48 hours	48 hours (28 days for chlorinated samples)
Nitrate + nitrite	P, G	200	g, c	Add H_2SO_4 to pH <2, refrigerate	1–2 days	28 days
Nitrite	P, G	100	g, c	Analyze as soon as possible; refrigerate	None	48 hours
Odor	G	500	g	Analyze as soon as possible; refrigerate	6 hours	N.S.††

Table continued on next page

Summary of Special Sampling and Handling Requirements* (continued)

Determination	Container†	Minimum Sample Size, mL	Sample Type‡	Preservation§	Maximum Storage Recommended	Regulatory**
Oil and grease	G, wide-mouth calibrated	1,000	g	Add HCl or H_2SO_4 to pH <2, refrigerate	28 days	28 days
Organic compounds						
Base/neutrals & acids	G(S) amber	1,000	g, c	Refrigerate	7 days	7 days until extraction; 40 days after extraction
Methylene blue active substances (MBAS)	P, G	250	g, c	Refrigerate	48 hours	N.S.††
Pesticides*	G(S), PTFE-lined cap	1,000	g, c	Refrigerate; add 1,000 mg ascorbic acid/L if residual chlorine present	7 days	7 days until extraction; 40 days after extraction
Phenols	P, G, PTFE-lined cap	500	g, c	Refrigerate; add H_2SO_4 to pH <2	*	28 days until extraction

Table continued on next page

Water Quality

Summary of Special Sampling and Handling Requirements* (continued)

Determination	Container†	Minimum Sample Size, mL	Sample Type‡	Preservation§	Maximum Storage Recommended	Regulatory**
Purgeables* by purge and trap	G, PTFE-lined cap	2 × 40	g	Refrigerate; add HCl to pH <2; add 1,000 mg ascorbic acid/L if residual chlorine present	7 days	14 days
Organic, Kjeldahl*	P, G	500	g, c	Refrigerate, add H_2SO_4 to pH <2	7 days	28 days
Oxygen, dissolved	G, BOD bottle	300	g			
Electrode				Analyze immediately	0.25 hour	0.25 hour
Winkler				Titration may be delayed after acidification	8 hours	8 hours
Ozone	G	1,000	g	Analyze immediately	0.25 hour	N.S.††
pH	P, G	50	g	Analyze immediately	0.25 hour	0.25 hour
Phosphate	G(A)	100	g	For dissolved phosphate filter, analyze immediately; refrigerate	48 hours	N.S.††
Phosphorus, total	P, G	100	g, c	Add H_2SO_4 to pH <2 and refrigerate	28 days	
Salinity	G, wax seal	240	g	Analyze immediately or use wax seal	6 months	N.S.††
Silica	P (PTFE) or quartz	200	g, c	Refrigerate; do not freeze	28 days	28 days
Sulfate	P, G	100	g, c	Refrigerate	28 days	28 days

Table continued on next page

Summary of Special Sampling and Handling Requirements* (continued)

Determination	Container[†]	Minimum Sample Size, mL	Sample Type[‡]	Preservation[§]	Maximum Storage Recommended	Regulatory**
Sludge digester gas	G, gas bottle	—	g	—	N.S.[††]	
Solids	P, G	200	g, c	Refrigerate	7 days	2–7 days
Sulfide	P, G	100	g, c	Refrigerate; add 4 drops 2N zinc acetate/100 mL; add NaOH to pH >9	28 days	7 days
Temperature	P, G	—	g	Analyze immediately	0.25 hour	0.25 hour
Turbidity	P, G	100	g, c	Analyze same day; store in dark up to 24 hours; refrigerate	24 hours	48 hours

Source: Standard Methods for the Examination of Water and Wastewater, *20th edition, 1998; APHA, AWWA, and WEF.*

* For determinations not listed, use glass or plastic containers; preferably refrigerate during storage and analyze as soon as possible.

† P = plastic (polyethylene [PTFE] or equivalent); G = glass; G(A) or P(A) = rinsed with 1 + 1 HNO_3; G(B) = glass, borosilicate; G(S) = glass, rinsed with organic solvents or baked.

‡ g = grab; c = composite.

§ Refrigerate = storage at 4°C ± 2°C, in the dark; analyze immediately = usually within 15 minutes of sample collection.

** See USEPA 40 CFR Part 100–149 for possible differences regarding container and preservation requirements.

†† N.S. = not stated.

‡‡ stat = no storage allowed; analyze immediately.

Water Quality

Bacteriological Sampling Guidelines

- Use only the bottles provided by the lab that are specifically for coliform sampling.

- Do not rinse sample bottles. Sample bottles contain a chemical that destroys any residual chlorine in the water. The residual chlorine would otherwise kill any bacteria in the sample, yielding an incorrect result.

- Keep sample bottles unopened until the moment of filling. The bottles are sterile.

- Make sure the faucet has no aerator and no swivel.

- Flush the faucet for 2 to 5 minutes to clear any stagnant water from the service line.

- Hold the bottle near the base; do not handle the stopper or cap and neck of the bottle.

- When flushing is complete, without changing the flow, gently fill the bottle without rinsing. Leave an air space at the top.

- Replace the cap or stopper immediately.

- Using a separate sample, test for free chlorine residual and record the result.

- Label the bottle, being sure to include the date and time the sample was taken, and package it for delivery to the lab.

Coliform Samples Required per Population Served

Population Served	Minimum Number of Samples per Month	Population Served	Minimum Number of Samples per Month
25 to 1,000*	1	59,001 to 70,000	70
1,001 to 2,500	2	70,001 to 83,000	80
2,501 to 3,300	3	83,001 to 96,000	90
3,301 to 4,100	4	96,001 to 130,000	100
4,101 to 4,900	5	130,001 to 220,000	120
4,901 to 5,800	6	220,001 to 320,000	150
5,801 to 6,700	7	320,001 to 450,000	180
6,701 to 7,600	8	450,001 to 600,000	210
7,601 to 8,500	9	600,001 to 780,000	240
8,501 to 12,900	10	780,001 to 970,000	270
12,901 to 17,200	15	970,001 to 1,230,000	300
17,201 to 21,500	20	1,230,001 to 1,520,000	330
21,501 to 25,000	25	1,520,001 to 1,850,000	360
25,001 to 33,000	30	1,850,001 to 2,270,000	390
33,001 to 41,000	40	2,270,001 to 3,020,000	420
41,001 to 50,000	50	3,020,001 to 3,960,000	450
50,001 to 59,000	60	More than 3,960,000	480

*Includes public water systems that have at least 15 service connections, but serve <25 people.

Water Quality

TYPICAL CUSTOMER COMPLAINTS AND CORRECTIVE ACTIONS

Inner wheel indicates categories.

Middle wheel indicates descriptors.

Outer wheel indicates reference standards.

*Presence confirmed in water.

†Distribution system has not been positively confirmed to be the source of these compounds.

Source: AwwaRF, Distribution Generated Taste-and-Odor Phenomena.

Distribution System Taste-and-Odor Wheel

Sensory Descriptions, Sources, and Possible Corrective Actions

Category	Possible Source	Possible Corrective Action
Sweet	Open reservoirs, biological activity	Cover reservoir. Breakpoint-chlorinate. Clean regularly.
Salty	Cross-connections	Eliminate cross-connections. Survey for potential cross-connections and install appropriate backflow-prevention devices.
	Treatment breakage at reverse osmosis membrane	
	Well contamination	
Sour/Acidic	Hot-water systems	Raise water temperature in water heaters. Flush water heaters annually. Replace sacrificial anodes in water heaters.
	Poorly maintained POU device	Replace POU device filters more regularly.
	Cross-connections	Eliminate cross-connections. Survey for potential cross-connections and install appropriate backflow-prevention devices.
Bitter	High pH from corrosion of cement lining	Fill, hold, and flush to waste newly installed or relined cement-lined pipelines left to stand. Periodically flush.
	Metal leaching	Stabilize water by adding polyphosphates or adjust pH.
	Cross-connections (e.g., backflow from carbonation)	Eliminate cross-connection. Survey for potential cross-connections and install appropriate backflow-prevention devices.
Mouthfeel/ Nosefeel	Galvanized or copper pipes, corrosion, rust, and other corrosion by-products	Add stabilizing agents such as polyphosphates, or adjust pH.
	New plumbing and water coolers	Fill, hold, and flush new plumbing.
	Cross-connections or backflow from carbonation	Eliminate cross-connection. Survey for potential cross-connections and install appropriate backflow-prevention devices.

Table continued on next page

Water Quality

113

Sensory Descriptions, Sources, and Possible Corrective Actions (continued)

Category	Possible Source	Possible Corrective Action
Earthy/Musty/Moldy	Open reservoirs, biological activity	Cover open treated-water reservoirs. Chlorinate open treated-water reservoirs. Install aeration systems.
	Dead ends or low-flow sections of system	Perform routine flushing of distribution system dead ends. Eliminate dead ends by looping mains.
	Loss of chlorine and unmasking of background odors	Maintain detectable residual at all times.
	Biomethylation of halogenated phenols	Treat with chlorine dioxide.
	Cross-connections	Eliminate cross-connections. Survey for potential cross-connections and install appropriate backflow-prevention devices.
Chlorinated	New main disinfection	For superchlorinated mains, dechlorinate and discharge.
	Residual boosting	Optimize weight ratio of chlorine to ammonia-nitrogen for chloraminated water. (Make sure the water is where it should be with respect to the breakpoint curve.)
	Blending	Eliminate breakpoint-chlorination occurrence in blended water. Reduce blend percentage or residuals of chlorinated water compared to chloraminated water.
	Continuing reactions	Chloraminate distribution system.
	Disinfection by-product (DBP) formation	Optimize treatment.

Table continued on next page

Sensory Descriptions, Sources, and Possible Corrective Actions (continued)

Category	Possible Source	Possible Corrective Action
Grassy/Hay/Straw/Woody	Open reservoirs, biological activity	Cover reservoirs. Treat reservoirs with chlorine. Utilize selective withdrawal from reservoirs. Blend with other sources.
Marshy/Swampy/Septic/Sulfurous	Poorly circulated reservoir	Aerate (or aerate hypolimnion of) source-water reservoir. Change withdrawal levels from reservoirs.
	Poorly maintained POU devices	Replace POU device filters more often.
	Poorly maintained hot-water systems	Flush hot-water systems annually.
	Poorly maintained sedimentation basins and/or dirty or stagnant filters in treatment plants	Clean sedimentation basins frequently. Backwash filters before returning to service.
	Dead-end mains	Loop or periodically flush dead-end mains.
	Oxidation of polysulfides	Breakpoint-chlorinate, followed by chloramination.
	Cross-connections or backflow	Eliminate cross-connections. Survey for potential cross-connections and install appropriate backflow-prevention devices.
Fragrant/Vegetable/Fruity/Flowery	Open reservoirs, biological activity	Aerate source-water reservoirs. Breakpoint-chlorinate, followed by chloramination of distribution system.
	Continuing reactions	Adjust chloramine-to-ammonia ratios.
Fishy	Open reservoirs, biological activity	Aerate source-water reservoirs. Preoxidate source waters before treatment.

Table continued on next page

Water Quality

115

Sensory Descriptions, Sources, and Possible Corrective Actions (continued)

Category	Possible Source	Possible Corrective Action
Medicinal/ Phenolic	Postchlorination of biofilms	Convert to chloramines. Maintain high chloramine residuals.
	Slow kinetics and DBP formation	Optimize treatment.
	Reservoir and tank linings, chlorination of phenolic linings	Use coatings approved by a recognized authority. Allow proper curing times on new coatings. Fill and hold, followed by flushing newly coated tanks and pipelines.
	Cross-connections or backflow	Eliminate cross-connections. Survey for potential cross-connections and install appropriate backflow-prevention devices.
Chemical/ Hydrocarbon/ Miscellaneous	Chlorine dioxide/carpet, volatilization in indoor air	Add ferrous salts or reduced sulfur compounds to remove chlorite at the treatment plant or substitute chloramines for chlorine in the distribution system.
	New plumbing and pipe materials	Use plumbing and fixtures approved by a recognized authority. Fill, hold, and flush plumbing.
	Plastic pipes, lubricants, new mains, asphaltic coatings, paint, epoxy/solvent-based linings	Use lubricants approved by a recognized authority. Allow proper cure times for new linings. Fill, hold, and flush newly lined mains.
	Permeation of buried plastic pipe	Replace plastic pipelines or remove contaminated soils surrounding pipelines.
	Old coal-tar linings on cast-iron mains	Gently flush to refresh water quality in mains. (Aggressive flushing can damage the coal tar that lines the mains.)

Source: AwwaRF, Distribution Generated Taste-and-Odor Phenomena.

Corrosion Properties of Materials Frequently Used in Water Distribution Systems

Plumbing Material	Corrosion Resistance	Primary Contaminants From Pipe
Asbestos cement, concrete, cement linings	Good corrosion resistance. Immune to electrolysis. Aggressive (soft) waters can leach calcium from cement; polyphosphate sequestering agents can deplete the calcium and substantially soften the pipe.	Asbestos fibers; increase in pH, aluminum, and calcium
Brass	Good overall resistance. Different types of brass respond differently to water chemistry; subject to dezincification by waters of pH >8.3 with high ratio of chloride to carbonate hardness. Conditions causing mechanical failure may not directly correspond to those promoting contaminant leaching.	Lead, copper, zinc
Cast or ductile iron (unlined)	Can be subject to surface erosion by aggressive waters and tuberculation in poorly buffered waters.	Iron, resulting in turbidity and red-water complaints
Copper	Good overall corrosion resistance; subject to corrosive attack from high flow velocities, soft water, chlorine, dissolved oxygen, low pH, and high inorganic carbon levels (alkalinities). May be prone to "pitting" failures.	Copper and possibly iron, zinc, tin, antimony, arsenic, cadmium, and lead from associated pipes and solder

Table continued on next page

Corrosion Properties of Materials Frequently Used in Water Distribution Systems (continued)

Plumbing Material	Corrosion Resistance	Primary Contaminants From Pipe
Galvanized iron or steel	Subject to galvanic corrosion of zinc by aggressive waters, especially of low hardness. Corrosion is accelerated by contact with copper materials; corrosion is accelerated at higher temperatures as in hot-water systems; corrosion is affected by the workmanship of the pipe and galvanized coating.	Zinc and iron; cadmium and lead (impurities in galvanizing process)
Lead	Corrodes in soft water with pH <8, and in hard waters with high inorganic carbon levels (alkalinities) and pH below ~7.5 or above ~8.5.	Lead
Mild steel	Subject to uniform corrosion. Affected primarily by high dissolved oxygen and chlorine levels, and poorly buffered waters.	Iron, resulting in turbidity and red-water complaints
Plastic	Resistant to corrosion.	Some pipes that contain metals in plasticizers, notably lead

Typical Customer Complaints Caused by Corrosion

Customer Complaint	Possible Cause
Red water, red or black particles, presence of reddish-brown staining on fixtures and laundry	Corrosion of iron pipes, old galvanized pipe, iron in source water
Bluish stains in sinks and tubs	Corrosion of copper lines
Black water lines	Sulfide corrosion of copper or iron
Foul taste or odors; fine, suspended bluish particles; orange, aqua, or black gelatinous deposits	By-products from microbial activity
Loss of pressure	Excessive scaling, tubercules building up, leak in system from pitting or other types of corrosion
Lack of hot water	Buildup of mineral deposits in hot-water heater system (can be reduced by setting thermostat to under 140°F or softening)
Short service life of household plumbing	Rapid deterioration of pipes from pitting and other types of corrosion
White or green-tinted fine particles in aerators and strainers	Deteriorated hot-water heater dip tube
White particles and cloudiness in ice cubes	Hard water

Hardness Classification Scale

Hardness Range, mg/L as $CaCO_3$	Hardness Description
0–75	Soft
75–150	Moderately hard
150–300	Hard
>300	Very hard

Alkalinity Relationships

Result of Titration	Hydroxide Alkalinity	Carbonate Alkalinity	Bicarbonate Alkalinity
$P^* = 0$	0	0	$MO^\dagger$
$P < \frac{1}{2} MO$	0	2P	$MO - 2P$
$P = \frac{1}{2} MO$	0	2P	0
$P > \frac{1}{2} MO$	$2P - MO$	$2(MO - P)$	0
$P - MO$	MO	0	0

*P = phenolphthalein.
† MO = methyl orange.

Leaking Faucets

Faucets With Threads

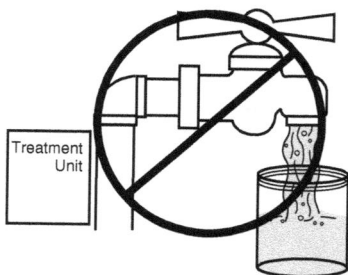

Faucets Connected to Home Treatment Units

Drinking Fountains

Types of Faucets That Should *Not* Be Used for Sampling

HEALTH EFFECTS

Adverse Effects of Secondary Contaminants

Contaminant	Suggested Levels	Adverse Effects
Aluminum	0.05–0.2 mg/L	Discoloration of water
Chloride	250 mg/L	Taste, corrosion of pipes
Color	15 color units	Aesthetic
Copper	1 mg/L	Taste, staining of porcelain
Corrosivity	Noncorrosive	Aesthetic and health related (corrosive water can leach lead from pipes into drinking water)
Fluoride	2.0 mg/L	Brownish discoloration of teeth
Foaming agents	0.5 mg/L	Aesthetic
Iron	0.3 mg/L	Taste, staining of laundry
Manganese	0.05 mg/L	Taste, staining of laundry
Odor	3 (threshold odor number)	Aesthetic
pH	6.5–8.5	Water is too corrosive
Silver	0.1 mg/L	Discoloration of the skin (argyria)
Sulfate	250 mg/L	Taste, laxative effects
Total dissolved solids (TDS)	500 mg/L	Taste and possible relation between low hardness and cardiovascular disease, also an indicator of corrosivity (related to lead levels in water), can damage plumbing and limit effectiveness of detergents
Zinc	5 mg/L	Taste

NOTE: Copper and fluoride appear on both Primary and Secondary Standards lists. The effects of each contaminant at the lower levels found on the Secondary list are aesthetic only. At higher concentrations, each can cause adverse health reactions and are therefore listed as Primary Standards. "Aesthetic" refers to effects of contaminants that may make water look, taste, or smell unpleasant, yet are not necessarily harmful to health.

Potential Waterborne Disease-Causing Organisms

Organism	Major Disease	Primary Source
	Bacteria	
Salmonella typhi	Typhoid fever	Human feces
Salmonella paratyphi	Paratyphoid fever	Human feces
Other *Salmonella* spp.	Gastroenteritis (salmonellosis)	Human/animal feces
Shigella	Bacillary dysentery	Human feces
Vibrio cholerae	Cholera	Human feces, coastal water
Pathogenic *Escherichia coli*	Gastroenteritis	Human/animal feces
Yersinia enterocolitica	Gastroenteritis	Human/animal feces
Campylobacter jejuni	Gastroenteritis	Human/animal feces
Legionella pneumophila	Legionnaires' disease, Pontiac fever	Warm water
Mycobacterium avium intracellulare	Pulmonary disease	Human/animal feces, soil, water
Pseudomonas aeruginosa	Dermatitis	Natural waters
Aeromonas hydrophila	Gastroenteritis	Natural waters
Helicobacter pylori	Peptic ulcers	Saliva, human feces suspected
	Enteric Viruses	
Poliovirus	Poliomyelitis	Human feces
Coxsackievirus	Upper respiratory disease	Human feces

Table continued on next page

Potential Waterborne Disease-Causing Organisms (continued)

Organism	Major Disease	Primary Source
Echovirus	Upper respiratory disease	Human feces
Rotavirus	Gastroenteritis	Human feces
Norwalk virus and other calciviruses	Gastroenteritis	Human feces
Hepatitis A virus	Infectious hepatitis	Human feces
Hepatitis E virus	Hepatitis	Human feces
Astrovirus	Gastroenteritis	Human feces
Enteric adenoviruses	Gastroenteritis	Human feces
Protozoa and Other Organisms		
Giardia lamblia	Giardiasis (gastroenteritis)	Human and animal feces
Cryptosporidium parvum	Cryptosporidiosis (gastroenteritis)	Human and animal feces
Entamoeba histolytica	Amoebic dysentery	Human feces
Cyclospora cayatanensis	Gastroenteritis	Human feces
Microspora	Gastroenteritis	Human feces
Acanthamoeba	Eye infection	Soil and water
Toxoplasma gondii	Flu-like symptoms	Cats
Naegleria fowleri	Primary amoebic meningoencephalitis	Soil and water
Blue-green algae	Gastroenteritis, liver damage, nervous system damage	Natural waters
Fungi	Respiratory allergies	Air, water suspected

Waterborne Diseases

Waterborne Disease	Causative Organism	Source of Organism in Water	Symptom/Outcome
Gastroenteritis	*Salmonella* (bacteria)	Animal or human feces	Acute diarrhea and vomiting—rarely fatal
Typhoid	*Salmonella typhosa* (bacteria)	Human feces	Inflamed intestine, enlarged spleen, high temperature—fatal
Dysentery	*Shigella*	Human feces	Diarrhea—rarely fatal
Cholera	*Vibrio cholerae* (bacteria)	Human feces	Vomiting, severe diarrhea, rapid dehydration, mineral loss—high mortality
Infectious hepatitis	Virus	Human feces, shellfish grown in polluted waters	Yellowed skin, enlarged liver, abdominal pain; lasts as long as 4 months—low mortality
Amoebic dysentery	*Entamoeba histolytica* (protozoa)	Human feces	Mild diarrhea, chronic dysentery—rarely fatal
Giardiasis	*Giardia lamblia* (protozoa)	Wild animal feces suspected	Diarrhea, cramps, nausea, general weakness; lasts 1 week to 30 weeks—not fatal
Cryptosporidiosis	*Cryptosporidium*	Human and animal feces	Diarrhea, abdominal pain, vomiting, low-grade fever—rarely fatal

Water Treatment

The types and concentrations of contaminants found in groundwater and surface water— including organic and inorganic substances, radionuclides, and disease-causing organisms— determine the treatment necessary to produce safe drinking water and to comply with federal standards. Over the years, conventional water treatment processes have been continually improved, and newer technologies, such as membrane filtration and advanced oxidation, have been put in place and are working well.

KEY FORMULAS FOR WATER TREATMENT _____

Jar Testing

$$\text{dosage, mg/L} = \frac{\text{stock, mL} \times 1,000 \text{ mg/g} \times \text{concentration, g/L}}{\text{sample size, mL}}$$

$$\text{g/L} = \frac{\text{mg/L} \times 1,000 \text{ mL}}{\text{mL} \times 1,000 \text{ mL}}$$

alum reacting, mg/L =

$$\frac{1.0 \text{ mg/L alum} \times \text{raw alkalinity, mg/L} - \text{alkalinity present, mg/L}}{0.45 \text{ mg/L alkalinity}}$$

alkalinity dosage, mg/L = total alum, mg/L – alum reacting, mg/L

$$\text{dilute solution, mg/L} = \frac{\text{mg of alum dosage} \times 1,000 \text{ mL/L}}{1.0 \text{ mL/L}}$$

$$g = \frac{\text{mg} \times 1.0 \text{ g}}{1,000 \text{ mg/L}}$$

$$\text{mg/L} = \frac{\text{g} \times 1,000 \text{ mg/L}}{1.0 \text{ g}}$$

Sedimentation Tanks and Clarifiers

circumference, ft = 3.141 (π) × diameter, ft

$$\text{solids loading rate, lb/day/ft}^2 = \frac{\text{solids into clarifier, lb/day}}{\text{surface area, ft}^2}$$

sludge solids, lb = flow, gal × 8.34 lb/gal × sludge, %

raw sludge pumping, gpm =

$$\frac{\text{settlable solids, mL/L} \times \text{plant flow, gpm}}{1,000 \text{ mL/L}}$$

sludge volume index, mg/L =

$$\frac{\text{settled sludge volume, mL/L} \times 1,000 \text{ mg/g}}{\text{suspended matter, mg/L}}$$

$$mg/L = \frac{mL \times 1,000,000}{mL \text{ sample}}$$

Hydraulic Cross-Check Formulas

$$\text{surface loading rate, gpd/ft}^2 = \frac{\text{total flow, gpd}}{\text{surface area, ft}^2}$$

design data: $800 - 1,200$ gpd/ft^2

$$\text{detention time, hr} = \frac{\text{volume/gal} \times 24 \text{ hr/day}}{\text{total 24-hr flow, gpd}}$$

design data: 1–4 hr; average 2.5 hr

$$\text{flow, gpd} = \frac{\text{volume, gal} \times 24 \text{ hr/day}}{\text{detention time, hr}}$$

$$\text{weir overflow rate, gpd/length ft} = \frac{\text{flow, gpd}}{\text{weir length, ft}}$$

design data: 10,000–40,000 gpd/length ft;
average 20,000 gpd/length ft

Filtration

$$\text{filtration rate, gpm} = \text{filter area, ft}^2 \times \text{gpm/ft}^2$$

$$\text{filtration rate, gpm/ft}^2 = \frac{\text{flow rate, gpm}}{\text{filter area, ft}^2}$$

$$\text{filtration rate, gpd} = \text{filter area, ft}^2 \times \text{gpm/ft}^2 \times 1,440 \text{ min/day}$$

$$\text{filter backwash rate} = \frac{\text{flow, gpm}}{\text{filter area, ft}^2}$$

$$\text{filter backwash rate} = \frac{\text{inches of water rise}}{\text{min}}$$

$$\text{backwash pumping rate, gpm} =$$
$$\text{filter area, ft}^2 \times \text{backwash rate, gpm/ft}^2$$

backwash volume, gal =
filter area, ft^2 × backwash rate, gpm/ft^2 × time, min

$$\text{backwash rate, gpm/ft}^2 = \frac{\text{backwash volume, gpm}}{\text{filter area, ft}^2}$$

backwash, gpm =
filter area, ft^2 × height, rise/fall/drop, ft min × 7.48 gal/ft^3

rate of rise, gpm/ft^2 = height, rise/fall/drop, ft min × 7.48 gal/ft^3

rate of rise, gpm/ft^2 = time, min × height, ft × 7.48 gal/ft^3

Ion Exchange

calcium hardness as mg/L $CaCO_3$ = 2.5 × calcium, mg/L

magnesium hardness as mg/L $CaCO_3$ = 4.1 × magnesium, mg/L

total hardness = calcium + magnesium hardness as $CaCO_3$

$$\text{gpg} = \frac{\text{total hardness, mg/L}}{17.1 \text{ mg/L/gr}}$$

total exchange capacity, kilograins =
resin capacity, $kilograins/ft^3$ × vol., ft^3

total gr capacity = kilograins × 1,000

gal of soft water per service run =

$$\frac{\text{total exchange capacity} \times \text{kilograins} \times 1,000}{\text{total hardness as CaCO}_3\text{, gpg}}$$

$$\text{bypass water, gpd} = \frac{\text{flow, gpd} \times \text{effluent hardness, gpg}}{\text{influent hardness, gpg}}$$

$$\text{bypass water, \%} = \frac{\text{discharge hardness, mg/L}}{\text{intitial hardness, mg/L}} \times 100$$

$$\text{salt, lb} = \frac{\text{capacity, gr} \times \text{salt, lb}}{1,000 \text{ gr}}$$

$$\text{brine, gal} = \frac{\text{salt needed, lb}}{\text{salt, lb/gal}}$$

$$\text{hardness removed, gr} =$$

$$\frac{\text{influent hardness, mg/L} - \text{effluent hardness, mg/L}}{17.1 \text{ mg/L/gr}}$$

$$\text{percent of soft water bypass} =$$

$$\frac{\text{blended discharge hardness, mg/L}}{\text{initial hardness, mg/L}} \times 100$$

$$\text{gpm bypass} = \frac{\text{percent bypass}}{100} \times \text{total flow, gpm}$$

$$\text{total flow through softener, gpm} =$$
$$\text{total flow, gpm} - \text{bypass flow, gpm}$$

Lime–Soda Ash Softening

$$\text{lb hardness removed} =$$
$$\text{mgd} \times \text{dosage, mg/L} \times \frac{\text{soda ash} - \text{mol. wt.}}{\text{calcium carbonate mol. wt.}} \times 8.34 \text{ lb/gal}$$

Disinfection—*CT* Calculation

$$CT = \text{disinfectant residual concentration, mg/L} \times \text{time, min}$$

NOTE: Contact your state health department for current CT tables.

CT VALUES FOR VARIOUS TYPES OF CONTAMINANTS

CT Values for Inactivation of *Giardia* Cysts by Free Chlorine at 0.5°C or Lower

Chlorine Concentration, mg/L	pH ≤ 6 Log Inactivation						pH = 6.5 Log Inactivation						pH = 7.0 Log Inactivation						pH = 7.5 Log Inactivation					
	0.5	1.0	1.5	2.0	2.5	3.0	0.5	1.0	1.5	2.0	2.5	3.0	0.5	1.0	1.5	2.0	2.5	3.0	0.5	1.0	1.5	2.0	2.5	3.0
≤ 0.4	23	46	69	91	114	137	27	54	82	109	136	163	33	65	98	130	163	195	40	79	119	158	198	237
0.6	24	47	71	94	118	141	28	56	84	112	140	168	33	67	100	133	167	200	40	80	120	159	199	239
0.8	24	48	73	97	121	145	29	57	86	115	143	172	34	68	103	137	171	205	41	82	123	164	205	246
1	25	49	74	99	123	148	29	59	88	117	147	176	35	70	105	140	175	210	42	84	127	169	211	253
1.2	25	51	76	101	127	152	30	60	90	120	150	180	36	72	108	143	179	215	43	86	130	173	216	259
1.4	26	52	78	103	129	155	31	61	92	123	153	184	37	74	111	147	184	221	44	89	133	177	222	266
1.6	26	52	79	105	131	157	32	63	95	126	158	189	38	75	113	151	188	226	46	91	137	182	228	273
1.8	27	54	81	108	135	162	32	64	97	129	161	193	39	77	116	154	193	231	47	93	140	186	233	279
2	28	55	83	110	138	165	33	66	99	131	164	197	39	79	118	157	197	236	48	95	143	191	238	286
2.2	28	56	85	113	141	169	34	67	101	134	168	201	40	81	121	161	202	242	50	99	149	198	248	297
2.4	29	57	86	115	143	172	34	68	103	137	171	205	41	82	124	165	206	247	50	99	149	199	248	298
2.6	29	58	88	117	146	175	35	70	105	139	174	209	42	84	126	168	210	252	51	101	152	203	253	304
2.8	30	59	89	119	148	178	36	71	107	142	178	213	43	86	129	171	214	257	52	103	155	207	258	310
3	30	60	91	121	151	181	36	72	109	145	181	217	44	87	131	174	218	261	53	105	158	211	263	316

Table continued on next page

CT Values for Inactivation of *Giardia* Cysts by Free Chlorine at 0.5°C or Lower (continued)

Chlorine Concentration, mg/L	pH = 8.0 Log Inactivation						pH = 8.5 Log Inactivation						pH = 9.0 Log Inactivation					
	0.5	1.0	1.5	2.0	2.5	3.0	0.5	1.0	1.5	2.0	2.5	3.0	0.5	1.0	1.5	2.0	2.5	3.0
≤0.4	46	92	139	185	231	277	55	110	165	219	274	329	65	130	195	260	325	390
0.6	48	95	143	191	238	286	57	114	171	228	285	342	68	136	204	271	339	407
0.8	49	98	148	197	246	295	59	118	177	236	295	354	70	141	211	281	352	422
1	51	101	152	203	253	304	61	122	183	243	304	365	73	146	219	291	364	437
1.2	52	104	157	209	261	313	63	125	188	251	313	376	75	150	226	301	376	451
1.4	54	107	161	214	268	321	65	129	194	258	323	387	77	155	232	309	387	464
1.6	55	110	165	219	274	329	66	132	199	265	331	397	80	159	239	318	398	477
1.8	56	113	169	225	282	338	68	136	204	271	339	407	82	163	245	326	408	489
2	58	115	173	231	288	346	70	139	209	278	348	417	83	167	250	333	417	500
2.2	59	118	177	235	294	353	71	142	213	284	355	426	85	170	256	341	426	511
2.4	60	120	181	241	301	361	73	145	218	290	363	435	87	174	261	348	435	522
2.6	61	123	184	245	307	368	74	148	222	296	370	444	89	178	267	355	444	533
2.8	63	125	188	250	313	375	75	151	226	301	377	452	91	181	272	362	453	543
3	64	127	191	255	318	382	77	153	230	307	383	460	92	184	276	368	460	552

Source: AWWA (1991).
CT = disinfectant residual concentration, mg/L × time, min.
Note: Contact your state health department to verify applicability.

CT Values for Inactivation of *Giardia* Cysts by Free Chlorine at 5°C

Chlorine Concentration, mg/L	pH ≤ 6 Log Inactivation						pH = 6.5 Log Inactivation						pH = 7.0 Log Inactivation						pH = 7.5 Log Inactivation					
	0.5	1.0	1.5	2.0	2.5	3.0	0.5	1.0	1.5	2.0	2.5	3.0	0.5	1.0	1.5	2.0	2.5	3.0	0.5	1.0	1.5	2.0	2.5	3.0
≤ 0.4	16	32	49	65	81	97	20	39	59	78	98	117	23	46	70	93	116	139	28	55	83	111	138	166
0.6	17	33	50	67	83	100	20	40	60	80	100	120	24	48	72	95	119	143	29	57	86	114	143	171
0.8	17	34	52	69	86	103	20	41	61	81	102	122	24	49	73	97	122	146	29	58	88	117	146	175
1	18	35	53	70	88	105	21	42	63	83	104	125	25	50	75	99	124	149	30	60	90	119	149	179
1.2	18	36	54	71	89	107	21	42	64	85	106	127	25	51	76	101	127	152	31	61	92	122	153	183
1.4	18	36	55	73	91	109	22	43	65	87	108	130	26	52	78	103	129	155	31	62	94	125	156	187
1.6	19	37	56	74	93	111	22	44	66	88	110	132	26	53	79	105	132	158	32	64	96	128	160	192
1.8	19	38	57	76	95	114	23	45	68	90	113	135	27	54	81	108	135	162	33	65	98	131	163	196
2	19	39	58	77	97	116	23	46	69	92	115	138	28	55	83	110	138	165	33	67	100	133	167	200
2.2	20	39	59	79	98	118	23	47	70	93	117	140	28	56	85	113	141	169	34	68	102	136	170	204
2.4	20	40	60	80	100	120	24	48	72	95	119	143	29	57	86	115	143	172	35	70	105	139	174	209
2.6	20	41	61	81	102	122	24	49	73	97	122	146	29	58	88	117	146	175	36	71	107	142	178	213
2.8	21	41	62	83	103	124	25	49	74	99	123	148	30	59	89	119	148	178	36	72	109	145	181	217
3	21	42	63	84	105	126	25	50	76	101	126	151	30	61	91	121	152	182	37	74	111	147	184	221

Table continued on next page

CT Values for Inactivation of *Giardia* Cysts by Free Chlorine at 5°C (continued)

Chlorine Concentration, mg/L	pH = 8.0 Log Inactivation						pH = 8.5 Log Inactivation						pH = 9.0 Log Inactivation					
	0.5	1.0	1.5	2.0	2.5	3.0	0.5	1.0	1.5	2.0	2.5	3.0	0.5	1.0	1.5	2.0	2.5	3.0
≤0.4	33	66	99	132	165	198	39	79	118	157	197	236	47	93	140	186	233	279
0.6	34	68	102	136	170	204	41	81	122	163	203	244	49	97	146	194	243	291
0.8	35	70	105	140	175	210	42	84	126	168	210	252	50	100	151	201	251	301
1	36	72	108	144	180	216	43	87	130	173	217	260	52	104	156	208	260	312
1.2	37	74	111	147	184	221	45	89	134	178	223	267	53	107	160	213	267	320
1.4	38	76	114	151	189	227	46	91	137	183	228	274	55	110	165	219	274	329
1.6	39	77	116	155	193	232	47	94	141	187	234	281	56	112	169	225	281	337
1.8	40	79	119	159	198	238	48	96	144	191	239	287	58	115	173	230	288	345
2	41	81	122	162	203	243	49	98	147	196	245	294	59	118	177	235	294	353
2.2	41	83	124	165	207	248	50	100	150	200	250	300	60	120	181	241	301	361
2.4	42	84	127	169	211	253	51	102	153	204	255	306	61	123	184	245	307	368
2.6	43	86	129	172	215	258	52	104	156	208	260	312	63	125	188	250	313	375
2.8	44	88	132	175	219	263	53	106	159	212	265	318	64	127	191	255	318	382
3	45	89	134	179	223	268	54	108	162	216	270	324	65	130	195	259	324	389

Source: AWWA (1991).

CT = disinfectant residual concentration, mg/L × time, min.

NOTE: Contact your state health department to verify applicability.

CT Values for Inactivation of *Giardia* Cysts by Free Chlorine at 10°C

Chlorine Concentration, mg/L	pH ≤ 6 Log Inactivation						pH = 6.5 Log Inactivation						pH = 7.0 Log Inactivation						pH = 7.5 Log Inactivation					
	0.5	1.0	1.5	2.0	2.5	3.0	0.5	1.0	1.5	2.0	2.5	3.0	0.5	1.0	1.5	2.0	2.5	3.0	0.5	1.0	1.5	2.0	2.5	3.0
≤ 0.4	12	24	37	49	61	73	15	29	44	59	73	88	17	35	52	69	87	104	21	42	63	83	104	125
0.6	13	25	38	50	63	75	15	30	45	60	75	90	18	36	54	71	89	107	21	43	64	85	107	128
0.8	13	26	39	52	65	78	15	31	46	61	77	92	18	37	55	73	92	110	22	44	66	87	109	131
1	13	26	40	53	66	79	16	31	47	63	78	94	19	37	56	75	93	112	22	45	67	89	112	134
1.2	13	27	40	53	67	80	16	32	48	63	79	95	19	38	57	76	95	114	23	46	69	91	114	137
1.4	14	27	41	55	68	82	16	33	49	65	82	98	19	39	58	77	97	116	23	47	70	93	117	140
1.6	14	28	42	55	69	83	17	33	50	66	83	99	20	40	60	79	99	119	24	48	72	96	120	144
1.8	14	29	43	57	72	86	17	34	51	67	84	101	20	41	61	81	102	122	25	49	74	98	123	147
2	15	29	44	58	73	87	17	35	52	69	87	104	21	41	62	83	103	124	25	50	75	100	125	150
2.2	15	30	45	59	74	89	18	35	53	70	88	105	21	42	64	85	106	127	26	51	77	102	128	153
2.4	15	30	45	60	75	90	18	36	54	71	89	107	22	43	65	86	108	129	26	52	79	105	131	157
2.6	15	31	46	61	77	92	18	37	55	73	92	110	22	44	66	87	109	131	27	53	80	107	133	160
2.8	16	31	47	62	78	93	19	37	56	74	93	111	22	45	67	89	112	134	27	54	82	109	136	163
3	16	32	48	63	79	95	19	38	57	75	94	113	23	46	69	91	114	137	28	55	83	111	138	166

Table continued on next page

134

CT Values for Inactivation of *Giardia* Cysts by Free Chlorine at 10°C (continued)

Chlorine Concentration, mg/L	pH = 8.0 Log Inactivation						pH = 8.5 Log Inactivation						pH = 9.0 Log Inactivation					
	0.5	1.0	1.5	2.0	2.5	3.0	0.5	1.0	1.5	2.0	2.5	3.0	0.5	1.0	1.5	2.0	2.5	3.0
≤0.4	25	50	75	99	124	149	30	59	89	118	148	177	35	70	105	139	174	209
0.6	26	51	77	102	128	153	31	61	92	122	153	183	36	73	109	145	182	218
0.8	26	53	79	105	132	158	32	63	95	126	158	189	38	75	113	151	188	226
1	27	54	81	108	135	162	33	65	98	130	163	195	39	78	117	156	195	234
1.2	28	55	83	111	138	166	33	67	100	133	167	200	40	80	120	160	200	240
1.4	28	57	85	113	142	170	34	69	103	137	172	206	41	82	124	165	206	247
1.6	29	58	87	116	145	174	35	70	106	141	176	211	42	84	127	169	211	253
1.8	30	60	90	119	149	179	36	72	108	143	179	215	43	86	130	173	216	259
2	30	61	91	121	152	182	37	74	111	147	184	221	44	88	133	177	221	265
2.2	31	62	93	124	155	186	38	75	113	150	188	225	45	90	136	181	226	271
2.4	32	63	95	127	158	190	38	77	115	153	192	230	46	92	138	184	230	276
2.6	32	65	97	129	162	194	39	78	117	156	195	234	47	94	141	187	234	281
2.8	33	66	99	131	164	197	40	80	120	159	199	239	48	96	144	191	239	287
3	34	67	101	134	168	201	41	81	122	162	203	243	49	97	146	195	243	292

Source: AWWA (1991).
CT = disinfectant residual concentration, mg/L × time, min.
NOTE: Contact your state health department to verify applicability.

CT Values for Inactivation of *Giardia* Cysts by Free Chlorine at 15°C

Chlorine Concentration, mg/L	pH ≤ 6 Log Inactivation						pH = 6.5 Log Inactivation						pH = 7.0 Log Inactivation						pH = 7.5 Log Inactivation					
	0.5	1.0	1.5	2.0	2.5	3.0	0.5	1.0	1.5	2.0	2.5	3.0	0.5	1.0	1.5	2.0	2.5	3.0	0.5	1.0	1.5	2.0	2.5	3.0
≤0.4	8	16	25	33	41	49	10	20	30	39	49	59	12	23	35	47	58	70	14	28	42	55	69	83
0.6	8	17	25	33	42	50	10	20	30	40	50	60	12	24	36	48	60	72	14	29	43	57	72	86
0.8	9	17	26	35	43	52	10	20	31	41	51	61	12	24	37	49	61	73	15	29	44	59	73	88
1	9	18	27	35	44	53	11	21	32	42	53	63	13	25	38	50	63	75	15	30	45	60	75	90
1.2	9	18	27	36	45	54	11	21	32	43	53	64	13	25	38	51	63	76	15	31	46	61	77	92
1.4	9	18	28	37	46	55	11	22	33	43	54	65	13	26	39	52	65	78	16	31	47	63	78	94
1.6	9	19	28	37	47	56	11	22	33	44	55	66	13	26	40	53	66	79	16	32	48	64	80	96
1.8	10	19	29	38	48	57	11	23	34	45	57	68	14	27	41	54	68	81	16	33	49	65	82	98
2	10	19	29	39	48	58	12	23	35	46	58	69	14	28	42	55	69	83	17	33	50	67	83	100
2.2	10	20	30	39	49	59	12	23	35	47	58	70	14	28	43	57	71	85	17	34	51	68	85	102
2.4	10	20	30	40	50	60	12	24	36	48	60	72	14	29	43	57	72	86	18	35	53	70	88	105
2.6	10	20	31	41	51	61	12	24	37	49	61	73	15	29	44	59	73	88	18	36	54	71	89	107
2.8	10	21	31	41	52	62	12	25	37	49	62	74	15	30	45	59	74	89	18	36	55	73	91	109
3	11	21	32	42	53	63	13	25	38	51	63	76	15	30	46	61	76	91	19	37	56	74	93	111

Table continued on next page

CT Values for Inactivation of *Giardia* Cysts by Free Chlorine at 15°C (continued)

Chlorine Concentration, mg/L	pH = 8.0 Log Inactivation						pH = 8.5 Log Inactivation						pH = 9.0 Log Inactivation					
	0.5	1.0	1.5	2.0	2.5	3.0	0.5	1.0	1.5	2.0	2.5	3.0	0.5	1.0	1.5	2.0	2.5	3.0
≤0.4	17	33	50	66	83	99	20	39	59	79	98	118	23	47	70	93	117	140
0.6	17	34	51	68	85	102	20	41	61	81	102	122	24	49	73	97	122	146
0.8	18	35	53	70	88	105	21	42	63	84	105	126	25	50	76	101	126	151
1	18	36	54	72	90	108	22	43	65	87	108	130	26	52	78	104	130	156
1.2	19	37	56	74	93	111	22	45	67	89	112	134	27	53	80	107	133	160
1.4	19	38	57	76	95	114	23	46	69	91	114	137	28	55	83	110	138	165
1.6	19	39	58	77	97	116	24	47	71	94	118	141	28	56	85	113	141	169
1.8	20	40	60	79	99	119	24	48	72	96	120	144	29	58	87	115	144	173
2	20	41	61	81	102	122	25	49	74	98	123	147	30	59	89	118	148	177
2.2	21	41	62	83	103	124	25	50	75	100	125	150	30	60	91	121	151	181
2.4	21	42	64	85	106	127	26	51	77	102	128	153	31	61	92	123	153	184
2.6	22	43	65	86	108	129	26	52	78	104	130	156	31	63	94	125	157	188
2.8	22	44	66	88	110	132	27	53	80	106	133	159	32	64	96	127	159	191
3	22	45	67	89	112	134	27	54	81	108	135	162	33	65	98	130	163	195

Source: AWWA (1991).

CT = disinfectant residual concentration, mg/L × time, min.

Note: Contact your state health department to verify applicability.

CT Values for Inactivation of *Giardia* Cysts by Free Chlorine at 20°C

Chlorine Concentration, mg/L	pH ≤ 6 Log Inactivation						pH = 6.5 Log Inactivation						pH = 7.0 Log Inactivation						pH = 7.5 Log Inactivation					
	0.5	1.0	1.5	2.0	2.5	3.0	0.5	1.0	1.5	2.0	2.5	3.0	0.5	1.0	1.5	2.0	2.5	3.0	0.5	1.0	1.5	2.0	2.5	3.0
≤0.4	6	12	18	24	30	36	7	15	22	29	37	44	9	17	26	35	43	52	10	21	31	41	52	62
0.6	6	13	19	25	32	38	8	15	23	30	38	45	9	18	27	36	45	54	11	21	32	43	53	64
0.8	7	13	20	26	33	39	8	15	23	31	38	46	9	18	28	37	46	55	11	22	33	44	55	66
1	7	13	20	26	33	39	8	16	24	31	39	47	9	19	28	37	47	56	11	22	34	45	56	67
1.2	7	13	20	27	33	40	8	16	24	32	40	48	10	19	29	38	48	57	12	23	35	46	58	69
1.4	7	14	21	27	34	41	8	16	25	33	41	49	10	19	29	39	48	58	12	23	35	47	58	70
1.6	7	14	21	28	35	42	8	17	25	33	42	50	10	20	30	39	49	59	12	24	36	48	60	72
1.8	7	14	22	29	36	43	9	17	26	34	43	51	10	20	31	41	51	61	12	25	37	49	62	74
2	7	15	22	29	37	44	9	17	26	35	43	52	10	21	31	41	52	62	13	25	38	50	63	75
2.2	7	15	22	29	37	44	9	18	27	35	44	53	11	21	32	42	53	63	13	26	39	51	64	77
2.4	8	15	23	30	38	45	9	18	27	36	45	54	11	22	33	43	54	65	13	26	39	52	65	78
2.6	8	15	23	31	38	46	9	18	28	37	46	55	11	22	33	44	55	66	13	27	40	53	67	80
2.8	8	16	24	31	39	47	9	19	28	37	47	56	11	22	34	45	56	67	14	27	41	54	68	81
3	8	16	24	31	39	47	10	19	29	38	48	57	11	23	34	45	57	68	14	28	42	55	69	83

Table continued on next page

CT Values for Inactivation of *Giardia* Cysts by Free Chlorine at 20°C (continued)

Chlorine Concentration, mg/L	pH = 8.0 Log Inactivation						pH = 8.5 Log Inactivation						pH = 9.0 Log Inactivation					
	0.5	1.0	1.5	2.0	2.5	3.0	0.5	1.0	1.5	2.0	2.5	3.0	0.5	1.0	1.5	2.0	2.5	3.0
≤0.4	12	25	37	49	62	74	15	30	45	59	74	89	18	35	53	70	88	105
0.6	13	26	39	51	64	77	15	31	46	61	77	92	18	36	55	73	91	109
0.8	13	26	40	53	66	79	16	32	48	63	79	95	19	38	57	75	94	113
1	14	27	41	54	68	81	16	33	49	65	82	98	20	39	59	78	98	117
1.2	14	28	42	55	69	83	17	33	50	67	83	100	20	40	60	80	100	120
1.4	14	28	43	57	71	85	17	34	52	69	86	103	21	41	62	82	103	123
1.6	15	29	44	58	73	87	18	35	53	70	88	105	21	42	63	84	105	126
1.8	15	30	45	59	74	89	18	36	54	72	90	108	22	43	65	86	108	129
2	15	30	46	61	76	91	18	37	55	73	92	110	22	44	66	88	110	132
2.2	16	31	47	62	78	93	19	38	57	75	94	113	23	45	68	90	113	135
2.4	16	32	48	63	79	95	19	38	58	77	96	115	23	46	69	92	115	138
2.6	16	32	49	65	81	97	20	39	59	78	98	117	24	47	71	94	118	141
2.8	17	33	50	66	83	99	20	40	60	79	99	119	24	48	72	95	119	143
3	17	34	51	67	84	101	20	41	61	81	102	122	24	49	73	97	122	146

Source: AWWA (1991).

CT = disinfectant residual concentration, mg/L × time, min.

Note: Contact your state health department to verify applicability.

CT Values for Inactivation of *Giardia* Cysts by Free Chlorine at 25°C

Chlorine Concentration, mg/L	pH ≤ 6 Log Inactivation						pH = 6.5 Log Inactivation						pH = 7.0 Log Inactivation						pH = 7.5 Log Inactivation					
	0.5	1.0	1.5	2.0	2.5	3.0	0.5	1.0	1.5	2.0	2.5	3.0	0.5	1.0	1.5	2.0	2.5	3.0	0.5	1.0	1.5	2.0	2.5	3.0
≤0.4	4	8	12	16	20	24	5	10	15	19	24	29	6	12	18	23	29	35	7	14	21	28	35	42
0.6	4	8	13	17	21	25	5	10	15	20	25	30	6	12	18	24	30	36	7	14	22	29	36	43
0.8	4	9	13	17	22	26	5	10	16	21	26	31	6	12	19	25	31	37	7	15	22	29	37	44
1	4	9	13	17	22	26	5	10	16	21	26	31	6	12	19	25	31	37	8	15	23	30	38	45
1.2	5	9	14	18	23	27	5	11	16	21	27	32	6	13	19	25	32	38	8	15	23	31	38	46
1.4	5	9	14	18	23	27	6	11	17	22	28	33	7	13	20	26	33	39	8	16	24	31	39	47
1.6	5	9	14	19	23	28	6	11	17	22	28	33	7	13	20	27	33	40	8	16	24	32	40	48
1.8	5	10	15	19	24	29	6	11	17	23	28	34	7	14	21	27	34	41	8	16	25	33	41	49
2	5	10	15	19	24	29	6	12	18	23	29	35	7	14	21	27	34	41	8	17	25	33	41	50
2.2	5	10	15	20	25	30	6	12	18	23	29	35	7	14	21	28	35	42	9	17	26	34	42	51
2.4	5	10	15	20	25	30	6	12	18	24	30	36	7	14	22	28	35	42	9	17	26	34	43	52
2.6	5	10	16	21	26	31	6	12	19	25	31	37	7	15	22	29	36	43	9	18	26	35	44	53
2.8	5	10	16	21	26	31	6	12	19	25	31	37	8	15	23	30	37	44	9	18	27	36	45	54
3	5	11	16	21	27	32	6	13	19	25	32	38	8	15	23	31	38	46	9	18	28	37	46	55

Table continued on next page

CT Values for Inactivation of *Giardia* Cysts by Free Chlorine at 25°C (continued)

Chlorine Concentration, mg/L	pH = 8.0 Log Inactivation						pH = 8.5 Log Inactivation						pH = 9.0 Log Inactivation					
	0.5	1.0	1.5	2.0	2.5	3.0	0.5	1.0	1.5	2.0	2.5	3.0	0.5	1.0	1.5	2.0	2.5	3.0
≤0.4	8	17	25	33	42	50	10	20	30	39	49	59	12	23	35	47	58	70
0.6	9	17	26	34	43	51	10	20	31	41	51	61	12	24	37	49	61	73
0.8	9	18	27	35	44	53	11	21	32	42	53	63	13	25	38	50	63	75
1	9	18	27	36	45	54	11	22	33	43	54	65	13	26	39	52	65	78
1.2	9	18	28	37	46	55	11	22	34	45	56	67	13	27	40	53	67	80
1.4	10	19	29	38	48	57	12	23	35	46	58	69	14	27	41	55	68	82
1.6	10	19	29	39	48	58	12	23	35	47	58	70	14	28	42	56	70	84
1.8	10	20	30	40	50	60	12	24	36	48	60	72	14	29	43	57	72	86
2	10	20	31	41	51	61	12	25	37	49	62	74	15	29	44	59	73	88
2.2	10	21	31	41	52	62	13	25	38	50	63	75	15	30	45	60	75	90
2.4	11	21	32	42	53	63	13	26	39	51	64	77	15	31	46	61	77	92
2.6	11	22	33	43	54	65	13	26	39	52	65	78	16	31	47	63	78	94
2.8	11	22	33	44	55	66	13	27	40	53	67	80	16	32	48	64	80	96
3	11	22	34	45	56	67	14	27	41	54	68	81	16	32	49	65	81	97

Source: AWWA (1991).

CT = disinfectant residual concentration, mg/L × time, min.

Note: Contact your state health department to verify applicability.

CT Values for Inactivation of Viruses by Free Chlorine, pH 6.0–9.0

Inactivation (log)	Temperature, °C																									
	0.5	1	2	3	4	5	6	7	8	9	10	11	12	13	14	15	16	17	18	19	20	21	22	23	24	25
2	6.0	5.8	5.3	4.9	4.4	4.0	3.8	3.6	3.4	3.2	3.0	2.8	2.6	2.4	2.2	2.0	1.8	1.6	1.4	1.2	1.0	1.0	1.0	1.0	1.0	1.0
3	9.0	8.7	8.0	7.3	6.7	6.0	5.6	5.2	4.8	4.4	4.0	3.8	3.6	3.4	3.2	3.0	2.8	2.6	2.4	2.2	2.0	1.8	1.6	1.4	1.2	1.0
4	12.0	11.6	10.7	9.8	8.9	8.0	7.6	7.2	6.8	6.4	6.0	5.6	5.2	4.8	4.4	4.0	3.8	3.6	3.4	3.2	3.0	2.8	2.6	2.4	2.2	2.0

Source: AWWA (1991). Modified by linear interpolation between 5°C increments.

CT Values for Inactivation of *Giardia* Cysts by Chlorine Dioxide, pH 6.0–9.0

Inactivation (log)	Temperature, °C																								
	1	2	3	4	5	6	7	8	9	10	11	12	13	14	15	16	17	18	19	20	21	22	23	24	25
0.5	10.0	8.6	7.2	5.7	4.3	4.2	4.2	4.1	4.1	4.0	3.8	3.7	3.5	3.4	3.2	3.1	2.9	2.8	2.6	2.5	2.4	2.3	2.2	2.1	2.0
1	21.0	17.9	14.9	11.8	8.7	8.5	8.3	8.1	7.9	7.7	7.4	7.1	6.9	6.6	6.3	6.0	5.8	5.5	5.3	5.0	4.7	4.5	4.2	4.0	3.7
1.5	32.0	27.3	22.5	17.8	13.0	12.8	12.6	12.4	12.2	12.0	11.6	11.2	10.8	10.4	10.0	9.5	9.0	8.5	8.0	7.5	7.1	6.7	6.3	5.9	5.5
2	42.0	35.8	29.5	23.3	17.0	16.6	16.2	15.8	15.4	15.0	14.6	14.2	13.8	13.4	13.0	12.4	11.8	11.2	10.6	10.0	9.5	8.9	8.4	7.8	7.3
2.5	52.0	44.5	37.0	29.5	22.0	21.4	20.8	20.2	19.6	19.0	18.4	17.8	17.2	16.6	16.0	15.4	14.8	14.2	13.6	13.0	12.2	11.4	10.6	9.8	9.0
3	63.0	53.8	44.5	35.3	26.0	25.4	24.8	24.2	23.6	23.0	22.2	21.4	20.6	19.8	19.0	18.2	17.4	16.6	15.8	15.0	14.2	13.4	12.6	11.8	11.0

Source: AWWA (1991). Modified by linear interpolation between 5°C increments.

CT Values for Inactivation of Viruses by Chlorine Dioxide, pH 6.0–9.0

Inactivation (log)	Temperature, °C																								
	1	2	3	4	5	6	7	8	9	10	11	12	13	14	15	16	17	18	19	20	21	22	23	24	25
2	8.4	7.7	7.0	6.3	5.6	5.3	5.0	4.8	4.5	4.2	3.9	3.6	3.4	3.1	2.8	2.7	2.5	2.4	2.2	2.1	2.0	1.8	1.7	1.5	1.4
3	25.6	23.5	21.4	19.2	17.1	16.2	15.4	14.5	13.7	12.8	12.0	11.1	10.3	9.4	8.6	8.2	7.7	7.3	6.8	6.4	6.0	5.6	5.1	4.7	4.3
4	50.1	45.9	41.8	37.6	33.4	31.7	30.1	28.4	26.8	25.1	23.4	21.7	20.1	18.4	16.7	15.9	15.0	14.2	13.3	12.5	11.7	10.9	10.0	9.2	8.4

Source: AWWA (1991). Modified by linear interpolation between 5°C increments.

CT Values for Inactivation of *Giardia* Cysts by Chloramine, pH 6.0–9.0

Inactivation (log)	Temperature, °C																								
	1	2	3	4	5	6	7	8	9	10	11	12	13	14	15	16	17	18	19	20	21	22	23	24	25
0.5	635	568	500	433	365	354	343	332	321	310	298	286	274	262	250	237	224	211	198	185	173	161	149	137	125
1	1,270	1,136	1,003	869	735	711	687	663	639	615	592	569	546	523	500	474	448	422	396	370	346	322	298	274	250
1.5	1,900	1,700	1,500	1,300	1,100	1,066	1,032	998	964	930	894	858	822	786	750	710	670	630	590	550	515	480	445	410	375
2	2,535	2,269	2,003	1,736	1,470	1,422	1,374	1,326	1,278	1,230	1,184	1,138	1,092	1,046	1,000	947	894	841	788	735	688	641	594	547	500
2.5	3,170	2,835	2,500	2,165	1,830	1,772	1,714	1,656	1,598	1,540	1,482	1,424	1,366	1,308	1,250	1,183	1,116	1,049	982	915	857	799	741	683	625
3	3,800	3,400	3,000	2,600	2,200	2,130	2,060	1,990	1,920	1,850	1,780	1,710	1,640	1,570	1,500	1,420	1,340	1,260	1,180	1,100	1,030	960	890	820	750

Source: AWWA (1991), Modified by linear interpolation between 5°C increments.

CT Values for Inactivation of Viruses by Chloramine

Inactivation (log)	Temperature, °C																								
	1	2	3	4	5	6	7	8	9	10	11	12	13	14	15	16	17	18	19	20	21	22	23	24	25
2	1,243	1,147	1,050	954	857	814	771	729	686	643	600	557	514	471	428	407	385	364	342	321	300	278	257	235	214
3	2,063	1,903	1,743	1,583	1,423	1,352	1,281	1,209	1,138	1,067	996	925	854	783	712	676	641	605	570	534	498	463	427	392	356
4	2,883	2,659	2,436	2,212	1,988	1,889	1,789	1,690	1,590	1,491	1,392	1,292	1,193	1,093	994	944	895	845	796	746	696	646	597	547	497

Source: AWWA (1991), Modified by linear interpolation between 5°C increments.

CT Values for Inactivation of Giardia Cysts by Ozone

Inactivation (log)	Temperature, °C																								
	1	2	3	4	5	6	7	8	9	10	11	12	13	14	15	16	17	18	19	20	21	22	23	24	25
0.5	0.48	0.44	0.40	0.36	0.32	0.30	0.28	0.27	0.25	0.23	0.22	0.20	0.19	0.17	0.16	0.15	0.14	0.14	0.13	0.12	0.11	0.10	0.10	0.09	0.08
1	0.97	0.89	0.80	0.72	0.63	0.60	0.57	0.54	0.51	0.48	0.45	0.42	0.38	0.35	0.32	0.30	0.29	0.27	0.26	0.24	0.22	0.21	0.19	0.18	0.16
1.5	1.50	1.36	1.23	1.09	0.95	0.90	0.86	0.81	0.77	0.72	0.67	0.62	0.58	0.53	0.48	0.46	0.43	0.41	0.38	0.36	0.34	0.31	0.29	0.26	0.24
2	1.90	1.75	1.60	1.45	1.30	1.23	1.16	1.09	1.02	0.95	0.89	0.82	0.76	0.69	0.63	0.60	0.57	0.54	0.51	0.48	0.45	0.42	0.38	0.35	0.32
2.5	2.40	2.20	2.00	1.80	1.60	1.52	1.44	1.36	1.28	1.20	1.12	1.04	0.95	0.87	0.79	0.75	0.71	0.68	0.64	0.60	0.56	0.52	0.48	0.44	0.40
3	2.90	2.65	2.40	2.15	1.90	1.81	1.71	1.62	1.52	1.43	1.33	1.24	1.14	1.05	0.95	0.90	0.86	0.81	0.77	0.72	0.67	0.62	0.58	0.53	0.48

Source: AWWA (1991), Modified by linear interpolation between 5°C increments.

CT Values for Inactivation of Viruses by Ozone

Inactivation (log)	Temperature, °C																								
	1	2	3	4	5	6	7	8	9	10	11	12	13	14	15	16	17	18	19	20	21	22	23	24	25
2	0.90	0.83	0.75	0.68	0.60	0.58	0.56	0.54	0.52	0.50	0.46	0.42	0.38	0.34	0.30	0.29	0.28	0.27	0.26	0.25	0.23	0.21	0.19	0.17	0.15
3	1.40	1.28	1.15	1.03	0.90	0.88	0.86	0.84	0.82	0.80	0.74	0.68	0.62	0.56	0.50	0.48	0.46	0.44	0.42	0.40	0.37	0.34	0.31	0.28	0.25
4	1.80	1.65	1.50	1.35	1.20	1.16	1.12	1.08	1.04	1.00	0.92	0.84	0.76	0.68	0.60	0.58	0.56	0.54	0.52	0.50	0.46	0.42	0.38	0.34	0.30

Source: AWWA (1991), Modified by linear interpolation between 5°C increments.

CT Values for Inactivation of Viruses by UV*

Log Inactivation,	
2.0	**3.0**
21	36

Source: AWWA (1991), Modified by linear interpolation between 5°C increments.
* UV inactivation of HAV. Units of CT values are mW-sec/cm². *CT* values include a safety factor of 3.

UV Dose Requirements for *Cryptosporidium*, *Giardia lamblia*, and Virus Inactivation Credit

Log Credit	*Cryptosporidium* UV dose, *mJ/cm²*	*Giardia lamblia* UV dose, *mJ/cm²*	Virus UV dose, *mJ/cm²*
0.5	1.6	1.5	39
1.0	2.5	2.1	58
1.5	3.9	3.0	79
2.0	5.8	5.2	100
2.5	8.5	7.7	121
3.0	12	11	143
3.5	NA	NA	163
4.0	NA	NA	186

Source: Federal Register, Vol 68, No. 154 (2003).
NA = not applicable.

Water Treatment

TYPICAL LOADING FILTRATION RATES
FOR VARIOUS TYPES OF FILTERS

Type of Filter	Common Filtration Rate
Slow sand	0.016 to 0.16 gpm/ft^2
Rapid sand	2 gpm/ft^2
Dual media (coal, sand)	2 to 4 gpm/ft^2
Multimedia (coal, sand, and garnet; or coal, sand, and ilmenite)	5 to 10 gpm/ft^2

$$\text{surface overflow rate} = \frac{\text{flow, gpd}}{\text{tank surface, ft}^2}$$

Typical Grain Sizes for Different Applications

	Effective Size, *mm*	Total Depth, *m*
A. Common US Practice After Coagulation and Settling		
1. Sand alone	0.45–0.55	0.6–0.7
2. Dual media Add anthracite (0.1 to 0.7 of bed)	0.9–1.1	0.6–0.9
3. Triple media Add garnet (0.1 m)	0.2–0.3	0.7–1.0
B. US Practice for Direct Filtration		
1. Practice not well established. With seasonal diatom blooms, use coarser top size.		
2. Dual-media or deep monomedium, 1.5-mm effective size.		
C. US Practice for Fe and Mn Filtration		
1. Dual media similar to A.2 above		
2. Single medium	<0.8	0.6–0.9
D. Coarse Single-Medium Filters Washed With Air and Water Simultaneously		
1. For coagulated and settled water	0.9–1.0	0.9–1.2
2. For direct filtration	1.4–1.6	1–2
3. For Fe and Mn removal	1–2	1.5–3

	Ionic Range	Molecular Range	Macro Range	Micro Particle Range	Macro Particle Range
Size, μm	0.001	0.01	0.1	1.0	10	100	1,000
Approximate Molecular	100 1,000	20,000 100,000	500,000			

Relative Size of Various Materials in Water

- Aqueous Salts
- Metal Ions
- Molecules
- Viruses
- Humic Acids
- Natural Organic Matter
- Clays
- Colloids
- Asbestos Fibers
- Bacteria
- Cysts
- Silt
- Algae
- Suspended
- Sand

Processes

- Reverse Osmosis Pervaporation
- Nanofiltration
- Electrodialysis
- Ultrafiltration
- Microfiltration
- Conventional Filtration
- Activated Carbon
- Coagulation
- Sand Activated Carbon (grains)

Size Ranges of Membrane Processes and Contaminants

147

1. Conventional Sand   2. Dual Media   3. Triple (mixed) Media

4. Deep Bed Filter   5. Upflow Filter

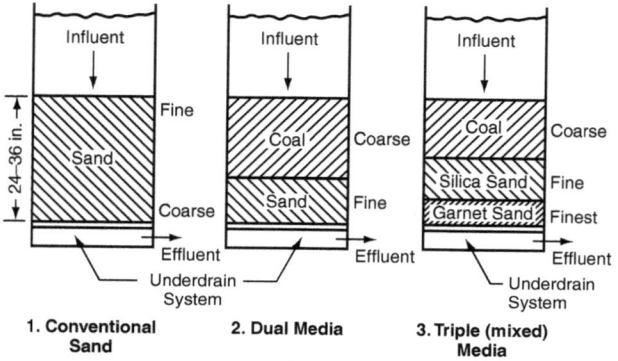

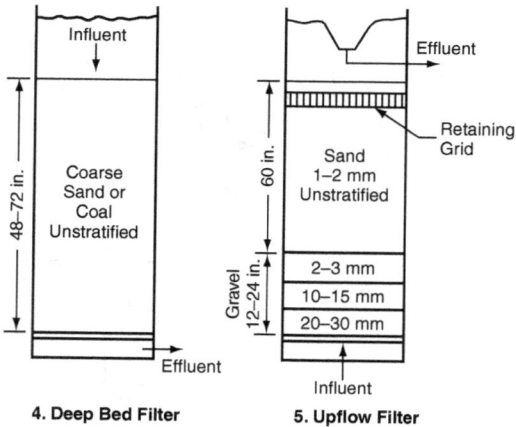

NOTE: Diagrams 2 and 3 are washed with fluidization; diagrams 4 and 5 are washed with air-plus-water without fluidization.

Schematic Diagrams of Filter Configurations for Rapid Filtration

General Effectiveness of Water Treatment Processes for Removal of Soluble Contaminants

Contaminant Category	Aeration and Stripping	Coagulation, Sedimentation, or DAF, Filtration	Precoat Filtration	Lime Softening	Chemical Oxidation and Disinfection	Nanofiltration	Reverse Osmosis	Electrodialysis/ED Reversal	Anion	Cation	Granular Activated Carbon	Powdered Activated Carbon	Activated Alumina
				Primary Contaminants			Membrane Process			Ion Exchange		Adsorption		
Inorganics													
Antimony							X*	X					
Arsenic (+3)		XO†		XO			X	X	X				X
Arsenic (+5)		X		X			X	X	X				X
Barium				X			X	X		X			
Beryllium		X		X			X	X					
Cadmium		X		X			X	X		X			
Chromium (+3)		X		X			X	X		X			
Chromium (+6)							X	X	X				
Cyanide					X								

Table continued on next page

Water Treatment

149

General Effectiveness of Water Treatment Processes for Removal of Soluble Contaminants (continued)

Contaminant Category	Aeration and Stripping	Coagulation, Sedimentation or DAF, Filtration	Precoat Filtration	Lime Softening	Chemical Oxidation and Disinfection	Nanofiltration	Reverse Osmosis	Electrodialysis/ ED Reversal	Anion	Cation	Granular Activated Carbon	Powdered Activated Carbon	Activated Alumina
Fluoride				X			X	X					X
Lead‡								X					
Mercury (inorganic)				X			X	X					
Nickel				X			X	X		X			
Nitrate							X	X	X				
Nitrite							X	X	X				
Selenium (+4)		X					X	X	X				X
Selenium (+6)							X	X	X				X
Thallium							X	X					X
Organic Contaminants													
Volatile organics	X										X	X	
Synthetic organics							X				X	X	

Table continued on next page

General Effectiveness of Water Treatment Processes for Removal of Soluble Contaminants (continued)

Contaminant Category	Aeration and Stripping	Coagulation, Sedimentation or DAF, Filtration	Precoat Filtration	Lime Softening	Chemical Oxidation and Disinfection	Membrane Process			Ion Exchange		Adsorption		
						Nanofiltration	Reverse Osmosis	Electrodialysis/ ED Reversal	Anion	Cation	Granular Activated Carbon	Powdered Activated Carbon	Activated Alumina
Pesticides/herbicides						X	X				X	X	
Dissolved organic carbon		X				X	X				X	X	
Radionuclides													
Radium (226 + 228)				X			X	X		X			
Uranium							X	X	X				
Secondary Contaminants and Constituents Causing Aesthetic Problems													
Hardness				X		X	X	X		X			
Iron		XO	XO	XO						X			
Manganese		XO	XO	X						X			
Total dissolved solids							X	X					

Table continued on next page

General Effectiveness of Water Treatment Processes for Removal of Soluble Contaminants (continued)

Contaminant Category	Aeration and Stripping	Coagulation, Sedimentation or DAF, Filtration	Precoat Filtration	Lime Softening	Chemical Oxidation and Disinfection	Nanofiltration	Reverse Osmosis	Electrodialysis/ED Reversal	Anion	Cation	Granular Activated Carbon	Powdered Activated Carbon	Activated Alumina
						Membrane Process			**Ion Exchange**		**Adsorption**		
Chloride							X	X					
Sulfate						X	X	X					
Zinc				X			X	X		X			
Color		X			X	X	X				X	X	
Taste and odor	X				X						X	X	

* X = appropriate process for this contaminant.
† X0 = appropriate when oxidation used in conjunction with this process.
‡ Lead is generally a product of corrosion and is controlled by corrosion control treatment rather than removed by water treatment processes.

Chemical Additions

Chemicals used for treating water include bromine, iodine, ozone (alone or in combination with other chemicals), potassium, permanganate, chlorine dioxide, chlorine and chlorine compounds, and oxygen. In some cases, the choice of chemicals used is dictated by the ability of the chemicals to perform secondary functions such as controlling biological growth in pipelines and basins, controlling taste and odors, removing color, reducing some organic compounds, aiding in flocculation, or oxidizing iron and manganese.

Details for Commonly Used Water Treatment Chemicals

Chemical Name and Formula	Common or Trade Name	Shipping Containers	Suitable Storage Materials	Available Forms/ Descriptions	Density	Solubility, lb/gal	Commercial Strength, %	Additional Characteristics and Properties
Activated carbon, powdered carbon	Aqua Nuchor, Hydrodarco, Herite	Bags, bulk	Dry: iron, steel; wet: rubber and silicon linings, type 316 stainless steel	Black granules, powder	15–30 lb/ft³	Insoluble (suspension used)		1 lb/gal suspension used for storage and handling
Aluminum oxide, Al_2O_3	Activated alumina	Bags, drums	Iron, steel	Powder granules (up to 1 in. in diameter)		Insoluble	100	
Aluminum sulfate, $Al_2(SO_4)_3 \cdot 14H_2O$ (dry)	Alum, filter alum, sulfate of alumina	100- to 200-lb bags, 300- to 400-lb barrels, bulk (carloads), tank truck, 228–36 tank car	Dry: iron, steel; wet: stainless steel, rubber, plastic	Ivory colored; powder, granule, lump	38–45 lb/ft³ 60–63 lb/ft³ 62–67 lb/ft³	6.2 (60°F)	17 as Al_2O_3 dry	pH of 1% solution: 3.4
Aluminum sulfate (liquid)	50% alum	Tank cars and tank trucks	FRP, PE, type 316 stainless steel, rubber linings	Liquid	11.2 lb/gal	—	8.5 as Al_2O_3	Freezing point: –4°F
Ammonium aluminum sulfate, $Al_2(SO_4)_3(NH_4)_2$-$SO_4 \cdot 24H_2O$	Ammonia alum, crystal alum	100-lb bags, barrels, bulk	FRP, PE, type 316 stainless steel, rubber linings	Colorless crystals or white powder	65–75 lb/ft³	0.3 (32°F)	99	pH of 1% solution: 3.5
Ammonium hydroxide, NH_4OH	Ammonia water, ammonium hydrate, aqua ammonia	Carboys, 750-lb drums, bulk	Glass lining, steel, iron, FRP, PE	Colorless liquid	7.48 lb/gal	Complete	29.4 (NH_3) max. 26°Bé	pH 14; Freezing point: –107°F

Table continued on next page

154

Details for Commonly Used Water Treatment Chemicals (continued)

Chemical Name and Formula	Common or Trade Name	Shipping Containers	Suitable Storage Materials	Available Forms/ Descriptions	Density	Solubility, lb/gal	Commercial Strength, %	Additional Characteristics and Properties
Ammonium silicofluoride, $(NH_4)_2SiF_6$	Ammonium fluorosilicate	100- and 400-lb drums	Steel, iron, FRP, PE	White crystals	65–70 lb/ft³	1.7 (63°F)	100	White, free-flowing solid
Ammonium sulfate, $(NH_4)_2SO_4$	Sulfate of ammonia	50- and 100-lb bags, 725-lb drums	FRP; PE; ceramic and rubber linings; iron (dry)	White or brown crystals	70 lb/ft³	6.3 (68°F)	>99	Cakes in dry feed; add $CaSO_4$ for free flow
Anhydrous ammonia, NH_3	Ammonia	50-, 100-, 150-lb cylinders; bulk tank cars; and trucks	Shipping containers	Colorless gas	38.6 lb/ft³	3.9 (32°F) 3.1 (60°F)	99.9+ (NH_3)	
Bentonite	Colloidal clay, volclay, wilkinite	100-lb bags, bulk	Iron, steel, FRP, PE	Powder, pellet, mixed sizes	60 lb/ft³	Insoluble (colloidal solution used)		Free flowing, nonabrasive
Calcium fluoride, CaF_2	Fluorspar	Bags, drums, barrels, hopper cars, trucks	Steel, iron, FRP, PE	Powder		Very slight	85 (CaF_2), less than 5 (SiO_2)	
Calcium hydroxide, $Ca(OH)_2$	Hydrated lime, slaked lime	50-lb bags, bulk	FRP, PE, iron, steel, rubber lining	White powder, light, dense	28–36 lb/ft³	0.14 (68°F) 0.12 (90°F)	85 to 99 ($Ca(OH)_2$) 63 to 73 (CaO)	Hopper agitation required for dry feed of light form
Calcium hypochlorite, $Ca(OCl)_2 \cdot 4H_2O$	HTH, perchloron, pittchlor	5-lb cans; 100-, 300-, 800-lb drums	Glass, plastic, and rubber linings; FRP; PE	White granule, powder, tablet	52.5 lb/ft³	1.5 at 25°C	65 (available Cl_2)	1 to 3 (available Cl_2 solution used)

Table continued on next page

Chemical Additions

155

Details for Commonly Used Water Treatment Chemicals (continued)

Chemical Name and Formula	Common or Trade Name	Shipping Containers	Suitable Storage Materials	Available Forms/ Descriptions	Density	Solubility, lb/gal	Commercial Strength, %	Additional Characteristics and Properties
Calcium oxide, CaO	Burnt lime, chemical lime, quicklime, unslaked lime	80- and 100-lb bags, bulk	FRP, PE, iron, steel, rubber linings	Lump, pebble, granule	35–71 lb/ft³	Slaked to form hydrated lime	75 to 99 (CaO)	pH of saturated solution, on detention time temperature amount of water critical for efficient slaking
Carbon dioxide, liquid CO_2	Carbonic anhydride	Bulk	Carbon steel (dry); type 316 stainless steel (solution)	Liquid		0.012 at 77°F	99.5	Solution is acid
Chlorinated lime, CaO, $2CaOCl_2 \cdot 3H_2O$	Bleaching powder, chloride of lime	100-, 300-, 800-lb drums	Glass and rubber linings, FRP, PE	White powder	48 lb/ft³		25–37 (available Cl_2)	Deteriorates
Chlorine, Cl_2	Chlorine gas, liquid chlorine	100-, 150-lb cylinders; 1-ton tanks; 16-, 30-, 55-ton tank cars	Shipping containers	Greenish-yellow; liquefied gas under pressure	91.7 lb/ft³	0.07 (60°F) 0.04 (100°F)	99.8 (Cl_2)	Forms HCl and HOCl when mixed with water
Chlorine dioxide, ClO_2	Chlorine dioxide	Generated as used	Glass, PVC, and rubber linings; FRP, PE	Greenish-yellow gas		0.02 (30 μ)	26.3 (available Cl_2)	Explosive under certain conditions
Copper sulfate, $CuSO_4 \cdot 5H_2O$	Blue vitriol, blue stone	100-lb bags, 450-lb barrels, drums	FRP, PE, silicon lining, iron, stainless steel	Crystal, lump, powder	75–90 lb/ft³ 73–80 lb/ft³ 60–64 lb/ft³	1.6 (32°F) 2.2 (68°F) 2.6 (86°F)	99 ($CuSO_4$)	pH of 25% solution: approx. 3.0

Table continued on next page

Details for Commonly Used Water Treatment Chemicals (continued)

Chemical Name and Formula	Common or Trade Name	Shipping Containers	Suitable Storage Materials	Available Forms/Descriptions	Density	Solubility, lb/gal	Commercial Strength, %	Additional Characteristics and Properties
Disodium phosphate, anhydrous $Na_2HPO_4 \cdot 12H_2O$	Basic sodium phosphate, DSP, secondary sodium phosphate	100- and 300-lb drums, 50- and 100-lb bags	Cast iron, steel, FRP, PE	White crystal, granular or powder	60–64 lb/ft³	0.4 (32°F) 6.4 (86°F)	64.3 (PO_4) 48 (P_2O_5)	Precipitates Ca, Mg; pH of 1% solution: 9.1; solubility is 11 g/100 g at 77°F
Ferric chloride, $FeCl_3$ (33% to 45% solution)	Ferrichlor, iron chloride	55-gal drums, bulk	Glass, PVC, and rubber linings; FRP; PE	Dark brown syrupy liquid	11.9 lb/gal (40%)	Complete	37–45 ($FeCl_3$) 20–21 (Fe)	
Ferric chloride, $FeCl_3 \cdot 6H_2O$	Crystal ferric chloride	300-lb barrels	Keep in original containers	Yellow-brown lump			59–61 ($FeCl_3$) 20–21 (Fe)	Hygroscopic (store lumps and powder in tight container), no dry feed; optimum pH 4.0–11.0
Ferric chloride $FeCl_3$	Anhydrous ferric chloride	500-lb casks; 100-, 300-, 400-lb kegs; 65-, 135-, 250-lb drums	Keep in original containers	Greenish-black powder or crystals	175 lb/ft³		98 ($FeCl_3$) 34 (Fe)	
Ferric sulfate, $Fe_2(SO_4)_3 \cdot 9H_2O$	Ferrifloc, ferrisul	100- to 175-lb bags, 400- to 425-lb drums	Glass, plastic, and rubber linings; FRP; PE; type 316 stainless steel	Red-brown powder 70 or granule 72	60–70 lb/ft³	Soluble in 2 to 4 parts cold water	90–94 [as $Fe(SO_4)_3$] 25 to 26 (Fe)	Mildly hygroscopic coagulant at pH 3.5–11.0
Ferrous sulfate, $FeSO_4 \cdot 7H_2O$	Copperas, green vitriol	Bags, barrels, bulk	Glass, plastic, and rubber linings; FRP; PE; type 316 stainless steel	Green crystal, granule, lump	63–66 lb/ft³		55 ($FeSO_4$) 20 (Fe)	Hygroscopic; cakes in storage; optimum pH 8.5–11.0

Table continued on next page

Chemical Additions

Details for Commonly Used Water Treatment Chemicals (continued)

Chemical Name and Formula	Common or Trade Name	Shipping Containers	Suitable Storage Materials	Available Forms/ Descriptions	Density	Solubility, lb/gal	Commercial Strength, %	Additional Characteristics and Properties
Fluorosilicic acid, H_2SiF_6	Fluorosilicic acid	Rubber-lined drums, trucks, or railroad tank cars	Rubber-lined steel, PE	Liquid	~0.4	Approx. 1.2 (68°F)	35 (approx.)	Corrosive, etches glass
Hydrogen fluoride, HF	Hydrofluoric acid	Steel drums, tank cars	Steel, FRP, PE	Liquid			70 (HF)	Below 60%, steel cannot be used
Oxygen, liquid	LOX	Dewars, cylinders, truck and rail tankers	Steel	Pale blue liquid	9.52 lb/gal at 68°F and 1 atm	3.16% by volume at 77°F	99.5	Prevent LOX from contacting grease, oil, asphalt, or other combustibles.
Ozone, O_3	Ozone	Generated at site of application		Colorless gas				
Phosphoric acid, H_3PO_4		PE drums, bulk	FRP, epoxy, rubber lining, polypro-pylene, type 316 stainless steel	Watery white liquid	13.1 lb/gal at 75% solution		75/80/85	Freezing points: 0.5°F at 75% 40.2°F at 80% 70.0°F at 85%
Polyaluminum chloride, $Al_{13}(OH)_{20}(SO_4)_2Cl_{15}$	SternPac	55-gal drums and bulk	FRP, PE, type 316 stainless steel, rubber linings	Pale amber liquid	10.0 lb/gal		10.3 (Al_2O_3)	Freezing point: −12°C
Potassium aluminum sulfate, K_2SO_4·$Al_2(SO_4)_3$·24H_2O	Potash alum, potassium alum	Bags, lead-lined bulk (carloads)	FRP, PE, ceramic and rubber linings	Lump, granule, powder	60–67 lb/ft³	0.5 (32°F) 1.0 (68°F) 1.4 (86°F)	10 to 11 (Al_2O_3)	Low, even solubility; pH of 1% solution: 3.5

Table continued on next page

Details for Commonly Used Water Treatment Chemicals (continued)

Chemical Name and Formula	Common or Trade Name	Shipping Containers	Suitable Storage Materials	Available Forms/ Descriptions	Density	Solubility, lb/gal	Commercial Strength, %	Additional Characteristics and Properties
Potassium permanganate, $KMnO_4$	Purple salt	Bulk, barrels, drums	Iron, steel, FRP, PE	Purple crystals	90–105 lb/ft^3	Infinite	100	Danger of explosion in contact organic matters
Pyrosodium sulfite $Na_2S_2O_5$	Sodium metabisulfite	Bags, drums, barrels	Iron, steel, FRP, PE	White crystalline powder		Complete in water	Dry 67 (SO_2), solution 33.3 (SO_2)	Sulfurous odor
Sodium aluminate, $Na_2OAl_2O_3$	Soda alum	100- to 150-lb bags, 250- to 440-lb drums, solution	Iron, FRP, PE, rubber, steel	Brown powder liquid (27°Bé)	50–60 lb/ft^3	3.0 (68°F) 3.3 (86°F)	70 to 80 (Na_2 Al_2O_4, min. 32 $Na_2Al_2O_4$	Hopper agitation required for dry feed; very hygroscopic
Sodium carbonate, Na_2CO_3	Soda ash	Bags, barrels, bulk (carloads), trucks	Iron, rubber lining, steel, FRP, PE	White powder, extra light, light, dense	31.2–56.2 lb/ft^3 (light); 56.2–68.7 lb/ft^3 (dense)	1.5 (68°F) 2.3 (86°F)	99.4 (Na_2CO_3) 57.9 (Na_2O)	Hopper agitation required for dry feed or light and extra-light forms; pH of 1% solution: 11.3
Sodium chloride, NaCl	Common salt, salt	Bags, barrels, bulk (carloads)	Bronze, FRP, PE, rubber lining	Rock, fine	50–60 lb/ft^3 58–78 lb/ft^3	2.9 (32°F) 3.0 (68°F)	98 (NaCl)	Absorbs moisture
Sodium chlorite, $NaClO_2$	ADOX dry	100-lb drums	Metals (avoid cellulose materials)	Light orange powder, flake or crystals	53–56 lb/ft^3	3.5 (68°F)	80 ($NaClO_2$) 30 (available Cl_2)	Generates ClO_2, at pH 3.0; explosive
Sodium fluoride, NaF	Fluoride	Bags, barrels, fiber drums, kegs	Iron, steel, FRP, PE	Nile blue or white powder, light, dense	50 lb/ft^3 75 lb/ft^3	0.35 (most temperatures)	90 to 95 (NaF)	pH of 4% solution: 6.6

Table continued on next page

Chemical Additions

159

Details for Commonly Used Water Treatment Chemicals (continued)

Chemical Name and Formula	Common or Trade Name	Shipping Containers	Suitable Storage Materials	Available Forms/ Descriptions	Density	Solubility, lb/gal	Commercial Strength, %	Additional Characteristics and Properties
Sodium fluorosilicate, Na_2SiF_6	Sodium silicofluoride	Bags, barrels, fiber drums	Cast iron, rubber linings, steel, FRP, PE	Nile blue or yellowish white powder	72 lb/ft³	0.03 (32°F) 0.06 (68°F) 0.12 (140°F)	99 (Na_2)	pH of 1% solution: 5.3
Sodium hexametaphosphate, $Na(PO_3)_6$	Calgon, glassy phosphate, vitreous phosphate	100-lb bags	Rubber linings, plastics, type 316 stainless steel	Crystal, flake, powder	47 lb/ft³	1.0–4.2	66 (P_2O_3 unadjusted)	pH of 0.25% solution: 6.0 to 8.3
Sodium hydroxide, NaOH	Caustic soda, soda lye	100- to 700-lb drums; bulk (trucks, tank cars)	Carbon steel, polypropylene, FRP, rubber lining	Flake, lump, liquid	95.5 lb/ft³; 12.8 lb/gal for 50% solution	2.4 (32°F) 4.4 (68°F) 4.8 (104°F)	98.9 (NaOH) 74–76 (NaO_2)	Solid, hygroscopic; pH of 1% solution: 12.9; freezing point of 50% solution: 53°F
Sodium hypochlorite, NaOCl	Sodium hypochlorite	5-, 13-, 50-gal carboys; 1,300- to 2,000-gal tank trucks	Ceramic, glass, plastic, and rubber linings; FRP, PE	Light yellow liquid			12–15 (available Cl_2)	Unstable
Sodium silicate, Na_2OSiO_2	Water glass	Drums, bulk (tank trucks, tank cars)	Cast iron, rubber linings, steel, FRP, PE	Opaque, viscous liquid		Complete	38–42°Bé	Variable ratio of Na_2O to SiO_2; pH of 1% solution: 12.3
Sodium sulfite, Na_2SO_3	Sulfite	Bags, drums, barrels	Cast iron, rubber linings, steel, FRP, PE	White crystalline powder	80–90 lb/ft³	Complete in water	23 (SO_2)	Sulfurous taste and odor

Table continued on next page

Details for Commonly Used Water Treatment Chemicals (continued)

Chemical Name and Formula	Common or Trade Name	Shipping Containers	Suitable Storage Materials	Available Forms/ Descriptions	Density	Solubility, lb/gal	Commercial Strength, %	Additional Characteristics and Properties
Sulfur dioxide, SO_2	Sulfurous acid anhydride	100- to 150-lb steel cylinders, ton containers, tank cars, tank trucks	Shipping container	Colorless gas		20% at 32°F, complete in water	99 (SO_2)	Irritating gas
Sulfuric acid, H_2SO_4	Oil of vitriol, vitriol	Bottles, carboys, drums, trucks, tank cars	FRP; PE; porcelain, glass, and rubber linings	Solution	81.4 lb/ft³ (59.3°Bé)	Complete	77 (59.3°Bé)	Approx. pH of 0.5% solution: 1.2
Tetrasodium pyrophosphate, $Na_4P_2O_7 \cdot 10H_2O$	Alkaline sodium, pyrophosphate, TSPP	125-lb kegs, 200-lb bags, 300-lb barrels	Cast iron, steel, plastics	White powder	68 lb/ft³	0.6 (80°F) 3.3 (212°F)	53 (P_2O_3)	pH of 1% solution: 10.8
Tricalcium phosphate	Fluorex	Bags, drums, bulk, barrels	Cast iron, steel, plastics	Granular	Variable	Insoluble		Also available as white powder
Trisodium phosphate, $Na_3PO_4 \cdot 12H_2O$	Normal sodium phosphate, tertiary sodium phosphate, TSP	125-lb kegs, 200-lb bags, 325-lb barrels	Cast iron, steel, plastics	Crystal—course, medium, standard	56 lb/ft³ 58 lb/ft³ 61 lb/ft³	0.1 (32°F) 13.0 (158°F)	19 (P_2O_3)	pH of 1% solution: 11.9

Chemical Additions

Process Use of Common Water Treatment Chemicals

Process	Chemicals	ANSI/AWWA Standard
Organics adsorption oxidation	Activated carbon, granular	B604
	Chlorine dioxide	None
	Ozone	None
	Potassium permanganate	B603
pH adjustment, stabilization, and corrosion control	Calcium carbonate	None
	Calcium chloride	B550
	Calcium hydroxide	B202
	Calcium oxide	B202
	Carbon dioxide	B510
	Disodium phosphate	B505
	Hydrochloric acid	B300
	Monosodium phosphate	B504
	Phosphoric acid	None
	Potassium hydroxide	B511
	Sodium carbonate	None
	Sodium hexametaphosphate	B502
	Sodium hydroxide	B501
	Sodium polyphosphate	B502
	Sodium silicate	None
	Sodium tripolyphosphate	B503
	Sulfuric acid	None
Softening	Calcium hydroxide	B202
	Calcium oxide	B202
	Sodium carbonate	B201
	Sodium chloride	B200
	Sodium hydroxide	B501
Taste and odor control	Activated carbon, granular	B604
	Activated carbon, powdered	B600

Table continued on next page

Process Use of Common Water Treatment Chemicals (continued)

Process	Chemicals	ANSI/AWWA Standard
	Chlorine	B301
	Chlorine dioxide (sodium chlorite + Cl_2)	B303
	Copper sulfate	B602
	Ozone	None
	Potassium permanganate	B603
Coagulants and coagulant aids	Aluminous sulfate	B403
	Bentonite	None
	Calcium carbonate	None
	Calcium hydroxide	None
	Calcium oxide	None
	EPI-DMA polyamines	B452
	Ferric chloride	B407
	Ferric sulfate	B406
	Ferrous sulfate	B402
	Polyaluminum chloride	B408
	PolyDADMAC	B451
	Polymers	B502
	Sodium aluminate	B405
	Sodium silicate	B404
Dechlorination	Activated carbon, granular	B604
	Ion-exchange resins	None
	Sodium bisulfite	None
	Sodium metabisulfite	B601
	Sulfur dioxide	B512
Disinfection and chlorination	Anhydrous ammonia	None
	Ammonium hydroxide	None
	Ammonium sulfate	B302
	Calcium hypochlorite	B300

Chemical Additions

Table continued on next page

Process Use of Common Water Treatment Chemicals (continued)

Process	Chemicals	ANSI/AWWA Standard
	Chlorinated lime	None
	Chlorine	B301
	Chlorine dioxide (sodium chlorite + Cl_2)	None
	Ozone	None
	Sodium chlorite	B303
	Sodium hypochlorite	B300
Fluoridation and fluoride adjustment	Activated alumina (aluminum oxide)	None
	Calcium fluoride	None
	Fluorosilicic acid	B703
	Hydrogen fluoride	None
	Sodium fluoride	B704
	Sodium fluorosilicate	B702
Mineral oxidation	Chlorine	B301
	Chlorine dioxide (sodium chlorite + Cl_2)	None
	Ozone	None
	Potassium permanganate	B603

KEY FORMULAS FOR CHEMICAL ADDITIONS _____

Gas Chlorine Feed, lb/day

$$\text{lb/day} = \text{flow, mgd} \times \text{concentration, mg/L} \times 8.34 \text{ lb/gal}$$

$$\text{dosage, mg/L} = \frac{\text{lb/day}}{\text{mgd} \times 8.34 \text{ lb/gal}}$$

65% HTH Feed, lb/day—Calcium Hypochlorite

$$\text{HTH, lb/day} =$$

$$\frac{\text{flow, mgd} \times \text{concentration, mg/L} \times 8.34 \text{ lb/gal}}{0.65}$$

$$\text{dosage, mg/L} = \frac{\text{lb/day} \times 0.65}{\text{mgd} \times 8.34 \text{ lb/gal}}$$

$$\text{lb, 65\% HTH} = \frac{\text{gal of water} \times 8.34 \text{ lb/gal} \times \% \text{ solution}}{0.65}$$

5¼%–12.5% Liquid Chlorine—Sodium Hypochlorite

$$\text{lb/gal} = \frac{\text{solution percentage}}{100} \times 8.34 \text{ lb/gal} \times \text{sp gr}$$

$$\text{gpd} = \frac{\text{volume, mgd} \times \text{concentration, mg/L} \times 8.34 \text{ lb/gal}}{\text{lb/gal}}$$

Chlorine Dosage, Demand, and Residual

$$\begin{array}{ccc}
\text{\% purity} & & \text{calculated dosage} \\
\text{of chemical} & \times \begin{array}{c}\text{actual dosage}\\\text{required for}\end{array} = & \text{of 100\% pure} \\
\text{actually used} & \text{chemical used} & \text{chemical required}
\end{array}$$

$$\text{dosage, mg/L} = \text{demand, mg/L} + \text{residual, mg/L}$$

$$\text{demand, mg/L} = \text{dosage, mg/L} - \text{residual, mg/L}$$

$$\text{residual, mg/L} = \text{dosage, mg/L} - \text{demand, mg/L}$$

Chemical Additions

Fluoridation

$$\text{feed, lb/day} = \text{mgd} \times \frac{\text{mg/L}}{\dfrac{\%\text{ purity}}{100} \times \dfrac{\%\text{ flouride}}{100}} \times 8.34 \text{ lb/gal} \times \text{sp gr}$$

adjusted feed, lb/day =

$$\text{mgd} \times \frac{\text{desired, mg/L} - \text{existing, mg/L}}{\dfrac{\%\text{ purity}}{100} \times \dfrac{\%\text{ flouride}}{100}} \times 8.34 \text{ lb/gal} \times \text{sp gr}$$

$$\text{dosage, mg/L} = \frac{\text{feed, lb/day} \times \dfrac{\%\text{ purity}}{100} \times \dfrac{\%\text{ flouride}}{100}}{\text{mgd} \times 8.34 \text{ lb/gal} \times \text{sp gr}}$$

Strength of Solutions—Chemical Feed Pumps

$$\text{gpd} = \frac{\text{required feed, lb/day}}{\text{dry lb/gal}} = \frac{\text{mgd} \times \text{mg/L} \times 8.34 \text{ lb/gal}}{\text{dry lb/gal}}$$

Strength of Solutions—Chemical Feed Rate

$$\text{gpd} = \frac{\text{feed, mL/min} \times 1,440 \text{ min/day}}{1,000 \text{ mL/L} \times 3.785 \text{ L/gal}}$$

$$\text{gpm} = \frac{\text{feed, ML/min}}{3,785 \text{ mL/gal}}$$

$$\text{mL/min} = \frac{\text{gpd} \times 1,000 \text{ mL/L} \times 3.785 \text{ L/gal}}{1,440 \text{ min/day}}$$

$$\text{mL/min} = \text{gpm} \times 3,785 \text{ mL/gal}$$

$$\text{lb/gal} = \frac{\%\text{ solution}}{100} \times 8.34 \text{ lb/gal} \times \text{sp gr}$$

$$\text{lb chemical} = \text{sp gr} \times 8.34 \text{ lb/gal} \times \text{gal of solution}$$

$$\text{sp gr} = \frac{8.34 \text{ lb/gal} + \text{chemical weight, lb/gal}}{8.34 \text{ lb/gal}}$$

$$\text{sp gr, lb/gal} = \text{sp gr} \times 8.34 \text{ lb/gal} - 8.34 \text{ lb/gal}$$

$$\text{percent of chemical in solution} = \frac{\text{dry chemical, lb}}{\text{dry weight chemical, lb + water, lb}} \times 100$$

Sizing Feed Pumps

$$\begin{array}{c}\text{well pump}\\\text{output rate,}\\\text{gpm}\end{array} \times \begin{array}{c}\text{required}\\\text{dosage,}\\\text{ppm}\end{array} \times 1{,}440 \div \begin{array}{c}\text{solution}\\\text{strength,}\\\text{ppm}\end{array} = \begin{array}{c}\text{feed pump}\\\text{output in}\\\text{gpd}\end{array}$$

Dilutions of Chemicals for Lab Reagents or Chemicals

$$\frac{\begin{array}{c}\text{concentration of}\\\text{chemical wanted}\end{array} \times \begin{array}{c}\text{volume of}\\\text{chemical wanted}\end{array}}{\text{concentration of the stock solution}} = \begin{array}{c}\text{volume of chemical}\\\text{to add}\end{array}$$

$$\frac{20 \text{ ntu in } 250 \text{ mL}}{4{,}000 \text{ ntu}} = 1.25 \text{ mL}$$

Milligrams-per-Liter to Pounds-per-Day Conversions

$$\begin{array}{c}\text{feed rate,}\\\text{lb/day}\end{array} = \begin{array}{c}\text{dosage,}\\\text{ppm}\end{array} \times \begin{array}{c}\text{flow rate,}\\\text{mgd}\end{array} \times \begin{array}{c}\text{conversion factor,}\\\text{8.34 lb/gal}\end{array}$$

$$\begin{array}{c}\text{feed rate,}\\\text{lb/day}\end{array} = \begin{array}{c}\text{dosage,}\\\text{mg/L}\end{array} \times \begin{array}{c}\text{flow rate,}\\\text{mgd}\end{array} \times \begin{array}{c}\text{conversion factor,}\\\text{8.34 lb/gal}\end{array}$$

$$\begin{array}{c}\text{chlorine}\\\text{weight, lb}\end{array} = \begin{array}{c}\text{dosage,}\\\text{mg/L}\end{array} \times \begin{array}{c}\text{volume of}\\\text{container,}\\\text{mil gal}\end{array} \times \begin{array}{c}\text{conversion factor,}\\\text{8.34 lb/gal}\end{array}$$

SOLVING FOR THE UNKNOWN VALUE _____

Dosage—Example 1

The chlorine dosage rate at a water treatment plant is 2 mg/L. The flow rate at the plant is 700,000 gpd. How many pounds per day of chlorine are required?

$$\text{mg/L} \times \text{mgd} \times 8.34 \text{ lb/gal} = \text{lb/day}$$

Convert 700,000 gal to million gallons by moving the decimal point six places to the left.

$$700{,}000 \text{ gpd} = 0.7 \text{ mgd}$$
$$2 \text{ mg/L} \times 0.7 \text{ mgd} \times 8.34 = 11.68 \text{ lb/day}$$

Dosage—Example 2

During coagulation, 100 lb/day of alum was fed into a flow of 0.6 mgd. What is the alum dosage in milligrams per liter?

$$\text{mg/L} \times \text{mgd} \times 8.34 \text{ lb/gal} = \text{lb/day}$$
$$\text{mg/L} \times 0.6 \text{ mgd} \times 8.34 \text{ lb/gal} = 100 \text{ lb/day}$$
$$\text{mg/L} = \frac{100}{0.6 \times 8.34} = 20 \text{ mg/L}$$

Dosage—Example 3

The chlorine demand of a water is 6 mg/L. The desired chlorine residual is 0.2 mg/L. How many pounds of chlorine will be required daily to chlorinate a flow of 8 mgd?

$$\text{chlorine dosage} = \text{chlorine demand} + \text{chlorine residual}$$
$$\text{dosage} = 6 \text{ mg/L} + 0.2 \text{ mg/L} = 6.2 \text{ mg/L}$$
$$6.2 \text{ mg/L} \times 8 \text{ mgd} \times 8.34 \text{ lb/gal} = 414 \text{ lb/day chlorine}$$

Wells—Example

How many gallons of 5.25% sodium hypochlorite will be needed to disinfect a well with an 18-in.-diameter casing, 200 feet under the water table, at a dose of 100 mg/L?

$$\text{hypochlorite} = 5.25\%$$
$$\text{chlorine dose} = 100 \text{ mg/L}$$
$$\text{casing} = 18 \text{ in.}$$
$$\text{depth in water} = 200 \text{ ft}$$

1. Find the volume of water in the well in gallons.

$$\text{well volume, gal} = \pi r^2 \times \text{water depth, ft} \times 7.48 \text{ gal/ft}^3$$

$$= \frac{\pi r^2 \times 18 \text{ in.}^2 \times 200 \text{ ft} \times 7.48 \text{ gal/ft}^3}{144 \text{ in.}^2/\text{ft}^2}$$

$$= 2{,}642 \text{ gal}$$

2. Determine the pounds of chlorine needed.

$$\text{chlorine, lb} = \text{volume, mil gal} \times \text{dose, mg/L} \times 8.34 \text{ lb/gal}$$
$$= 0.002642 \text{ mil gal} \times 100 \text{ mg/L} \times 8.34 \text{ lb/gal}$$
$$= 2.2 \text{ lb chlorine}$$

3. Calculate the gallons of 5.25% sodium hypochlorite solution needed.

$$\begin{array}{l}\text{sodium}\\ \text{hypochlorite} =\\ \text{solution, gal}\end{array} \frac{\text{chlorine, lb} \times 100\%}{8.34 \text{ lb/gal} \times \text{hypochlorite, \%}}$$

$$= \frac{2.2 \text{ lb} \times 100\%}{8.34 \text{ lb/gal} \times 5.25\%}$$

$$= 5.0 \text{ gal}$$

Five gallons of 5.25% sodium hypochlorite should do the job.

Mains—Example

A section of an old 8-in. water main has been replaced and a 350-ft section of pipe needs to be disinfected. An initial chlorine dose of 400 mg/L is expected to maintain a chlorine residual of over 300 mg/L during the 3-hour disinfection period. How many gallons of 5.25% sodium hypochlorite solution will be needed?

Known	Unknown
diameter of pipe = 8 in.	5.25% hypochlorite, gal
or 8 in./12 in. = 0.67 ft	
length of pipe = 350 ft	
chlorine dose = 400 mg/L	
hypochlorite = 5.25%	

1. Calculate the volume of water in the pipe in gallons.

$$\text{pipe volume, gal} = \pi r^2 \times \text{diameter, ft}^2 \times \text{length, ft} \times 7.48 \text{ gal/ft}^3$$

$$= \pi r^2 \times 0.67 \text{ ft}^2 \times 350 \text{ ft} \times 7.48 \text{ gal/ft}^3$$

$$= 923 \text{ gal water}$$

2. Determine the pounds of chlorine needed.

$$\text{chlorine, lb} = \text{volume, mil gal} \times \text{dose, mg/L} \times 8.34 \text{ lb/gal}$$

$$= 0.000923 \text{ mil gal} \times 400 \text{ mg/L} \times 8.34 \text{ lb/gal}$$

$$= 3.08 \text{ lb chlorine}$$

3. Calculate the gallons of 5.25% sodium hypochlorite solution needed.

$$\text{sodium hypochlorite solution, gal} = \frac{\text{chlorine, lb} \times 100\%}{8.34 \text{ lb/gal} \times \text{hypochlorite, \%}}$$

$$= \frac{3.08 \text{ lb} \times 100\%}{8.34 \text{ lb/gal} \times 5.25\%}$$

$$= 7.0 \text{ gal}$$

Seven gallons of 5.25% sodium hypochlorite should do the job.

Tanks—Example

An existing service storage reservoir has been taken out of service for inspection, maintenance, and repairs. The reservoir needs to be disinfected before being placed back online. The reservoir is 6 ft deep, 10 ft wide, and 25 ft long. An initial chlorine dose of 100 mg/L is expected to maintain a chlorine residual of over 500 mg/L during the 24-hour disinfection period. How many gallons of 5.25% sodium hypochlorite solution will be needed?

Known	Unknown
tank depth = 6 ft	5.25% hypochlorite, gal
tank width = 10 ft	
tank length = 25 ft	
chlorine dose = 100 mg/L	
hypochlorite = 5.25%	

1. Calculate the volume of water in the tank in gallons.

$$\text{tank volume, gal} = \text{length, ft} \times \text{width, ft} \times \text{depth, ft} \times 7.48 \text{ gal/ft}^3$$

$$= 25 \text{ ft} \times 10 \text{ ft} \times 6 \text{ ft} \times 7.48 \text{ gal/ft}^3$$

$$= 11,200 \text{ gal}$$

2. Determine the pounds of chlorine needed.

$$\text{chlorine, lb} = \text{volume water, mil gal} \times \text{chlorine dose, mg/L} \times 8.34 \text{ lb/gal}$$

$$= 0.01122 \text{ mil gal} \times 100 \text{ mg/L} \times 8.34 \text{ lb/gal}$$

$$= 9.36 \text{ lb chlorine}$$

3. Calculate the gallons of 5.25% sodium hypochlorite solution needed.

$$\text{sodium hypochlorite solution, gal} = \frac{\text{chlorine, lb} \times 100\%}{8.34 \text{ lb/gal} \times \text{hypochlorite \%}}$$

$$= \frac{9.36 \text{ lb} \times 100\%}{8.34 \text{ lb/gal} \times 5.25\%}$$

$$= 21.4 \text{ gal}$$

Twenty-two gallons of 5.25% sodium hypochlorite should do the job.

Chlorinator—Example 1

A deep-well turbine pump is connected to a hydropneumatic tank. Under normal operating heads, the pump delivers 500 gpm. If the desired chlorine dosage is 3.5 mg/L, what should be the setting on the rotameter for the chlorinator (in pounds of chlorine per 24 hours)?

Known	Unknown
pump flow = 500 gpm	rotameter setting,
chlorine dose = 3.5 mg/L	lb chlorine/24 hr

1. Convert pump flow to million gallons per day.

$$\text{flow, mgd} = \frac{500 \text{ gpm} \times 1 \text{ mil} \times 60 \text{ min/hr} \times 24 \text{ hr/day}}{1,000,000}$$

$$= 0.72 \text{ mgd}$$

2. Calculate the rotameter setting in pounds of chlorine per 24 hours.

$$\text{rotameter setting,} \atop \text{lb/day} = \text{flow, mgd} \times \text{dose, mg/L} \times 8.34 \text{ lb/gal}$$

$$= 0.72 \text{ mgd} \times 3.5 \text{ mg/L} \times 8.34 \text{ lb/gal}$$

$$= 21.0 \text{ lb chlorine/day}$$

$$= 21.0 \text{ lb chlorine/24 hr}$$

Chlorinator—Example 2

Using the results from the example on page 174 (a chlorinator setting of 21 lb/24 hr), how many pounds of chlorine would be used in one month if the pump hour meter shows the pump operates an average of 20 hours per day? The chlorinator operates only when the pump operates. How many 150-lb cylinders will be needed per month?

Chemical Additions

Known	Unknown
chlorinator setting = 21 lb/day	1. chlorine used, lb/month
pump operation = 20 hr/day	2. cylinders needed, no./month
chlorine cylinders = 150 lb/cylinder	

1. Calculate the chlorine used in pounds per month.

$$\text{chlorine used, lb/month} = \frac{\text{chlorine setting, lb/day} \times \text{operation, hr/day} \times 30 \text{ days/month}}{24 \text{ hr/day}}$$

$$= \frac{21 \text{ lb/day} \times 20 \text{ hr/day} \times 30 \text{ days/month}}{24 \text{ hr/day}}$$

$$= 525 \text{ lb/month}$$

2. Determine the number of 150-lb cylinders used per month.

$$\text{cylinders needed, no./month} = \frac{\text{chlorine used, lb/month}}{150 \text{ lb chlorine/cylinder}}$$

$$= \frac{525 \text{ lb chlorine/month}}{150 \text{ lb chlorine/cylinder}}$$

$$= 3.5 \text{ cylinders/month}$$

Chlorinator—Example 3

A deep-well turbine pump delivers 400 gpm throughout a 24-hour period. The weight of chlorine in a 150-lb cylinder was 123 lb at the start of the time period and 109 lb at the end of the 24 hours. What was the chlorine dose rate in milligrams per liter?

Known	Unknown
pump flow = 400 gpm	chlorine dose, mg/L
time period = 24 hours	
chlorine weight at start = 123 lb	
chlorine weight at end = 109 lb	

1. Convert flow of 400 gpm to million gallons per day.

$$\text{flow, mgd} = \frac{400 \text{ gal}}{\text{min}} \times \frac{60 \text{ min}}{\text{hr}} \times \frac{24 \text{ hr}}{\text{day}} \times \frac{1 \text{ MG}}{1,000,000}$$

$$= 0.576 \text{ mgd}$$

2. Calculate the chlorine dose rate in mg/L.

$$\text{chlorine dose, mg/L} = \frac{\text{chlorine used, lb/day}}{\text{flow, mgd} \times 8.34 \text{ lb/gal}}$$

$$= \frac{(123 \text{ lb} - 109 \text{ lb})/1 \text{ day}}{0.576 \text{ mgd} \times 8.34 \text{ lb/gal}}$$

$$= \frac{14 \text{ lb chlorine/day}}{0.576 \text{ mgd} \times 8.34 \text{ lb/gal}}$$

$$= \frac{2.9 \text{ lb chlorine}}{1 \text{ MG lb water}}$$

$$= 2.9 \text{ mg/L}$$

Hypochlorinator—Example

Water from a well is being treated by a hypochlorinator. If the hypochlorinator is set at a pumping rate of 50 gpd and uses a 3% available hypochlorite solution, what is the chlorine dose rate in milligrams per liter if the pump delivers 350 gpm?

Known	Unknown
hypochlorinator = 50 gpd	chlorine dose, mg/L
hypochlorite = 3%	
pump = 350 gpm	

1. Convert the pumping rate to million gallons per day.

$$\text{pumping rate, mgd} = \frac{350 \text{ gal}}{\text{min}} \times \frac{60 \text{ min}}{\text{hr}} \times \frac{24 \text{ hr}}{\text{day}} \times \frac{1 \text{ MG}}{1,000,000}$$

$$= 0.50 \text{ mgd}$$

2. Calculate the chlorine dose rate in pounds per day.

$$\text{chlorine dose, lb/day} = \frac{\text{flow, gpd} \times \text{hypochlorite, \%} \times 8.34 \text{ lb/gal}}{100\%}$$

$$= \frac{50 \text{ gpd} \times 3\% \times 8.34 \text{ lb/gal}}{100\%}$$

$$= 12.5 \text{ lb/day}$$

3. Calculate the chlorine dose in milligrams per liter.

$$\text{chlorine dose, mg/L} = \frac{\text{chlorine dose, lb/day}}{\text{flow, mgd} \times 8.34 \text{ lb/gal}}$$

$$= \frac{12.5 \text{ lb chlorine dose/day}}{0.50 \text{ mgd} \times 8.34 \text{ lb/gal}}$$

$$= 3 \text{ lb chlorine/mil lb water}$$

$$= 3 \text{ mg/L}$$

Chemical Dose—Example

Determine the chlorinator setting in pounds per 24 hr if a well pump delivers 300 gpm and the desired chlorine dose is 2.0 mg/L.

Known	Unknown
flow = 300 gpm	chlorinator setting,
chlorine dose = 2.0 mg/L	lb/24 hr

1. Convert the flow from gallons per minute to million gallons per day.

$$\text{flow, mgd} = \frac{\text{flow, gal}}{\text{min}} \times \frac{60 \text{ min}}{\text{hr}} \times \frac{24 \text{ hr}}{\text{day}} \times \frac{1 \text{ MG}}{1,000,000}$$

$$= \frac{300 \text{ gal}}{\text{min}} \times \frac{60 \text{ min}}{\text{hr}} \times \frac{24 \text{ hr}}{\text{day}} \times \frac{1 \text{ MG}}{1,000,000}$$

$$= 0.432 \text{ mgd}$$

NOTE: When multiplying an equation by 1 mil/1,000,000, do not change anything except the units. This is just like multiplying an equation by 12 in./ft or 60 min/hr; all that is being done is changing units.

2. Determine the chlorinator setting in pounds per 24 hours or pounds per day.

$$\frac{\text{chemical}}{\text{feed, lb/day}} = \text{flow, mgd} \times \text{dose, mg/L} \times 8.34 \text{ lb/gal}$$

$$= 0.432 \text{ mgd} \times 2.0 \text{ mg/L} \times 8.34 \text{ lb/gal}$$

$$= 7.2 \text{ lb/day}$$

Small Water Treatment Plants—Example

The optimum dose of liquid alum from the jar tests is 13 mg/L. Determine the setting on the liquid alum chemical feeder in gallons per day when the flow is 1.1 mgd. The liquid alum delivered to the plant contains 5.36 lb of alum per gallon of liquid solution.

Known	Unknown
alum dose = 13 mg/L	chemical feeder setting, gpd
flow = 1.1 mgd	
liquid alum = 5.36 lb/gal	

1. Calculate the chemical feeder setting in gallons per day.

$$\text{chemical feeder setting, gpd} = \frac{\text{flow, mgd} \times \text{alum dose, mg/L} \times 8.34 \text{ lb/gal}}{\text{liquid alum, lb/gal}}$$

$$= \frac{1.1 \text{ mgd} \times 13 \text{ mg/L} \times 8.34 \text{ lb/gal}}{5.36 \text{ lb/gal}}$$

$$= 22.2 \text{ gpd}$$

Amounts of Chemicals Required to Give Various Chlorine Concentrations in 100,000 gal (378.5 m³) of Water*

Desired Chlorine Concentration in Water,	Chlorine Required,		Sodium Hypochlorite Required						Calcium Hypochlorite Required	
			5% Available Chlorine,		10% Available Chlorine,		15% Available Chlorine,		65% Available Chlorine,	
mg/L	lb	(kg)	gal	(L)	gal	(L)	gal	(L)	lb	(kg)
2	1.7	(0.8)	3.9	(14.7)	2.0	(7.6)	1.3	(4.9)	2.6	(1.1)
10	8.3	(3.8)	19.4	(73.4)	9.9	(37.5)	6.7	(25.4)	12.8	(5.8)
50	42.0	(19.1)	97.0	(367.2)	49.6	(187.8)	33.4	(126.4)	64.0	(29.0)

* Amounts of sodium hypochlorite are based on concentrations of available chlorine by volume. For either sodium hypochlorite or calcium hypochlorite, extended or improper storage of chemicals may cause a loss of available chlorine.

Amounts of Chemicals Required to Give Various Chlorine Concentrations in 200 mg/L in Various Volumes of Water*

Volume of Water,		Chlorine Required,		Sodium Hypochlorite Required						Calcium Hypochlorite Required	
				5% Available Chlorine,		10% Available Chlorine,		15% Available Chlorine,		65% Available Chlorine,	
gal	(L)	lb	(kg)	gal	(L)	gal	(L)	gal	(L)	lb	(kg)
10	(37.9)	0.02	(9.1)	0.04	(0.15)	0.02	(0.08)	0.02	(0.08)	0.03	(13.6)
50	(189.3)	0.1	(45.4)	0.2	(0.76)	0.1	(0.38)	0.07	(0.26)	0.15	(68.0)
100	(378.5)	0.2	(90.7)	0.4	(1.51)	0.2	(.76)	0.15	(0.57)	0.3	(136.1)
200	(757.1)	0.4	(181.4)	0.8	(3.03)	0.4	(1.51)	0.3	(1.14)	0.6	(272.2)

* Amounts of sodium hypochlorite are based on concentrations of available chlorine by volume. For either sodium hypochlorite or calcium hypochlorite, extended or improper storage of chemicals may cause a loss of available chlorine.

Number of 5-g Calcium Hypochlorite Tablets Required for Dose of 25 mg/L*

Pipe Diameter,		Length of Pipe Section, *ft (m)*				
		≤13 (4.0)	18 (5.5)	20 (6.1)	30 (9.1)	40 (12.2)
in.	*(mm)*	Number of 5-g Calcium Hypochlorite Tablets				
4	(100)	1	1	1	1	1
6	(150)	1	1	1	2	2
8	(200)	1	2	2	3	4
10	(250)	2	3	3	4	5
12	(300)	3	4	4	6	7
16	(400)	4	6	7	10	13

* Based on 3.25-g available chlorine per tablet; any portion of tablet rounded to the next higher integer.

Chemical Additions

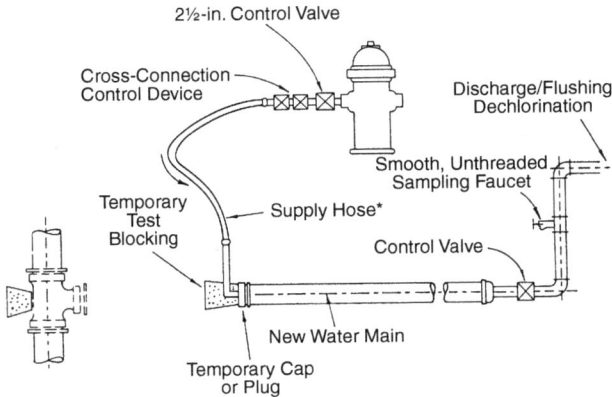

NOTE: This figure applies to pipes with diameters 4 in. (100 mm) through 12 in. (300 mm). All larger sizes must be handled on a case-by-case basis.

*Clean potable-water hose only. This hose must be removed during the hydrostatic pressure test.

Suggested Temporary Flushing/Testing Connection

179

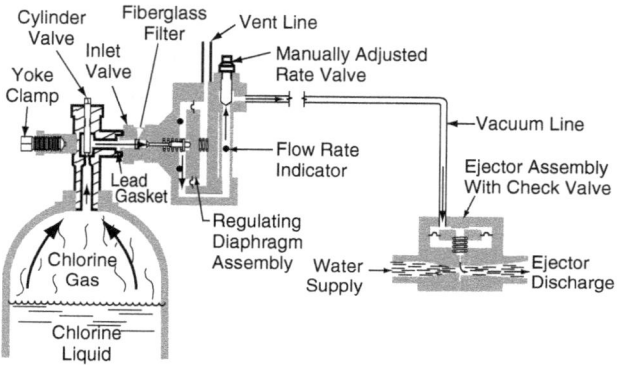

Gas Chlorinator

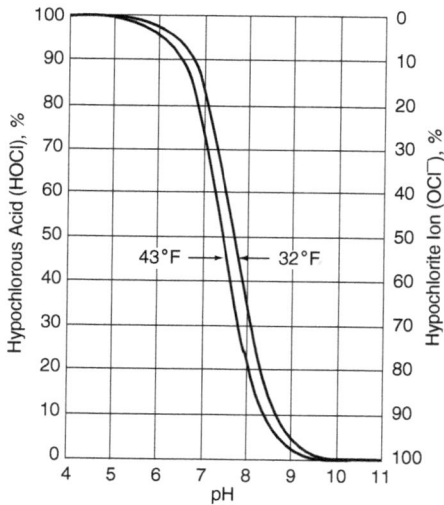

Relationship Among Hypochlorous Acid, Hypochlorite Ion, and pH

Chlorine Required to Produce 25-mg/L Concentration in 100 ft (30.5 m) of Pipe by Diameter

Pipe Diameter,		100% Chlorine,		1% Chlorine Solution,	
in.	*(mm)*	*lb*	*(g)*	*gal*	*(L)*
4	(100)	.013	(5.9)	.16	(0.6)
6	(150)	.030	(13.6)	.36	(1.4)
8	(200)	.054	(24.5)	.65	(2.5)
10	(250)	.085	(38.6)	1.02	(3.9)
12	(300)	.120	(54.4)	1.44	(5.4)
16	(400)	.217	(98.4)	2.60	(9.8)

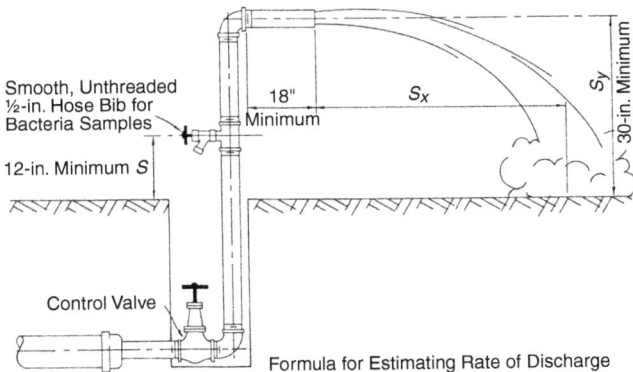

Smooth, Unthreaded ½-in. Hose Bib for Bacteria Samples

18" Minimum

S_x

S_y

30-in. Minimum

12-in. Minimum S

Control Valve

Formula for Estimating Rate of Discharge

$$Q = \frac{2.83\, d_2\, S_x}{\sqrt{S_y}}$$

Where:

Q = discharge in gallons per minute
d = inside diameter of discharge pipe
d, S_x, S_y = measured in inches

NOTE: This figure applies to pipes up to and including 8 in. (200 mm) in diameter.

Suggested Combination Blowoff and Sampling Tap

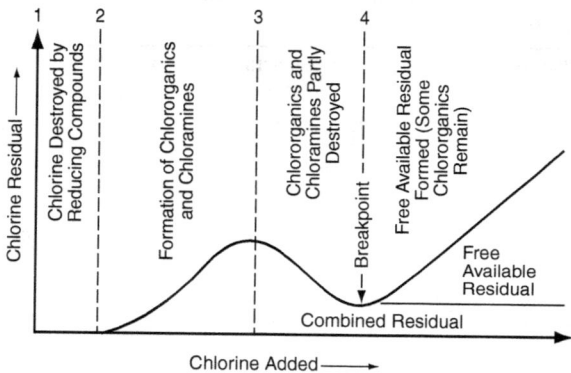

Breakpoint Chlorination Curve

The figure is labeled with the following vertical axis and regions:

- Chlorine Residual (vertical axis, upward)
- 1 — Chlorine Destroyed by Reducing Compounds
- 2 — Formation of Chlororganics and Chloramines
- 3 — Chlororganics and Chloramines Partly Destroyed
- 4 — Free Available Residual Formed (Some Chlororganics Remain)
- Breakpoint
- Free Available Residual
- Combined Residual
- Chlorine Added (horizontal axis)

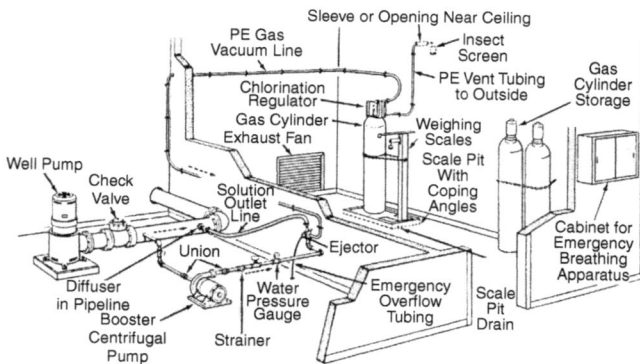

Typical Deep Well Chlorination System

Labels on the diagram:
- Sleeve or Opening Near Ceiling
- PE Gas Vacuum Line
- Insect Screen
- Chlorination Regulator
- PE Vent Tubing to Outside
- Gas Cylinder Storage
- Gas Cylinder Exhaust Fan
- Weighing Scales
- Scale Pit With Coping Angles
- Well Pump
- Check Valve
- Solution Outlet Line
- Ejector
- Cabinet for Emergency Breathing Apparatus
- Union
- Diffuser in Pipeline
- Booster Centrifugal Pump
- Water Pressure Gauge
- Strainer
- Emergency Overflow Tubing
- Scale Pit Drain

To Remote Chlorine Flowmeter

Diaphragm

O-Ring Seat

From Vacuum Regulator No. 1

From Vacuum Regulator No. 2

Toggle Assembly

Vacuum Tubing

Automatic Switchover Module

Remote Flowmeter

Vacuum Regulator No. 2

Vacuum Regulator No. 1

Vent

Vacuum Tubing

Vent

Gas Cylinder No. 1

Gas Cylinder No. 2

Ejector

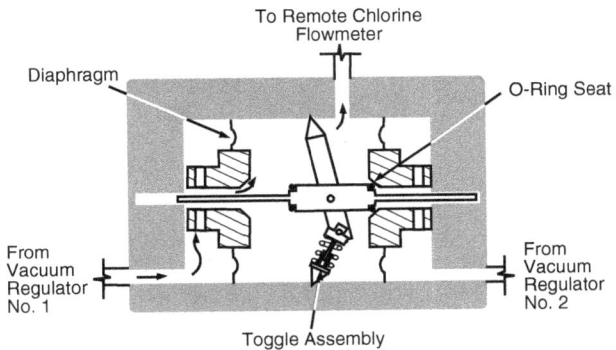

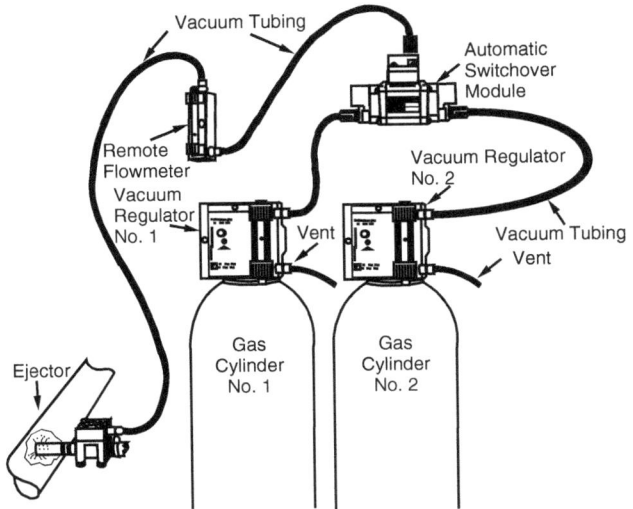

Typical Chlorinator Flow Diagrams

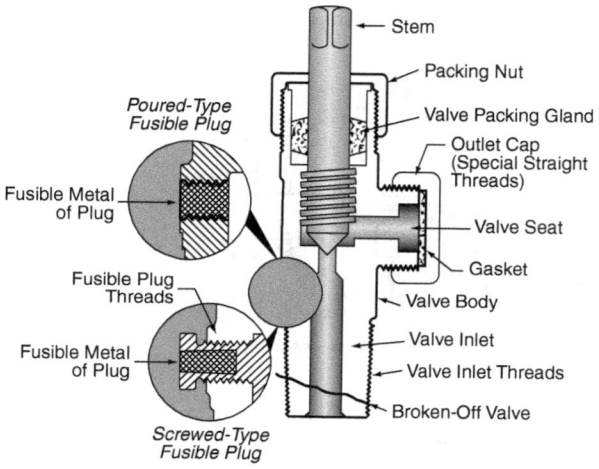

100–150-lb Cylinders

NOTE: Valve closes by turning clockwise; there are about 1¼ turns between wide-open and fully closed position. All threads are right-hand threads.

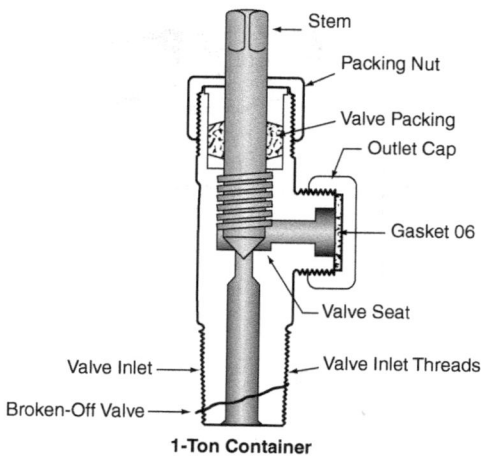

1-Ton Container

Courtesy of Chorine Specialties, Inc.

Standard Chlorine Cylinder Valves

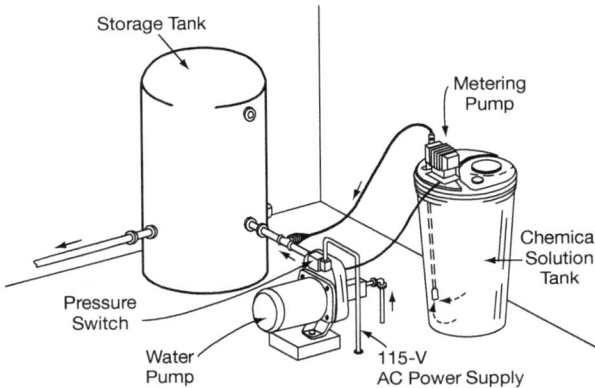

WARNING: When hazardous chemicals are pumped against positive pressure at point of application, use rigid pipe discharge line.

Courtesy of US Filter/Wallace & Tiernan.

Typical Hypochlorinator Installation

Factors for Converting Constituent Concentrations to Softening Chemical Dosages

Constituent Concentration	Conversion Factor* to Determine Required Amount of Lime (100% Pure CaO)	
	mg/L	*lb/MG*
Carbon dioxide, *mg/L as CO$_2$*	1.27	10.63
Bicarbonate alkalinity, *mg/L as CaCO$_3$*	0.56	4.67
Magnesium, *mg/L as Mg*	2.31	19.24
Excess lime, *mg/L as CaCO$_3$*	0.56	4.67

Constituent Concentration	Conversion Factor* to Determine Required Amount of Soda Ash (100% Pure Na$_2$CO$_3$)	
	mg/L	*lb/MG*
Noncarbonate hardness, *mg/L as CaCO$_3$*	1.06	8.83
Excess soda ash, *mg/L as CaCO$_3$*	1.06	8.83

* Multiply constituent concentration by conversion factor to determine softening chemical dosage in units noted.

Disposal of Heavily Chlorinated Water

Check with local sewer department for conditions of disposal to sanitary sewer, and with the state regulatory agency for conditions of disposal to natural drainage courses.

Chlorine residual of disposed water will be neutralized by treating with one of the chemicals listed in the table below.

Amounts of Chemicals Required to Neutralize Various Residual Chlorine Concentrations in 100,000 gal (378.5 m³) of Water

Residual Chlorine Concentration, mg/L	Chemical Required									
	Sulfur Dioxide (SO_2),		Sodium Bisulfite $(NaHSO_3)$,		Sodium Sulfite (Na_2SO_3),		Sodium Thiosulfate $(Na_2S_2O_3 \cdot 5H_2O)$,		Ascorbic Acid* $(C_6O_8H_6)$	
	lb	(kg)	lb	(kg)	lb	(kg)	lb	(kg)	lb	(kg)
1	0.8	(0.36)	1.2	(0.54)	1.4	(0.64)	1.2	(0.54)	2.1	(0.95)
2	1.7	(0.77)	2.5	(1.13)	2.9	(1.32)	2.4	(1.09)	4.2	(1.90)
10	8.3	(3.76)	12.5	(5.67)	14.6	(6.62)	12.0	(5.44)	20.9	(9.47)
50	41.7	(18.91)	62.6	(28.39)	73.0	(33.11)	60.0	(27.22)	104	(47.11)

Source: ANSI/AWWA Standard C651, Disinfecting Water Mains.
  * User should confirm required dosage with chemical supplier.

Amounts of Calcium Hypochlorite Granules to Be Placed at Beginning of Main and at Each 500-ft Interval

Pipe Diameter (d),		Calcium Hypochlorite Granules,	
in.	(mm)	oz	(g)
4	(100)	1.7	(48)
6	(150)	3.8	(108)
8	(200)	6.7	(190)
10	(250)	10.5	(298)
12	(300)	15.1	(428)
14 and larger	(350 and larger)	$D^2 \times 15.1$	$(D^2 \times 428)$
where D is the inside pipe diameter in feet D = $^d/12$			

ALUM PROPERTIES AND DOSAGES _____

Properties of Liquid Alum

Specific Gravity, g/mL	lb/gal	% Al$_2$O$_3$	Equivalent % Dry Alum*	Dry Alum per Gallon Solution, lb	Dry Alum per Liter Solution, g
1.0069	8.40	0.19	1.12	0.09	11.277
1.0140	8.46	0.39	2.29	0.19	23.221
1.0211	8.52	0.59	3.47	0.30	35.432
1.0284	8.58	0.80	4.71	0.40	48.438
1.0357	8.64	1.01	5.94	0.51	61.521
1.0432	8.70	1.22	7.18	0.62	74.902
1.0507	8.76	1.43	8.41	0.74	88.364
1.0584	8.83	1.64	9.65	0.85	102.136
1.0662	8.89	1.85	10.88	0.97	116.003
1.0741	8.96	2.07	12.18	1.09	130.825
1.0821	9.02	2.28	13.41	1.21	145.110
1.0902	9.09	2.50	14.71	1.34	160.368
1.0985	9.16	2.72	16.00	1.47	175.760
1.1069	9.23	2.93	17.24	1.59	190.830
1.1154	9.30	3.15	18.53	1.72	206.684
1.1240	9.37	3.38	19.88	1.86	223.451
1.1328	9.45	3.60	21.18	2.00	239.927
1.1417	9.52	3.82	22.47	2.14	256.540
1.1508	9.60	4.04	23.76	2.28	273.430
1.1600	9.67	4.27	25.12	2.43	291.392

Table continued on next page

Properties of Liquid Alum (continued)

Specific Gravity, g/mL	lb/gal	% Al$_2$O$_3$	Equivalent % Dry Alum*	Dry Alum per Gallon Solution, lb	Dry Alum per Liter Solution, g
1.1694	9.75	4.50	26.47	2.58	309.540
1.1789	9.83	4.73	27.82	2.74	327.970
1.1885	9.91	4.96	29.18	2.89	346.804
1.1983	9.99	5.19	30.53	3.05	365.841
1.2083	10.08	5.43	31.94	3.22	385.931
1.2185	10.16	5.67	33.35	3.39	406.370
1.2288	10.25	5.91	34.76	3.56	427.131
1.2393	10.34	6.16	36.24	3.74	449.122
1.2500	10.43	6.42	37.76	3.93	472.000
1.2609	10.52	6.67	39.24	4.12	494.777
1.2719	10.61	6.91	40.65	4.31	517.027
1.2832	10.70	7.16	42.12	4.51	540.484
1.2946	10.80	7.40	43.53	4.71	563.539
1.3063	10.89	7.66	45.06	4.91	588.619
1.3182	10.99	7.92	46.59	5.12	614.149
1.3303	11.09	8.19	48.18	5.34	640.938
1.3426	11.20	8.46	49.76	5.57	668.078
1.3551	11.30	8.74	51.41	5.81	696.657
1.3679	11.41	9.01	53.00	6.05	724.987

*17% Al$_2$O$_3$ in dry alum + 0.03% free Al$_2$O$_3$.

Alum Addition Required for Stock Solutions

Solution, %	Concentration, mg/L	mg of Alum Added to 1-L Flask
0.1	1,000	1,000
0.2	2,000	2,000
0.5	5,000	5,000
1.0	10,000	10,000
1.5	15,000	15,000
2.0	20,000	20,000

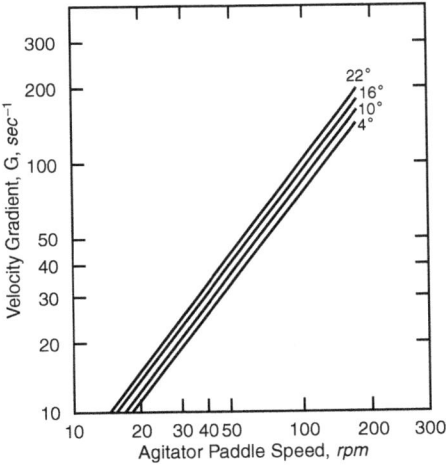

Velocity Gradient Versus rpm at Various Temperatures (°C) for a 2-L Square Beaker Using a Phipps and Bird Stirrer

JAR TESTING

Jar Testing Cheat Sheet*

mL of Solution Added to Jar Containing 2 L Raw Water	Concentration in Raw Water, *mg/L*
0.1	1
0.2	2
0.3	3
0.4	4
0.5	5
0.6	6
0.7	7
0.8	8
0.9	9
1	10
2	20
3	30
4	40
5	50
6	60
7	70
8	80
9	90
10	100

* Dissolve 20 g chemical in lab-pure water and dilute to 1 L (or 1,000 mL). The concentration of this solution will be 20 g/L, or 20 mg/mL, or 20,000 mg/L.
Concentration of solution added × liters of concentration added = concentration in raw water × 2 liters of raw water.

Determination of Jar Test Dosage for Selected Variables

Solution, %	Concentration, *mg/L*	mg/L Dosage per mL of Stock Solution Added	
		To 2-L Jar	To 1-L Jar
0.1	1,000	0.5	1
0.2	2,000	1.0	2
0.5	5,000	2.5	5
1.0	10,000	5.0	10
1.5	15,000	7.5	15
2.0	20,000	10.0	20

Required Flow and Openings to Flush Pipelines (40 psi [276 kPa] Residual Pressure in Water Main)*

Pipe Diameter,		Flow Required to Produce 2.5 ft/sec (approx.) Velocity in Main,		Size of Tap, in. (mm)			Number of 2½-in. (64-mm) Hydrant Outlets
				1 (25)	1½ (38)	2 (51)	
in.	(mm)	gpm	(L/sec)	Number of Taps on Pipe†			
4	(100)	100	(6.3)	1	—	—	1
6	(150)	200	(12.6)	—	1	—	1
8	(200)	400	(25.2)	—	2	1	1
10	(250)	600	(37.9)	—	3	2	1
12	(300)	900	(56.8)	—	—	2	2
16	(400)	1,600	(100.9)	—	—	4	2

* With a 40-psi (276-kPa) pressure in the main with the hydrant flowing to atmosphere, a 2½-in. (64-mm) hydrant outlet will discharge approximately 1,000 gpm (63.1 L/sec); and a 4½-in. (114-mm) hydrant outlet will discharge approximately 2,500 gpm (160 L/sec).

† Based on discharge through 5 ft (1.5 m) of galvanized iron pipe with one 90° elbow.

Chemical Additions

Distribution

With more than 54,000 community water systems
across the nation, consisting of wells, reservoirs,
storage facilities, drinking water treatment
plants, and transmission and distribution
water mains, the water transmission and
distribution system is one of the most complicated
infrastructures in the United States. The system
is made up of intricate linkages among wells,
reservoirs, pumps, pipes, valves, meters, and
myriad other components through which treated
water is moved from the source to homes, offices,
industries, and other consumers. The distribution
system must supply every customer with
a sufficient volume of water, at adequate pressure,
to satisfy the customer's safety, quality, and
aesthetic expectations.

PIPES

Common Pipe Materials

Pipe Material	Range of Diameter, *mm*	Period of Installation	ANSI/AWWA Standard	AWWA Manual	CSA Standard
Pit CI	75–1,500	1850s–1940s	C100*	—	—
Spun CI	75–1,500	1930s–1960s	C100*	—	—
DI	75–1,600	Since 1960s	C151	M41	—
Steel	>150	Since 1850s	C200	M11	Z245.1
PVC	100–1,200	Since 1970s	C900/C905	M23	B137.3
HDPE	100–1,575	Since 1980s	C901/C906	M55	B137.1
(A–C)	100–1,050	1930s to 1980s	—	—	—
CPP	250–3,660	Since 1940s	C300/C301/C302/C303	M9	—

Source: National Guide to Sustainable Municipal Infrastructures—InfraGuide, 2002. Deterioration and Inspection of Water Distribution Systems, Ottawa, Ontario.

  * British standard cast-iron pipe is also common in Canada.

Pit CI = pit cast iron
Spun CI = spun cast iron
DI = ductile iron
PVC = polyvinyl chloride
HDPE = high-density polyethylene
CPP = concrete pressure pipe

Comparison of Transmission and Distribution Pipeline Materials

Material	Common Sizes, Diameter		Normal Maximum Working Pressure		Advantages	Disadvantages
	in.	*(mm)*	*psi*	*(kPa)*		
Ductile iron (cement lined)	3–64	(76–1,625)	350	(2,413)	Durable, strong, high flexural strength, good corrosion resistance, lighter weight than cast iron, greater carrying capacity for same external diameter, easily tapped	Subject to general corrosion if installed unprotected in a corrosive environment
Concrete (reinforced)	12–168	(305–4,267)	250	(1,724)	Durable with low maintenance, good corrosion resistance, good flow characteristics, O-ring joints are easy to install, high external load capacity, minimal bedding and backfill requirements	Requires heavy lifting equipment for installation, may require special external protection in high-chloride soils
Concrete (prestressed)	16–144	(406–3,658)	50	(2,413)	Same as for reinforced concrete	Same as for reinforced concrete

Table continued on next page

Distribution

195

Comparison of Transmission and Distribution Pipeline Materials (continued)

Material	Common Sizes, Diameter		Normal Maximum Working Pressure		Advantages	Disadvantages
	in.	(mm)	psi	(kPa)		
Steel	4–120	(100–3,048)	High		Lightweight, easy to install, high tensile strength, low cost, good hydraulically when lined, adapted to locations where some movement may occur	Subject to electrolysis; external corrosion in acidic or alkaline soil; poor corrosion resistance unless properly lined, coated, and wrapped; air-and-vacuum relief valves imperative in large sizes; subject to tuberculation when unlined
Polyvinyl chloride	4–36	(100–914)	200	(1,379)	Lightweight, easy to install, excellent resistance to corrosion, good flow characteristics, high tensile strength and impact strength	Difficult to locate underground, requires special care during tapping, susceptible to damage during handling, requires special care in bedding

Comparison Between US Standard Cast-Iron Pipe Sizes and ISO Standard Sizes*

	US Standard			ISO Standard		
Nominal Size, in.	OD, in.	OD, mm	Nominal Size, mm	OD, in.	OD, mm
3	3.96	100.58	80	3.86	98
4	4.80	121.92	100	4.65	118
			125	5.67	144
6	6.90	175.26	150	6.69	170
8	9.05	229.87	200	8.74	222
10	11.10	281.94	250	10.79	274
12	13.20	335.28	300	12.83	326
14	15.30	388.62	350	14.88	378
16	17.40	441.96	400	16.89	429
18	19.50	495.30			
20	21.60	548.64	500	20.94	532
24	25.80	655.32	600	25.00	635

*ISO Standard is used in most countries other than the United States. Nominal size is the approximate inside diameter.

Distribution

Types of Plastic Pipe

It is very important that the correct primers and solvents be used on each type of pipe or the joints will not seal properly and the overall strength will be weakened.

Types	Characteristics
ABS (acrilonitrile butadiene styrene)	Strong, rigid, and resistant to a variety of acids and bases. Some solvents and chlorinated hydrocarbons may damage the pipe. Maximum usable temperature is 160°F (71°C) at low pressures. It is most common as a DWV pipe.
CPVC (chlorinated polyvinyl chloride)	Similar to PVC but designed specifically for piping water at up to 180°F (82°C). Pressure rating is 100 psi.
FRP (fiberglass-reinforced plastic) epoxy	A thermosetting plastic over fiberglass. Very high strength and excellent chemical resistance. Good to 220°F (105°C). Excellent for labs.
PB (polybutylene)	A flexible pipe for pressurized water systems, both hot and cold. *Only* compression and banded-type joints can be used.
PE (polyethylene)	A flexible pipe for pressurized water systems such as sprinklers. Not for hot water.
Polypropylene	Low pressure, lightweight material that is good up to 180°F (82°C). Highly resistant to acids, bases, and many solvents. Good for laboratory plumbing.
PVC (polyvinyl chloride)	Strong, rigid, and resistant to a variety of acids and bases. Some solvents and chlorinated hydrocarbons may damage the pipe. Can be used with water, gas, and drainage systems but *not* with hot-water systems.
PVDF (polyvinylidene fluoride)	Strong, very tough, and resistant to abrasions, acids, bases, solvents, and much more. Good to 280°F (138°C). Good in lab.

PVC Pipe Terminology

ANSI/AWWA C900 is a category of standard dimension ratio (SDR) pipe that is the same diameter as ductile-iron (DI) pipe. The following are all classified as C900. SDR/14 is Class 305, SDR/18 is Class 235, SDR/25 is Class 165.

The class signifies working pressure. SDR refers to a ratio of wall thickness to actual pipe outside diameter (OD). For example, SDR/18 pipe × 6.90 in. (the actual OD of 6-in. DI pipe) has a wall thickness of 6.90/18 = 0.38 in. Mechanical joints on C900 fittings are used with C900 pipe.

SDR/21 and SDR/26 have class designations that correspond to rated working pressure. The ratings incorporate a lower service factor (same as safety factor) than C900 pipe, which explains why SDR/21 and 26 list a higher class rating for a given wall thickness. SDR/21 is Class 200; SDR/26 is Class 160.

The SDR numbers relate to wall thickness SDR/21 × 6.63 in. (actual 6-in. steel pipe OD) has a wall thickness of 6.63/21 = 0.32 in.

Schedules 40 and 80 have the same diameter as steel pipe. The pressure ratings vary with the diameter of the pipe. The larger the diameter, the lower the rating.

Pipe Size, in.	Working Pressure		
	Schedule 40	Schedule 80	
	Socket	Socket	Threaded
½	600	850	420
¾	480	690	340
1	450	630	320
1¼	370	520	260
1½	330	471	240
2	300	425	210
2½	280	400	200
3	260	375	190
4	220	324	160
6	180	280	140

NOTE: ASTM D1785 is standard for Schedules 40, 80, and 120 PVC pipe. ASTM D2241 is standard for SDR series PVC pipe.

SDR/21, SDR/26, and all Schedule pipe can be used with Schedule 40 and Schedule 80 fittings because they conform to steel pipe dimensions.

All of the above, C900, SDR/21 and 26, and Schedule 40/80, can be used for both water and sewer lines.

SDR/35 and SDR/41 are used exclusively for sewer drain only. Their outside dimensions are different from SDR pressure pipe and are different from each other in sizes other than 4 in. and 6 in.

Flange Guide

Gasket and Machine Bolt Dimensions for 150-lb Flange

Pipe Size, in.	Bolts Needed	Machine Bolt Dimension, in.	Gasket Dimensions	
			Ring, in.	Full Face, in.
2	4	$\frac{5}{8} \times 2\frac{3}{4}$	$2\frac{3}{8} \times 4\frac{1}{8}$	$2\frac{3}{8} \times 6$
$2\frac{1}{2}$	4	$\frac{5}{8} \times 3$	$2\frac{7}{8} \times 4\frac{7}{8}$	$2\frac{7}{8} \times 7$
3	4	$\frac{5}{8} \times 3$	$3\frac{1}{2} \times 5\frac{3}{8}$	$3\frac{1}{2} \times 7\frac{1}{2}$
$3\frac{1}{2}$	8	$\frac{5}{8} \times 3$	$4 \times 6\frac{3}{8}$	$4 \times 8\frac{1}{2}$
4	8	$\frac{5}{8} \times 3$	$4\frac{1}{2} \times 6\frac{7}{8}$	$4\frac{1}{2} \times 9$
5	8	$\frac{3}{4} \times 3\frac{1}{4}$	$5\frac{9}{16} \times 7\frac{3}{4}$	$5\frac{9}{16} \times 10$
6	8	$\frac{3}{4} \times 3\frac{1}{4}$	$6\frac{5}{8} \times 8\frac{3}{4}$	$6\frac{5}{8} \times 11$
8	8	$\frac{3}{4} \times 3\frac{1}{2}$	$8\frac{5}{8} \times 11$	$8\frac{5}{8} \times 13\frac{1}{2}$
10	12	$\frac{7}{8} \times 3\frac{3}{4}$	$10\frac{3}{4} \times 13\frac{3}{8}$	$10\frac{3}{4} \times 16$
12	12	$\frac{7}{8} \times 4$	$12\frac{3}{4} \times 16\frac{1}{8}$	$12\frac{3}{4} \times 19$

Outside Diameter of Small Pipe and Tube

Type of Pipe or Tubing	Nominal Pipe Size, in.									
	¼	⅜	½	⅝	¾	1	1¼	1½	2
Copper and CTS-PE*	.375	.500	.625	.750	.875	1.125	1.375	1.625	2.125
Iron pipe and IPS-PE†	.540	.675	.840	—	1.050	1.315	1.660	1.900	2.375
Lead pipe‡									
Strong	—	—	—	1.010	1.156	1.428	—	—	—
Extra strong	—	—	.876	1.082	1.212	1.492	1.765	2.076	2.751
Double extra strong	—	—	1.012	1.335	1.596	—	—	—	—

Note: Polyethylene (PE) pipe is also available with the same inside diameter as iron pipe. The wall thickness varies with the pressure class, so the outside diameter (OD) is variable. See manufacturer's information or ANSI/AWWA Standard C901 for details.
* CTS-PE—Copper tubing size polyethylene. Tubing has the same OD as copper tube.
† IPS-PE—Iron pipe size polyethylene. Pipe has the same OD as iron pipe.
‡ The OD of lead pipe is approximate.

Distribution

Copper Tubing

Copper tubing is available in three types that are used by the water and wastewater industries:

    Type K = heavy wall

    Type L = medium wall

    Type M = light weight

All types have the same outside diameter.

Copper tubing is also available in a form called "soft" which will bend, and as "hard" which is rigid.

- Type K (soft) is primarily used where the pipe is to be buried, such as for water services lines.

- Type L (hard) is primarily used for interior water piping.

- Type M (hard) is primarily used for hot-water heating and drain lines.

- DWV (drain-waste-vent) is used for aboveground use in no-pressure applications.

Dimensions of Copper Tubing

Nominal Size, in.	Outside Diameter, in.	Inside Diameter, in.		
		Type K	Type L	Type M
¼	0.375	0.035	0.315	
⅜	0.500	0.402	0.430	
½	0.625	0.527	0.545	
¾	0.875	0.745	0.785	
1	1.125	0.995	1.025	
1¼	1.375	1.245	1.265	1.291
1½	1.625	1.481	1.505	1.527
2	2.125	1.959	1.985	2.009

Pipe Capacities

Pipe Capacity Comparison*

Main Size, in.	Smaller Pipe Size, in.							
	¾	1	2	3	4	6	8	10
2	13	6	1					
3	39	18	2	1				
4	84	39	6	2	1			
6	247	115	18	6	2	1		
8	530	247	39	13	6	2	1	
10	957	447	71	24	11	3	1	1
12		724	115	39	18	6	2	1
14		1,090	174	59	27	9	4	2
16			247	84	39	13	6	3
18			338	115	53	18	8	4
20			447	153	71	24	11	6

* Number of smaller pipes required to provide carrying capacity equal to a larger pipe.

Contents of Pipe

Pipe Diameter, in.	Inside Pipe Diameter, ft	Approximate Cubic-Feet-per-Foot Length	Approximate US Gallons-per-Foot Length
¾	.0625	.0031	.0230
1	.0833	.0055	.0408
1¼	.1042	.0086	.0638
1½	.1250	.0123	.0918
2	.1667	.0218	.1632
3	.2500	.0491	.3673
4	.3333	.0873	.6528
6	.5000	.1963	1.469
8	.6667	.3490	2.611
10	.8333	.5455	4.018
12	1.000	.7854	5.876
14	1.167	1.069	7.997
16	1.333	1.396	10.44
18	1.500	1.767	13.22
20	1.666	2.182	16.32
24	2.000	3.142	23.50

Symbols for Pipe Fittings

Flanged	Screwed	Bell & Spigot	Welded	Soldered	
—‖—	—┼—	—←—	—×—	—⊙—	Joint
					Elbow—45°
					Elbow—90°
⊖‖—	⊖┼—	⊖→—	⊖×—	⊖o—	Elbow—Turned Down
○‖—	○┼—	○→—	○×—	○o—	Elbow—Turned Up
					Elbow—Long Radius
					Side Outlet Elbow—Outlet Down
					Side Outlet Elbow—Outlet Up
					Double Branch Elbow
					Base Elbow
					Reducing Elbow
					Simple Sweep Tee
					Double Sweep Tee
‖—┼—‖	—┼—	→—┼—←	×—┼—×	o—┼—o	Tee
‖○‖	—⊙—	→○←	×○×	o⊖o	Tee—Outlet Down
‖○‖	—○—	→○←	×○×	o○o	Tee—Outlet Up
‖⊕‖	—⊕—	→⊕←			Side Outlet Tee—Outlet Down
‖○‖	—○—	→○←			Side Outlet Tee—Outlet Up
‖—┼—‖	—┼—	→—┼—←		⊖—┼—⊙	Cross

Flanged	Screwed	Bell & Spigot	Welded	Soldered	
					Reducer
					Concentric Reducer
					Lateral
					Globe Valve
					Angle Globe Valve
					Gate Valve
					Angle Gate Valve
					Check Valve
					Angle Check Valve
					Safety Valve
					Quick Opening Valve
					Float Operating Valve
					Motor Operating Gate Valve
					Motor Operating Globe Valve
					Stop Cock
					Expansion Joint Flange
					Reducing Flange
					Bushing
					Union
					Sleeve

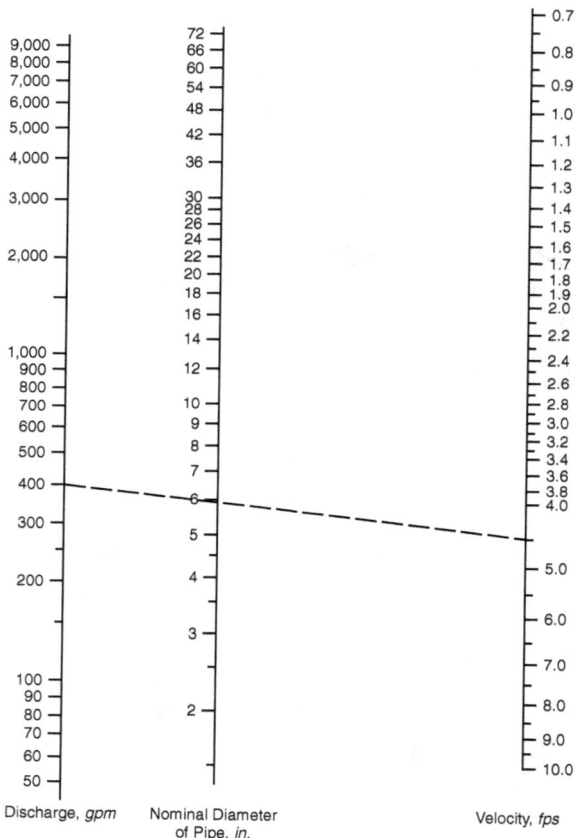

Discharge, *gpm*　　Nominal Diameter　　Velocity, *fps*
　　　　　　　　　　of Pipe, *in.*

EXAMPLE: A 6-in. pipe is flowing full at 400 gpm. Draw a line through the 400 and the 6 to intersect the velocity line at about 4.5.

Diagram for Calculating Approximate Velocity of Flow in Pipe

Pipe and Hose Threads Most Commonly Used in the United States

Taper Pipe Thread

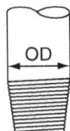

NPT   American standard taper pipe thread (old nomenclature IPT)

IPT   Iron pipe thread (old nomenclature). Widely used to designate all types of pipe thread (NPT, NPSM, NPSH).

Straight Pipe Thread

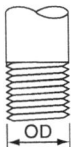

NPSM   American standard straight pipe thread for free mechanical joints (same as NPS)

IPS   Iron pipe straight thread "V" pattern. Same as NPSH.

Fire Hose Coupling Straight Thread

NST   Coarse thread used on most fire hydrants. Not interchangeable with any other threads. Also known as NH.

GHT   Garden hose thread. Only ¾ in. with 11½ threads per inch. Used for ½-in., ⅝-in., and ¾-in. hose. Not interchangeable with any other thread.

Compatibility of Pipe and Hose Threads

Size, in.	NST (Fire Hose)		NPSM		GHT	
	ODM*	TPI†	ODM*	TPI†	ODM*	TPI†
¾	—	—	1.0353	14	1.0625	1½
1	1.375	8	1.295	11½	—	—
1¼	1.6718	9	1.6399	11½	—	—
1½	1.990	9	1.8788	11½	—	—
2	2.5156	8	2.3528	8	—	—
2½	3.0686	7½	2.841	8	—	—
3	3.6239	6	3.470	8	—	—
3½	4.2439	6	3.970	8	—	—
4	5.0109	4	4.470	8	—	—
5	6.260	4	—	—	—	—
6	7.025	4	—	—	—	—

* ODM = maximum outside diameter in inches.
† TPI = threads per inch.

Thread Combinations That Are Compatible

Female Thread	Male Thread	Sealant That Should Be Used
NPT	NPT	Thread seal
NPSM	NPSM	Washer seal
	NPT	Washer seal
NPSH	NPSH	Washer seal
	NPSM	Washer seal
	NPT	Washer seal
GHT*	GHT*	Washer seal

Male Thread	Female Thread	Sealant That Should Be Used
NPT	NPT	Thread seal
	NPSM	Washer seal
	NPSH	Washer seal
NPSM	NPSM	Washer seal
	NPSH	Washer seal
NPSH	NPSH	Washer seal
GHT*	GHT*	Washer seal

* Not compatible with any other threads.

Locating Underground Pipe

Electronic equipment is now available at reasonable costs that allow all water systems to locate piping. Electronic locators are used for locating metallic water mains, service pipes, valve boxes, and access covers. The units will also locate metallic gas pipe and telephone and television cables, but it is generally best to let other utilities locate their own pipes and cables to ensure accuracy and avoid liability.

Ground-Probing Radar

Radar is now used for many purposes and would be very useful for all utilities if it could be easily used to detect all underground pipes and cables. Unfortunately, the units that are available now work only under certain soil conditions, require special expertise to interpret the results, and are expensive. Radar equipment may someday be developed for use by all water systems.

Metal Detectors

Units similar to military mine detectors have flat detection coils on the ends of their handles. When the coil is near a metal object that is relatively close to the ground surface, there is a change in audible tone or meter reading. Relatively inexpensive units can save time in locating metal access covers, valve boxes, and meter-pit covers that have been paved over, have grass growing over them, or are covered by snow.

Some valve boxes and meter pits are now made entirely of plastic. To make one of these detectable by an electronic locator, fasten a small piece of metal or a small magnet to the underside of the cover.

Magnetic Locators

A magnetic locator consists of a single unit that monitors the earth's magnetic field. When it is brought near any object containing iron or steel, there is an imbalance in the magnetic field, which the locator translates into a change in sound or meter reading. A unit will generally detect an 8-in. (200-mm) ductile-iron pipe 8 ft (2.5 m) deep. It will not detect noniron metals, such as aluminum cans, copper water service pipe, or cables.

Radio Transmission Units

Another type of locator uses a radio transmitter and receiver. Commonly called a *line tracer*, a *continuous metal locator*, or a *pipe and cable locator*, it consists of two units: a radio transmitter that sends out a signal and a receiver with a loop antenna that receives a maximum signal only in the plane of the loop. It will locate any continuous pipe or cable made of any type of metal. The transmitter introduces a signal into the metal, either by a direct connection or by placing the transmitter above the line.

Pipe and cable locators will locate copper and galvanized service pipes over a considerable distance. But when used for new ductile-iron pipe with rubber joints, the signal may travel for only a few pipe lengths because the signal does not conduct well from one pipe to the next.

Nonmetallic Pipe Locators

The best way of locating nonmetallic pipe is to bury a metallic tape or tracer wire in the ditch when the pipe is installed. If a metal tape is used, it is usually buried about 1.0 ft (0.3 m) below the surface so it will be easily detectable and will act as a visual warning to anyone excavating in the vicinity of the pipe. The tape or wire can be easily located with a pipe and cable locator by either direct or inductive signal.

Unfortunately, few installers have had the foresight to install tracers, so there are many water systems with nonmetallic pipe and no location records. One way of locating nonmetallic mains and services is to use a unit that uses a transmitter to send small shock waves through the water. The pipe is then located using a receiver that detects the vibration in the soil above the pipe. In most soils, pipe can be located at least 250 ft (76 m) from the transmitter, and may work over as long as 1.0 mile (1.5 km) under ideal conditions. This type of unit will usually not work well in dry, loose soil or very wet ground.

Color Codes for Pipes and Hydrants

Uniform Color Code Used for Identifying Buried Public Works Pipe and Cables

Red	Electric power lines
	Lighting cables
	Conduit
Yellow	Gas
	Oil
	Steam
	Petroleum
Orange	Communications cables
	Alarm cables
	Signal lines
Blue	Potable water
	Irrigation water
	Slurry lines
Green	Sewers
	Drain lines
Pink	Temporary survey markings
White	Proposed excavation

Color Scheme for Identifying the Capacity of Fire Hydrants

AWWA recommends a color scheme for painting hydrants to indicate their relative capacity.

The capacity is to be determined by flow measurements of individual hydrants taken at a period of ordinary demand. When initial pressures are over 40 psig at the hydrant under test, the rating is to be based on 20 psig residual pressure, observed at the nearest hydrant connected to the same main and when no water is being drawn. When initial pressures are less than 40 psig, residual pressures must be at least half that of the initial pressure.

The tops and caps of hydrants should be painted the colors indicated on page 214 to indicate the hydrant capacity. It is recommended that private hydrants be painted a color that will distinguish them from public hydrants.

Standard Hydrant Color Scheme to Indicate Flow Capacity

Hydrant Class	Color*	Usual Flow Capacity at 20 psig (140 kPa [gauge])† gpm	Usual Flow Capacity at 20 psig (140 kPa [gauge])† (L/min)	Hydrants That on Individual Tests Usually Have a Flow Capacity of:
AA	Light Blue	1,500	(5,680)	1,500 gpm or greater
A	Green	1,000 to 1,499	(3,785 to 5,675)	1,000 gpm or greater
B	Orange	500 to 999	(1,900 to 3,780)	500 to 1,000 gpm
C	Red	Less than 500	(Less than 1,900)	0 to 500 gpm

* As designed in Federal Standard 595B, General Services Administration, Specification Section, Washington, D.C.

† Capacities are to be rated by flow measurements of individual hydrants at a period of ordinary demand. See *ANSI/AWWA Standard for Dry-Barrel Fire Hydrants*, C502, for additional details.

Treatment Plant Color Coding

Pipeline Color Coding Used in Water Treatment Plants (continued)

Type of Line	Contents of Line	Color of Pipe
Water lines	Raw water	Olive green
	Settled or clarified water	Aqua
	Finished or potable water	Dark blue
Chemical lines	Alum or primary coagulant	Orange
	Ammonia	White
	Carbon slurry	Black
	Caustic	Yellow with green band
	Chlorine gas or solution	Yellow
	Fluoride	Light blue with red band
	Lime slurry	Light green
	Ozone	Yellow with orange band
	Phosphate compounds	Light green with red band
	Polymers or coagulant aids	Orange with green band
	Potassium permanganate	Violet
	Soda ash	Light green with orange band
	Sulfuric acid	Yellow with red band

Table continued on next page

	Sulfur dioxide	Light green with yellow band

Pipeline Color Coding Used in Water Treatment Plants (continued)

Type of Line	Contents of Line	Color of Pipe
Waste lines	Backwash waste	Light brown
	Sludge	Dark brown
	Sewer (sanitary or other)	Dark gray
Other lines	Compressed air	Dark green
	Gas	Red
	Other pipes	Light gray

Pipeline Color Coding Used in Wastewater Treatment Plants

Type of Line	Contents of Line	Color of Pipe
Sludge lines	Raw sludge	Brown with black bands
	Sludge recirculation or suction	Brown with yellow bands
	Sludge draw off	Brown with orange bands
	Sludge recirculation discharge	Brown
Gas lines	Sludge gas	Orange (or red)
	Natural gas	Orange (or red) with black bands
Water lines	Nonpotable water	Blue with black bands
	Potable water	Blue
	Water for heating digestors or buildings	Blue with a 6-in. (150-mm) red band spaced 30 in. (760 mm) apart
Other lines	Chlorine	Yellow
	Sulfur dioxide	Yellow with red bands
	Sewage (wastewater)	Gray
	Compressed air	Green

Source: Recommended Standards for Water Works and Recommended Standards for Wastewater Facilities (*The "Ten States Standards"*).
NOTE: It is recommended that the direction of flow and name of the contents be noted on all lines.

JOINTS AND ASSEMBLIES

A. Lap-Welded Slip Joint

May be welded inside or outside, or both inside and outside when required.

B. Single-Butt Welded Joint

C. Double-Butt Welded Joint

Butt Strap

D. Butt Strap Joint

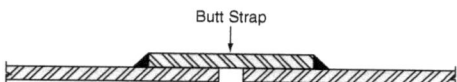

E. Fabricated Rubber Gasket Joint

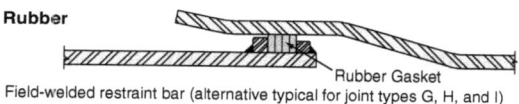

Rubber Gasket

Field-welded restraint bar (alternative typical for joint types G, H, and I)

F. Rolled-Groove Rubber Gasket Joint

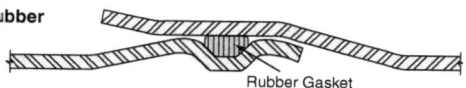

Rubber Gasket

G. Tied Rubber Gasket Joint

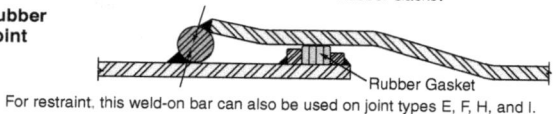

Rubber Gasket

For restraint, this weld-on bar can also be used on joint types E, F, H, and I.

H. Carnegie-Shape Rubber Gasket Joint

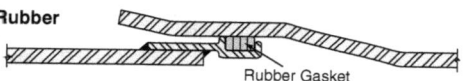

Rubber Gasket

I. Carnegie-Shape Rubber Gasket Joint With Weld-on Bell Ring

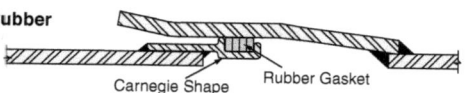

Carnegie Shape   Rubber Gasket

J. Sleeve Coupling

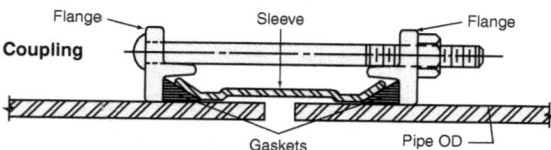

Flange   Sleeve   Flange

Gaskets   Pipe OD

Source: AWWA M41—Ductile Iron Pipe and Fittings.

Common Welded and Rubber-Gasketed Joints Used for Connecting Steel Pipe

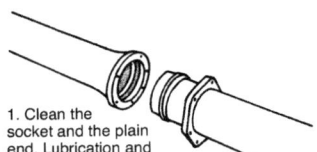

1. Clean the socket and the plain end. Lubrication and additional cleaning should be provided by brushing both the gasket and plain end with soapy water or an approved pipe lubricant meeting the requirements of ANSI/AWWA C111/A21.11, just prior to slipping the gasket onto the plain end for joint assembly. Place the gland on the plain end with the lip extension toward the plain end, followed by the gasket with the narrow edge of the gasket toward the plain end.

NOTE: In cold weather, it is preferred to warm the gasket to facilitate assembly of joint.

2. Insert the pipe into the socket and press the gasket firmly and evenly into the gasket recess. Keep the joint straight during assembly.

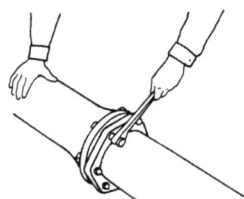

3. Push the gland toward the socket and center it around the pipe with the gland lip against the gasket. Insert bolts and hand-tighten nuts. Make deflection after joint assembly but before tightening bolts.

4. Tighten the bolts to the normal range of bolt torque as indicated in the Maximum Joint Deflection table on page 219 while at all times maintaining approximately the same distance between the gland and the face of the flange at all points around the socket. This can be accomplished by partially tightening the bottom bolt first, then the top bolt, next the bolts at either side, finally the remaining bolts. Repeat the process until all bolts are within the appropriate range of torque. In large sizes (30–48 in. [762–1,219 mm]), five or more repetitions may be required. The use of a torque-indicating wrench will facilitate this procedure.

Source: AWWA M41—Ductile Iron Pipe and Fittings.

Mechanical-Joint Assembly

Mechanical-Joint Bolt Torque

Joint Size,		Bolt Size,		Range of Torque,	
in.	*(mm)*	*in.*	*(mm)*	*ft-lb*	*(N·m)*
3	(76)	⁵⁄₈	(16)	45–60	(61–81)
4–24	(102–610)	³⁄₄	(19)	75–90	(102–122)
30–36	(762–914)	1	(25)	100–120	(136–163)
42–48	(1,067–1,219)	1¼	(32)	120–150	(163–203)

Source: AWWA M41—Ductile Iron Pipe and Fittings.

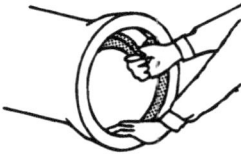

1. Thoroughly clean the groove and the bell socket of the pipe or fitting; also clean the plain end of the mating pipe. Using a gasket of the proper design for the joint to be assembled, make a small loop in the gasket and insert it in the socket, making sure the gasket faces the correct direction and that it is properly seated. For pipe sizes larger than 20 in., it may be necessary to make two loops in the gasket (6 and 12 o'clock). NOTE: In cold weather, it is necessary to warm the gasket to facilitate assembly of the joint.

2. Apply lubricant to the exposed surface of the gasket and plain end of the pipe in accordance with the pipe manufacturer's recommendations. Do not apply lubricant to the bell socket or the surface of the gasket in contact with the bell socket. Lubricant is furnished in sterile containers, and every effort should be made to protect against contamination of the container's contents.

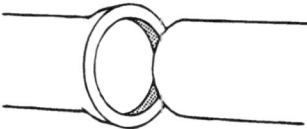

3. Be sure that the plain end is beveled per the manufacturer's recommendations; square or sharp edges may damage or dislodge the gasket and cause a leak. When pipe is cut in the field, bevel the plain end with a heavy file or grinder to remove all sharp edges. Push the plain end into the bell of the pipe. Keep the joint straight while pushing. Make deflection after the joint is assembled.

4. Small pipe can be pushed into the bell socket with a long bar. Large pipe requires additional power, such as a jack, lever puller, or backhoe. The supplier may provide a jack or lever puller on a rental basis. A timber header should be used between the pipe and the jack or backhoe bucket to avoid damage to the pipe.

Source: AWWA M41—Ductile Iron Pipe and Fittings.

Push-On-Joint Assembly

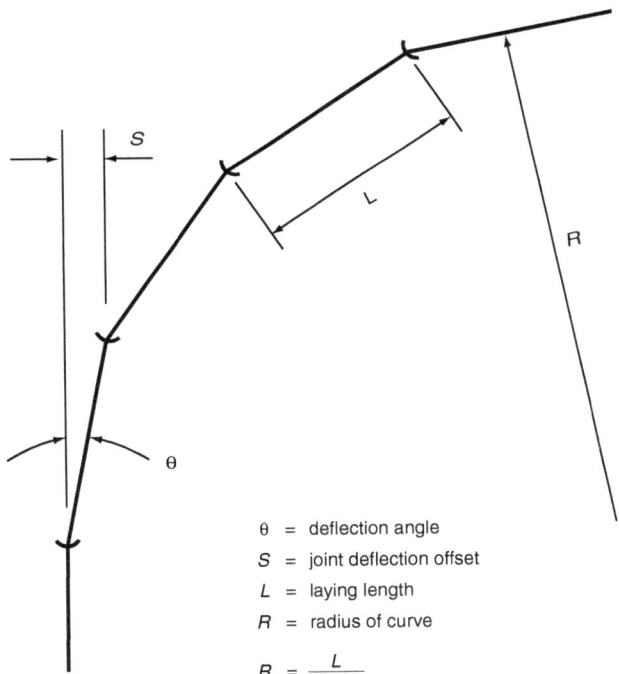

θ = deflection angle

S = joint deflection offset

L = laying length

R = radius of curve

$$R = \frac{L}{2 \tan \frac{\theta}{2}}$$

NOTE: See tables on the following pages for θ and S values.
Source: AWWA M41—Ductile Iron Pipe and Fittings.

Pipeline Curve Geometry

Maximum Joint Deflection* Full-Length Pipe—Push-on-Type Joint Pipe

Nominal Pipe Size,		Deflection Angle—θ,	Maximum Offset—S,† in. (m)				Approximate Radius of Curve—R† Produced by Succession of Joints, ft (m)			
in.	(mm)	degrees	L† = 18 ft (5.5 m)		L† = 20 ft (6 m)		L† = 18 ft (5.5 m)		L† = 20 ft (6 m)	
3	(76)	5	19	(0.48)	21	(0.53)	205	(62)	230	(70)
4	(102)	5	19	(0.48)	21	(0.53)	205	(62)	230	(70)
6	(152)	5	19	(0.48)	21	(0.53)	205	(62)	230	(70)
8	(203)	5	19	(0.48)	21	(0.53)	205	(62)	230	(70)
10	(254)	5	19	(0.48)	21	(0.53)	205	(62)	230	(70)
12	(305)	5	19	(0.48)	21	(0.53)	205	(62)	230	(70)
14	(356)	3*	11	(0.28)	12	(0.30)	340	(104)	380	(116)
16	(406)	3*	11	(0.28)	12	(0.30)	340	(104)	380	(116)
18	(457)	3*	11	(0.28)	12	(0.30)	340	(104)	380	(116)
20	(508)	3*	11	(0.28)	12	(0.30)	340	(104)	380	(116)
24	(610)	3*	11	(0.28)	12	(0.30)	340	(104)	380	(116)
30	(762)	3*	11	(0.28)	12	(0.30)	340	(104)	380	(116)
36	(914)	3*	11	(0.28)	12	(0.30)	340	(104)	380	(116)
42	(1,067)	3*	11	(0.28)	12	(0.30)	340	(104)	380	(116)
48	(1,219)	3*			12	(0.30)			380	(116)
54	(1,400)	3*			12	(0.30)			380	(116)
60	(1,500)	3*			12	(0.30)			380	(116)
64	(1,600)	3*			12	(0.30)			380	(116)

Source: AWWA M41—Ductile Iron Pipe and Fittings.
* For 14-in. and larger push-on joints, maximum deflection angle may be larger than shown above. Consult the manufacturer.
† See Pipeline Curve Geometry figure on page 217.

Maximum Joint Deflection Full-Length Pipe—Mechanical-Joint Pipe

Nominal Pipe Size,		Deflection Angle—θ,	Maximum Offset—S,* in. (m)				Approximate Radius of Curve—R* Produced by Succession of Joints, ft (m)			
			L* = 18 ft (5.5 m)		L* = 20 ft (6 m)		L* = 18 ft (5.5 m)		L* = 20 ft (6 m)	
in.	(mm)	degrees								
3	(76)	8-18	31	(0.79)	35	(0.89)	125	(38)	140	(43)
4	(102)	8-18	31	(0.79)	35	(0.89)	125	(38)	140	(43)
6	(152)	7-07	27	(0.69)	30	(0.76)	145	(44)	160	(49)
8	(203)	5-21	20	(0.51)	22	(0.56)	195	(59)	220	(67)
10	(254)	5-21	20	(0.51)	22	(0.56)	195	(59)	220	(67)
12	(305)	5-21	20	(0.51)	22	(0.56)	195	(59)	220	(67)
14	(356)	3-35	13.5	(0.34)	15	(0.38)	285	(87)	320	(98)
16	(406)	3-35	13.5	(0.34)	15	(0.38)	285	(87)	320	(98)
18	(457)	3-00	11	(0.28)	12	(0.30)	340	(104)	380	(116)
20	(508)	3-00	11	(0.28)	12	(0.30)	340	(104)	380	(116)
24	(610)	2-23	9	(0.23)	10	(0.25)	450	(137)	500	(152)

Source: AWWA M41—Ductile Iron Pipe and Fittings.
* See Pipeline Curve Geometry figure on page 217.

Distribution

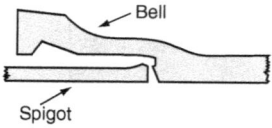

Spigot End Pushed Into Bell

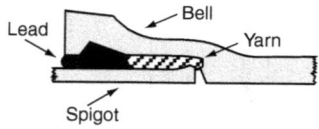

Completed

Leaded Bell-and-Spigot Joint

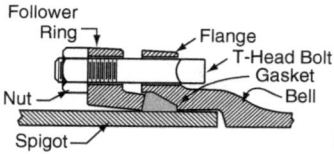

Mechanical Joint

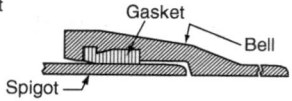

Push-on Joint for DI Pipe

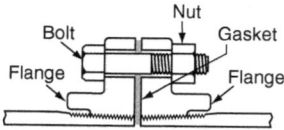

Flanged Joint

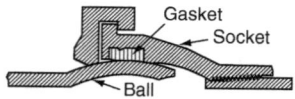

Boltless Ball Joint

Source: AWWA M41—Ductile Iron Pipe and Fittings.

Common Types of Ductile-Iron Pipe Joints

EXCAVATION AND TRENCHING

NOTE: Clays, silts, loams, or nonhomogeneous soils require shoring and bracing. The presence of groundwater requires special treatment.

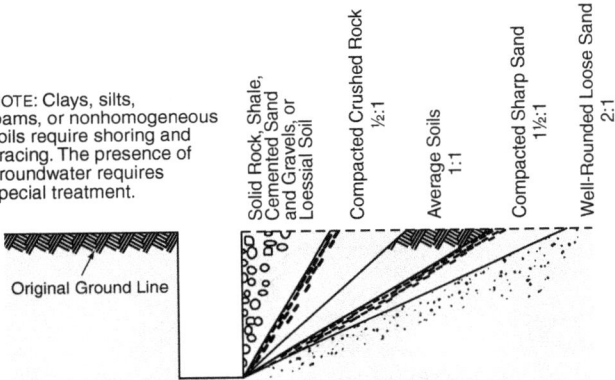

Solid Rock, Shale, Cemented Sand and Gravels, or Loessial Soil

Compacted Crushed Rock
½:1

Average Soils
1:1

Compacted Sharp Sand
1½:1

Well-Rounded Loose Sand
2:1

Original Ground Line

Approximate Angles of Repose of Various Types of Soil

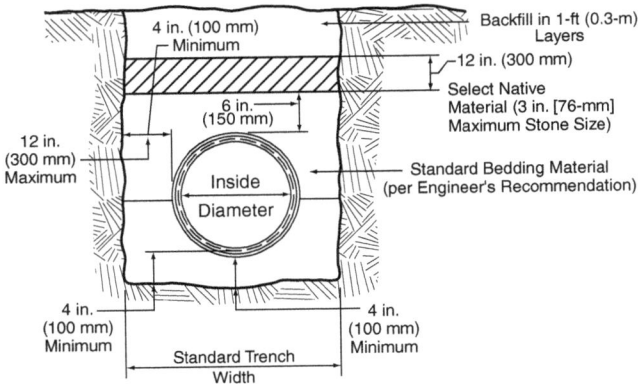

4 in. (100 mm) Minimum

Backfill in 1-ft (0.3-m) Layers

12 in. (300 mm)

6 in. (150 mm)

Select Native Material (3 in. [76-mm] Maximum Stone Size)

12 in. (300 mm) Maximum

Inside Diameter

Standard Bedding Material (per Engineer's Recommendation)

4 in. (100 mm) Minimum

4 in. (100 mm) Minimum

Standard Trench Width

Trench Conditions

Type 1* Flat-bottom trench† with loose backfill.

Type 2 Flat-bottom trench† with backfill lightly consolidated to centerline of pipe.

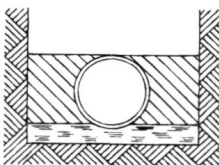

Type 3 Pipe bedded in 4-in. (100-mm) minimum of loose soil‡ with backfill lightly consolidated to top of pipe.

Type 4 Pipe bedded in sand, gravel, or crushed stone to depth of ⅛ pipe diameter, 4-in. (100-mm) minimum with backfill compacted to top of pipe. (Approximately 80 percent Standard Proctor, AASHTO§ T-99**)

Type 5 Pipe bedded to its centerline in compacted granular material, 4-in. (100-mm) minimum under pipe. Compacted granular or select material‡ to top of pipe. (Approximately 90 percent Standard Proctor, AASHTO§ T-99**)

* For nominal pipe sizes 14 in. (356 mm) and larger, consideration should be given to the use of laying conditions other than Type 1.
† "Flat-bottom" is defined as "undisturbed earth."
‡ Loose soil or select material is defined as "native soil excavated from the trench, free of rocks, foreign materials, and frozen earth."
§ Granular materials are defined per the AASHTO Soil Classification System (ASTM D3282) or the Unified Soil Classification System (ASTM D2487), with the exception that gravel bedding/backfill adjacent to the pipe is limited to 2 in. (50 mm) maximum particle size per ANSI/AWWA C600.
** AASHTO T-99 "Standard Method of Test for the Moisture-Density Relations of Soils Using a 5.5 lb (2.5 kg) Rammer and a 12 in. (305 mm) Drop." Available from the American Association of State Highway and Transportation Officials.

Laying Conditions for Ductile-Iron Pipe

Trench Widths

Recommended Trench Widths for Ductile-Iron Pipe

Nominal Pipe Size,		Trench Width,	
in.	*(mm)*	*in.*	*(m)*
3	(80)	27	(0.69)
4	(100)	28	(0.70)
6	(150)	30	(0.76)
8	(200)	32	(0.81)
10	(200)	34	(0.86)
12	(300)	36	(0.92)
14	(350)	38	(0.96)
16	(400)	40	(1.00)
18	(450)	42	(1.07)
20	(500)	44	(1.12)
24	(600)	48	(1.22)
30	(750)	54	(1.37)
36	(900)	60	(1.52)
42	(1,050)	66	(1.68)
48	(1,200)	72	(1.83)
54	(1,350)	78	(1.98)
60	(1,500)	84	(2.13)

Recommended Trench Widths for PVC Pipe

Pipe Diameter,		Trench Width			
		Minimum,		Maximum,	
in.	*(mm)*	*in.*	*(m)*	*in.*	*(m)*
4	(100)	18	(0.46)	29	(0.74)
6	(150)	18	(0.46)	31	(0.79)
8	(150)	21	(0.53)	33	(0.84)
10	(250) and greater	12	(0.31) greater than OD of pipe	24	(0.61) greater than OD of pipe

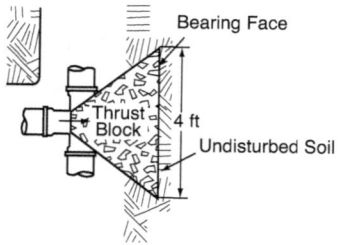

Top View

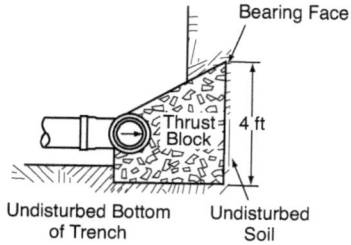

Side View

Courtesy of J-M Manufacturing Co., Inc.

Correctly Sized Thrust Block (see Thrust Anchoring table on page 225)

Thrust Anchoring for 11¼°, 22½°, 30°, and 45° Vertical Bends

Pipe Size Nominal Diameter, in.	Test Pressure, psi	Vertical Bend,* degrees	Volume of Concrete Blocking, ft³	Side of Cube, ft	Diameter of Shackle Rods, in.	Depth of Rods in Concrete, ft
4	300	11¼	8	2	¾	1.5
		22½	11	2.2	¾	2.0
		30	17	2.6	¾	2.0
		45	30	3.1	¾	2.0
6	300	11¼	11	2.2	¾	2.0
		22½	25	2.9	¾	2.0
		30	41	3.5	¾	2.0
		45	68	4.1	¾	2.0
8	300	11¼	16	2.5	¾	2.0
		22½	47	3.6	¾	2.0
		30	70	4.1	¾	2.5
		45	123	5.0	¾	2.0
12	250	11¼	32	3.2	¾	2.0
		22&	88	4.5	⅞	3.0
		30	132	5.1	⅞	3.0
		45	232	6.1	¾	2.5
16	225	11¼	70	4.1	⅞	3.0
		22½	184	5.7	1⅛	4.0
		30	275	6.5	1¼	4.0
		45	478	7.8	1⅛	4.0
20	200	11¼	91	4.5	⅞	3.0
		22½	225	6.1	1¼	4.0
		30	330	6.9	1⅜	4.5
		45	560	8.2	1¼	4.0
24	200	11¼	128	5.0	1	3.5
		22½	320	6.8	1⅜	4.5
		30	480	7.9	1⅝	5.5
		45	820	9.4	1⅜	4.5

*Four rods for 45° vertical bends; two rods for all others.

Distribution

TESTING, LEAKS, AND FLUSHING

The constructor may perform simultaneous pressure and leakage tests or perform separate pressure and leakage tests on the installed system at test durations and pressures specified in the System Test Methods table that follows. Tests shall be witnessed by the purchaser or the purchaser's agent, and the equipment used for the test shall be subject to the approval of the purchaser or the purchaser's agent.

System Test Methods

Procedure	Pressure	Duration of Test
Simultaneous pressure and leakage tests	150% of working pressure* at point of test, but not less than 125% of normal working pressure at highest elevation.[†]	2 hours
Separate pressure test	150% of working pressure* at point of test, but not less than 125% of normal working pressure at highest elevation.[†]	1 hour
Separate leakage test	150% of working pressure* of segment tested.[†]	2 hours

* Working pressure is the maximum anticipated sustained operating pressure.
[†] In no case shall the test pressure be allowed to exceed the design pressure for pipe, appurtenances, or thrust restraints.

Rapid loss of pressure will usually be due to an open valve, a cracked or broken pipe, or a joint that slipped out after it was made up. These leaks are usually relatively easy to locate by continuing to apply pressure until the water comes to the surface.

One possible cause of a slow drop of pressure is that all of the air was not completely removed before testing began. In this case, the amount of apparent leakage will usually be less on each subsequent repeat test.

A slow leak is often difficult to locate. If the initial test was on a long section of main, it will probably be necessary to locate a small leak by performing tests on small sections between valves to narrow down the location. Leak detection equipment can then be used to locate the leak. Another alternative is to continuously subject the main to the highest pressure that can be safely applied, and wait for the water to come to the surface.

Hydrostatic Testing of Pipe

Warning: The testing methods described in this section are specific for water-pressure testing. These procedures should not be applied for air-pressure testing because of the serious safety hazards involved.

Test pressure shall not be less than 1.25 times the working pressure at the highest point along the test section.

Test pressure shall not exceed pipe or thrust-restraint design pressures.

The hydrostatic test shall be of at least a 2-hour duration.

Test pressure shall not vary by more than ±5 psi (±34.5 kPa) for the duration of the test.

The test pressure shall not exceed the rated pressure of the valves when the pressure boundary of the test section includes closed, resilient-seated gate valves or butterfly valves.

After the pipe has been laid, all newly laid pipe or any valved section thereof shall be subjected to a hydrostatic pressure of at least 1.5 times the working pressure at the point of testing. Each valved section of pipe shall be slowly filled with water, and the specified test pressure (based on the elevation of the lowest point of the line or section under test and corrected to the elevation of the test gauge) shall be applied using a pump connected to the pipe. Valves shall not be operated in either the opened or closed direction at differential pressures above the rated pressure. The system should be allowed to stabilize at the test pressure before conducting the hydrostatic test.

Before applying the specified test pressure, air shall be expelled completely from the section of piping under test. If permanent air vents are not located at all high points, corporation cocks shall be installed at these points to expel the air as the line is filled with water. After the air has been expelled, the corporation cocks shall be closed and the test pressure applied. At the conclusion of the pressure test, the corporation cocks shall be removed and the pipe plugged or left in place as required by the specifications.

Any exposed pipe, fittings, valves, hydrants, and joints shall be examined carefully during the test. Any damaged or defective pipe, fittings, valves, hydrants, or joints that are discovered following the pressure test shall be repaired or replaced with reliable material, and the test shall be repeated until satisfactory results are obtained.

Testing allowance shall be defined as the quantity of makeup water that must be supplied into the newly laid pipe or any valved section thereof to maintain pressure within 5 psi (34.5 kPa) of the specified test pressure after the pipe has been filled with water and the air has been expelled. Testing allowance shall not be measured by a drop in pressure in a test section over a period of time.

Allowable Leakage or Makeup Water for Ductile or PVC Pipe Leakage Test

The constructor shall furnish the gauges and measuring device for the leakage test, pump, pipe, connections, and all other necessary apparatus, unless otherwise specified, and shall furnish the necessary assistance to conduct the test. The duration of each leakage test shall be 2 hours, unless otherwise specified. During the test, the pipeline shall be subjected to the pressure listed in the System Test Methods table above. Leakage shall be defined as the quantity of water that must be supplied into the pipe section being tested to maintain a pressure within 5 psi (34 kPa) of the specified leakage-test pressure after the pipe has been filled with water and the air in the pipeline has been expelled. No installation will be accepted if the leakage (makeup water) is greater than that determined by the formula:

$$Q = \frac{LD\sqrt{P}}{148,000}$$

Where:

Q = quantity of makeup water, in gph

L = length of pipe section being tested, in ft

D = nominal diameter of the pipe, in in.

P = average test pressure during the hydrostatic test, in psi (gauge)

In metric units,

$$Q_m = \frac{LD\sqrt{P}}{795,000}$$

Where:

Q_m = quantity of makeup water, in L/hr

L = length of pipe section being tested, in m

D = nominal diameter of the pipe, in mm

P = average test pressure during the leakage test, in kPa

See table on next page for allowable makeup water per 1,000 ft.

Allowable Leakage per 50 Joints of PVC Pipe, *gph*

Average Test Pressure,		Nominal Pipe Diameter, *in. (mm)*															
		4	6	8	10	12	14	16	18	20	24	30	36	42	48
psi	*(kPa)*	(100)	(150)	(200)	(250)	(300)	(350)	(400)	(450)	(500)	(610)	(760)	(915)	(1,070)	(1,220)
300	(2,070)	0.47	0.70	0.94	1.17	1.40	1.64	1.87	2.11	2.34	2.81	3.51	4.21	4.92	5.62
275	(1,900)	0.45	0.67	0.90	1.12	1.34	1.57	1.79	2.02	2.24	2.69	3.36	4.03	4.71	5.38
250	(1,720)	0.43	0.64	0.85	1.07	1.28	1.50	1.71	1.92	2.14	2.56	3.21	3.85	4.49	5.13
225	(1,550)	0.41	0.61	0.81	1.01	1.22	1.42	1.62	1.82	2.03	2.43	3.04	3.65	4.26	4.86
200	(1,380)	0.38	0.57	0.76	0.96	1.15	1.34	1.53	1.72	1.91	2.29	2.87	3.44	4.01	4.59
175	(1,210)	0.36	0.54	0.72	0.89	1.07	1.25	1.43	1.61	1.79	2.15	2.68	3.22	3.75	4.29
150	(1,030)	0.33	0.50	0.66	0.83	0.99	1.16	1.32	1.49	1.66	1.99	2.48	2.98	3.48	3.97
125	(860)	0.30	0.45	0.60	0.76	0.91	1.06	1.21	1.36	1.51	1.81	2.27	2.72	3.17	3.63
100	(690)	0.27	0.41	0.54	0.68	0.81	0.95	1.08	1.22	1.35	1.62	2.03	2.43	2.84	3.24
75	(520)	0.23	0.35	0.47	0.59	0.70	0.82	0.94	1.05	1.17	1.40	1.76	2.11	2.46	2.81
50	(340)	0.19	0.29	0.38	0.48	0.57	0.67	0.76	0.86	0.96	1.15	1.43	1.72	2.01	2.29

Source: ANSI/AWWA C605, Underground Installation of PVC Pipe.

* If the pipeline under test contains sections of various diameters, the allowable leakage will be the sum of the computed leakage for each size.
† To obtain leakage in liters per hour, multiply the values in the table by 3.72.

Hydrostatic Testing Allowance per 1,000 ft of Pipeline,* gph† Ductile Iron and PVC Pipe

Average Test Pressure, psi	Nominal Pipe Diameter, in.																	
	3	4	6	8	10	12	14	16	18	20	24	30	36	42	48	54	60	64
450	0.43	0.57	0.86	1.15	1.43	1.72	2.01	2.29	2.58	2.87	3.44	4.30	5.16	6.02	6.88	7.74	8.60	9.17
400	0.41	0.54	0.81	1.08	1.35	1.62	1.89	2.16	2.43	2.70	3.24	4.05	4.86	5.68	6.49	7.30	8.11	8.65
350	0.38	0.51	0.76	1.01	1.26	1.52	1.77	2.02	2.28	2.53	3.03	3.79	4.55	5.31	6.07	6.83	7.58	8.09
300	0.35	0.47	0.70	0.94	1.17	1.40	1.64	1.87	2.11	2.34	2.81	3.51	4.21	4.92	5.62	6.32	7.02	7.49
275	0.34	0.45	0.67	0.90	1.12	1.34	1.57	1.79	2.02	2.24	2.69	3.36	4.03	4.71	5.38	6.05	6.72	7.17
250	0.32	0.43	0.64	0.85	1.07	1.28	1.50	1.71	1.92	2.14	2.56	3.21	3.85	4.49	5.13	5.77	6.41	6.84
225	0.30	0.41	0.61	0.81	1.01	1.22	1.42	1.62	1.82	2.03	2.43	3.04	3.65	4.26	4.86	5.47	6.08	6.49
200	0.29	0.38	0.57	0.76	0.96	1.15	1.34	1.53	1.72	1.91	2.29	2.87	3.44	4.01	4.59	5.16	5.73	6.12
175	0.27	0.36	0.54	0.72	0.89	1.07	1.25	1.43	1.61	1.79	2.15	2.68	3.22	3.75	4.29	4.83	5.36	5.72
150	0.25	0.33	0.50	0.66	0.83	0.99	1.16	1.32	1.49	1.66	1.99	2.48	2.98	3.48	3.97	4.47	4.97	5.30
125	0.23	0.30	0.45	0.60	0.76	0.91	1.06	1.21	1.36	1.51	1.81	2.27	2.72	3.17	3.63	4.08	4.53	4.83
100	0.20	0.27	0.41	0.54	0.68	0.81	0.95	1.08	1.22	1.35	1.62	2.03	2.43	2.84	3.24	3.65	4.05	4.32

* If the pipeline under test contains sections of various diameters, the testing allowance will be the sum of the testing allowance for each size.
† Calculated on the basis of the equations on page 228–229.

Distribution

Water Waste From Leaks

The amount of water wasted at 40 psi (a low water pressure) and 100 psi (high pressure) would be:

Diameter of Leak, *in.*	Gallons of Water per Day at	
	40 psi	**100 psi**
$\frac{1}{8}$	2,550	3,700
$\frac{1}{16}$	650	950
$\frac{1}{32}$	160	230
$\frac{1}{64}$	40	60

Leak Losses for Circular Holes Under Different Pressures*

		Leak Losses, *gpm*									
		Water Pressure, *psi*									
Diameter of Hole, in.	Area of Hole, in.²	20	40	60	80	100	120	140	160	180	200
0.1	0.007	1.067	1.510	1.850	2.136	2.388	2.616	2.825	3.021	3.204	3.337
0.2	0.031	4.271	6.041	7.399	8.544	9.522	10.464	11.302	12.083	12.816	13.509
0.3	0.070	9.611	13.593	16.648	19.224	21.493	23.544	25.430	27.186	28.835	30.395
0.4	0.125	17.087	24.165	29.597	34.175	38.209	41.856	45.209	48.331	51.263	54.036
0.5	0.196	26.699	37.758	46.245	53.399	59.702	65.400	70.640	75.518	80.098	84.431
0.6	0.282	38.477	54.372	66.593	76.894	85.971	94.176	101.721	108.745	115.341	121.581
0.7	0.384	52.331	74.007	90.640	104.662	117.010	128.184	138.454	148.014	156.993	165.485
0.8	0.502	68.350	96.662	118.387	136.701	152.840	167.424	180.839	193.325	205.052	216.144
0.9	0.636	86.506	122.338	149.833	173.012	193.434	211.896	228.874	244.676	259.519	273.557
1.0	0.785	106.798	151.035	184.979	213.596	238.807	261.600	282.561	302.070	320.394	337.725
1.1	0.950	129.225	182.752	223.825	258.451	288.957	316.536	341.898	365.505	387.676	408.647
1.2	1.131	153.789	217.490	266.370	307.578	343.882	376.704	406.887	434.981	461.367	486.323
1.3	1.327	180.488	255.249	312.615	360.977	403.584	442.104	477.527	510.498	541.465	570.755
1.4	1.539	209.324	296.028	362.559	418.648	468.062	512.737	553.819	592.057	627.972	661.941
1.5	1.767	240.295	339.829	416.203	480.590	537.317	588.601	635.762	679.658	720.886	759.880
1.6	2.011	273.402	386.649	473.547	546.805	611.347	669.697	723.355	773.299	820.208	864.575
1.7	2.270	308.646	436.491	534.590	617.292	690.153	756.025	816.600	872.983	925.938	976.024
1.8	2.545	346.025	489.353	599.333	692.050	773.736	847.585	915.496	978.707	1,038.070	1,094.220
1.9	2.836	385.540	545.237	667.776	771.081	862.095	944.378	1,020.040	1,090.470	1,156.620	1,219.180
2.0	3.142	427.191	604.140	739.918	854.383	955.230	1,046.400	1,130.240	1,208.280	1,281.570	1,350.890

* Calculated using Greeley's formula (see equation on following page).

Distribution

Leak Losses for Joints and Cracks*

Area of Joint or Crack		Leak Losses, *gpm*									
Length, Width,		Water Pressure, *psi*									
in.	*in.*	20	40	60	80	100	120	140	160	180	200
1.0	1/32	3.2	4.5	5.5	6.4	7.1	7.8	8.4	9.0	9.6	10.1
1.0	1/16	6.4	9.0	11.0	12.7	14.2	15.6	16.9	18.0	19.1	20.1
1.0	1/8	12.7	18.0	22.1	25.5	28.5	31.2	33.7	36.0	38.2	40.3
1.0	1/4	25.5	36.0	44.1	51.0	57.0	62.4	67.4	72.1	76.5	80.6

*For leaks emitted from joints and cracked service pipes, an orifice coefficient of 0.60 is used in the following equation:

$$Q = (22.796)(A)(\sqrt{P})$$

Where: Q = flow, in gpm; A = area, in in.^{2}; P = pressure, in psi

For losses from such items as pipes or broken taps, assume an orifice coefficient of 0.80 and calculate flow in gallons per minute from Greeley's formula:

$$Q = \frac{43,767}{1,440} \times A \times \sqrt{P}$$

Where:

Q = flow, in gpm

A = the cross-sectional area of the leak, in in.^{2}

P = pressure, in psi

No pipe installation will be accepted if the amount of makeup water is greater than that determined by the following formula:

In inch-pound units,

$$L = \frac{SD\sqrt{P}}{133,200}$$

Where:

L = testing allowance (makeup water), in gph

S = length of pipe tested, in ft

D = nominal diameter of the pipe, in in.

P = average test pressure during the hydrostatic test, in psi (gauge)

In metric units,

$$L_m = \frac{SD\sqrt{P}}{715,317}$$

Where:

L_m = testing allowance (makeup water), in L/hr

S = length of pipe tested, in m

D = nominal diameter of the pipe, in mm

P = average test pressure during the hydrostatic test, in kPa

Pressure Testing the Main

After the trench has been at least partially backfilled, the new main must be tested to determine whether there are any leaks. The test may be performed one section at a time between valves, or the installer may wait and test the entire job at one time.

Many years ago, when only lead joints were used on cast-iron pipe, it was always assumed that there would be leaks. The only thing the installer could do was hope that it would not be too much. Now, an installation of all mechanical, push-on, or other rubber-gasket joints will have virtually no leakage unless something is defective.

Testing Procedure

The following general procedures apply to testing of all types of pipe:

- If poured concrete blocking was used, allow at least 5 days before testing. If high early strength concrete was used, this time can be shortened.

- Make sure the valves at all connections with existing mains are holding completely tight.

- Fill the new main with water and be sure that air has been released at all high points. If chlorine tablets have been installed in the pipe as work progressed, be sure to fill very slowly so the tablets will not be dislodged.

- Close all fire hydrant auxiliary valves.

- Connect a pressure pump to a corporation stop in the main. The pump must have a pressure gauge and connection to a small tank of makeup water.

- Apply partial pressure and again check that all air has been removed from the system. Allow the pipe to stand with pressure on it for at least 24 hours to stabilize.

- Pump up the pressure as specified in the applicable ANSI/AWWA standard. The minimum is usually 1.5 times the operating pressure, or 150 psi (1,030 kPa) for a period of 30 minutes.

- Examine the piping and fittings for visible leaks or air that was not previously released.

- Again pump up the pressure and wait for at least 2 hours.

- During the waiting period, pump up the pressure as required to maintain the minimum test pressure. Pump from a calibrated container and record the quantity of makeup water used.

- Compare the amount of leakage (the quantity of water required to bring the pressure back up) with the suggested maximum allowable leakage in the appropriate ANSI/AWWA standard.

Flushing and Disinfection

All new sections of water main must be thoroughly flushed, disinfected, and tested for bacteriological quality before the water can be used by customers.

New water mains and equipment must be disinfected with some form of chlorine. Two methods typically used are tablets and chlorine solution.

Number of 5-g Calcium Hypochlorite Tablets Required to Produce Chlorine Residual of 25 mg/L in 20-ft (6-m) Pipe Lengths

Pipe Diameter, in.	(mm)	Number of Tablets per Pipe Length
4	(100)	1
6	(150)	1
8	(200)	2
10	(250)	3
12	(300)	4
16	(400)	7

Tablet Method

With the tablet method, calcium High Test Hypochlorite (HTH) tablets are placed in the section of pipe and fire hydrant as the work progresses, and they will dissolve when the new pipe is filled with water. The tablets are usually glued to the top of the pipe with an epoxy resin, in sufficient quantities to produce a chlorine residual of 25 mg/L after they have dissolved. The number of tablets required for each 20-ft (6-m) pipe section is listed in the table above.

After the main has been filled with water, the chlorine solution should be maintained in the pipe for at least 24 hours. Because the tablets are placed only at the end of each pipe section, it is advisable to periodically bleed off a small amount of water at the end of the line to move the chlorine solution to new locations in the piping.

When tablets are used for disinfection, the velocity of the water filling the pipe must be kept below 1 ft/sec (0.3 m/sec), or the tablets will be dislodged and washed to the end of the pipeline. When the tablet method is used, workers must take particular care to keep the pipe clean during installation because the main cannot be flushed before it is disinfected. If it is anticipated that working conditions will make it difficult to keep the pipe clean, the tablet method should not be used so that the line may be flushed before being disinfected.

Hypochlorite Disinfection

Although chlorine gas may be used for disinfecting water mains, it requires special equipment and is dangerous to use, so is recommended for only large water main installations where it can be done under supervision of someone experienced with the equipment. Calcium hypochlorite and sodium hypochlorite (bleach) are generally used for disinfecting smaller mains.

A concentrated chlorine solution is usually injected through a corporation stop that has been installed close to the valve that connects to the existing water system. The chlorine is administered by either the continuous feed or the slug method.

In the continuous feed method, water is slowly admitted to the new pipeline where chlorine solution is forced in through the connection using a chemical feed booster pump. The water flow rate can be gauged by measuring the flow of water from a fire hydrant at the end of the line. The chemical feed rate should be such that it will produce a concentration of about 50 mg/L when mixed with the incoming water.

The feed should continue until a residual of at least 25 mg/L can be measured in the flow at the end of the line. The flow should then be stopped and the chlorine allowed to remain in the pipe for at least 24 hours. During this time, all valves and hydrants on the line should be operated to make sure they are also properly disinfected. The quantity of HTH required to produce a 50-mg/L concentration is listed in the table on page 239.

In the slug method, a long slug of water having a very high dose of chlorine is initially created, and then it is slowly moved through the pipeline. The concentration should be at least 300 mg/L, and the slug should be moved at a speed that will provide at least 3 hours of contact as it moves through the pipeline. Fire hydrants and side connections must be operated as the slug passes to make sure the chlorine reaches all parts of the piping. This method is primarily used for larger pipelines for which the continuous feed method is impractical.

At the end of the contact period, the chlorinated water should be flushed from the pipeline and disposed of in an environmentally responsible manner. The high chlorine concentration will probably kill grass, so the flow should be carried to a disposal site through hoses. State and local regulatory agencies should be contacted in advance to determine whether they have any special requirements that must be met. In some cases, they may require that the water be dechlorinated before it is released to a waterway.

Prior to placing the installed water line in service, the new pipe and all exposed sections and appurtenances of existing pipelines shall be cleaned and disinfected in accordance with ANSI/AWWA C651, unless otherwise specified. Pipelines shall be flushed following completion of disinfection procedures. Disposal or neutralization of disinfection water shall comply with applicable regulations. (Refer to Appendix B of ANSI/AWWA C651.)

Bacteriological Testing

Quantity of HTH Required to Produce 50 mg/L Chlorine Residual

Nominal Pipe Diameter,		Amount of Hypochlorite per 100 ft (30.5 m) of Pipe,	
in.	(mm)	lb	kg
4	(100)	0.04	0.018
6	(150)	0.09	0.04
8	(200)	0.17	0.08
10	(250)	026	0.12
12	(300)	0.38	0.17
14	(350)	0.51	0.23
16	(400)	0.67	0.30
18	(450)	0 85	0.39
20	(500)	1.05	0.47

After a new pipeline has been disinfected and flushed, it should be refilled with water from the distribution system and tested for bacteriological quality. This test takes at least 24 hours from the time of sampling. When planning the pipe installation, this time should be included. The tests must meet requirements of the state regulatory agency, and customers must not be allowed to use the water until the results of the testing have been received. The state agency should be contacted in advance for sample bottles and instructions on sampling procedures.

If the results of the sampling are reported as negative, it means that no coliform bacteria were present in the sample and the system has been adequately disinfected.

If the results are reported as positive, the agency will usually suggest resampling. If the results of the second set of samples are still positive, disinfection of the pipeline will have to be repeated, and more samples will have to be processed to make sure the pipeline has been properly disinfected.

Flow Rate and Number of Hydrant Outlets Required to Flush Pipelines

Pipe Diameter, in.	Flow Required to Produce Velocity of Approximately 2.5 ft/sec in Main, gpm	Number of 2½-in. (65-mm) Hydrant Outlets
4	100	1
6	200	1
8	400	1
10	600	1
12	900	2
16	1,600	2

Factors That Contribute to Water System Deterioration

Factor	Explanation
Physical	
Pipe material	Pipes made from different materials fail in different ways.
Pipe wall thickness	Corrosion will penetrate thinner walled pipe more quickly.
Pipe age	Effects of pipe degradation become more apparent over time.
Pipe vintage	Pipes made at a particular time and place may be more vulnerable to failure.
Pipe diameter	Small-diameter pipes are more susceptible to beam failure.
Type of joints	Some types of joints have experienced premature failure (e.g., leadite joints).
Thrust restraint	Inadequate restraint can increase longitudinal stresses.
Pipe lining and coating	Lined and coated pipes are less susceptible to corrosion.
Dissimilar metals	Dissimilar metals are susceptible to galvanic corrosion.
Pipe installation	Poor installation practices can damage pipes, making them vulnerable to failure.
Pipe manufacture	Defects in pipe walls produced by manufacturing errors can make pipes vulnerable to failure. This problem is most common in older pit-cast pipes.
Environmental	
Pipe bedding	Improper bedding may result in premature pipe failure.
Trench backfill	Some backfill materials are corrosive or frost susceptible.
Soil type	Some soils are corrosive; some soils experience significant volume changes in response to moisture changes, resulting in changes to pipe loading. Presence of hydrocarbons and solvents in soil may result in some pipe deterioration.

Table continued on next page

Distribution

241

Factors That Contribute to Water System Deterioration (continued)

Factor	Explanation
Environmental (continued)	
Groundwater	Some groundwater is aggressive toward certain pipe materials.
Climate	Climate influences frost penetration and soil moisture. Permafrost must be considered in the north.
Pipe location	Migration of road salt into soil can increase the rate of corrosion.
Disturbances	Underground disturbances in the immediate vicinity of an existing pipe can lead to actual damage or changes in the support and loading structure on the pipe.
Stray electrical currents	Stray currents cause electrolytic corrosion.
Seismic activity	Seismic activity can increase stresses on pipe and cause pressure surges.
Operational	
Internal water pressure, transient pressure	Changes to internal water pressure will change stresses acting on the pipe.
Leakage	Leakage erodes pipe bedding and increases soil moisture in the pipe zone.
Water quality	Some water is aggressive, promoting corrosion.
Flow velocity	Rate of internal corrosion is greater in unlined dead-end mains.
Backflow potential	Cross-connections with systems that do not contain potable water can contaminate water distribution system.
O&M practices	Poor practices can compromise structural integrity and water quality.

Source: National Guide to Sustainable Municipal Infrastructures—InfraGuide, 2002. Deterioration and Inspection of Water Distribution Systems, Ottawa, Ontario.

AWWA PIPE REPAIR CHECKLIST _____

Location:

Date:

- ☐ Notify other agencies (public works, department of transportation, police, mayor's office) and affected customers in advance, if possible, so they can prepare.
- ☐ Put up signs for public warnings and communication; notify public of work and hazards.
- ☐ Put up signs and barriers for traffic control.
- ☐ Use safety lights for night work if needed.
- ☐ Set up and follow valve lockout/tagout procedures.
- ☐ Locate, mark, and protect all existing utility lines in the vicinity including water, storm sewer, phone, cable, gas, and power lines. Call the local "dig safe" system to identify other utilities.
- ☐ Locate and mark nearby water grid isolation valves so they can be easily found and used.
- ☐ Determine if temporary service can be provided.
- ☐ Locate where dewatering/evacuation/runoff water will occur and mitigate erosion or property damage.
- ☐ Follow confined space protocols.
- ☐ Put excavation and trenching controls into place for safety.
- ☐ Install temporary diversions to control surface water runoff into trench.
- ☐ Provide for dewatering of excavations below the level of pipe invert.
- ☐ Check groundwater condition and effects wet weather will have on project.
- ☐ Control dust during excavation. Have coverings and water source ready.
- ☐ Isolate main section with isolation valves.
- ☐ Maintain positive pressure inline to reduce backflow and contamination.
- ☐ Expose break; thoroughly scrape and clean area. Inspect for rough spots and sharp edges and file smooth.
- ☐ Determine type of leak/break, beam, split, blowout, joint, or corrosion.

- [ ] Determine and obtain pipe material, outside diameter, and repair equipment and parts/clamps needed.
- [ ] Identify and store materials onsite in secure area.
- [ ] Maintain protective caps and plugs and coverings on pipe and fittings until material is needed.
- [ ] Disinfect hand tools, saws, tapping machines used in repair.
- [ ] Clean and disinfect parts and materials before installing into system.
- [ ] Make repairs.
- [ ] Flush, clean, disinfect new pipe repair. (Depending on the magnitude of the repair, ANSI/AWWA Standard C651 can be used as a guideline for disinfection).
- [ ] Complete repair and tighten or secure all installations.
- [ ] Apply any corrosion protection materials.
- [ ] Follow applicable ANSI/AWWA standard or manual for hydrostatic testing of line, depending on your pipe material.
- [ ] Test for coliform using ANSI/AWWA C651 standards and turbidity.
- [ ] Ensure proper disposal of chlorinated water using dechlorinating or neutralizing agent.
- [ ] Follow sequence for lockout/tagout of valves and hydrants.
- [ ] Flush hydrants when project is complete to remove debris, and introduce fresh chlorinated water.
- [ ] Sequence operations of valve openings to avoid negative pressures in system due to filling repaired section.
- [ ] Backfill and repair ground and surfaces to original conditions or better. Insure proper backfill and compaction of materials.
- [ ] Document new components, field conditions, potential causes of leak, location of project, and estimated cost. This information will be important when determining major line replacement needs and projects.
- [ ] Report results to proper authorities and get approval from inspectors or authority to return to service.
- [ ] Notify affected public of repair and service. Suggest they flush their home plumbing after the repair is completed.
- [ ] Complete the Field Data for Main Break Evaluation form on the following page.

Field Data for Main Break Evaluation

Date of Break _____ Time_____ A.M. _____P.M.

Type of Main _____

Size _____ Joint _____ Cover _____ ___ ft. ___ in.

Thickness at Point of Failure _____ in.

Nature of Break:

☐ Circumferential ☐ Longitudinal ☐ Circumferential & Longitudinal

☐ Blowout ☐ Joint ☐ Split at Corporation ☐ Sleeve ☐ Misc. _____

Apparent Cause of Break:

☐ Water Hammer (surge) ☐ Defective Pipe ☐ Deterioriation ☐ Corrosion

☐ Improper Bedding ☐ Excessive Operating Pressure ☐ Temp. Change

☐ Differential Settlement ☐ Contractor ☐ Misc. _____

Street Surface: ☐ Paved ☐ Unpaved   Traffic: ☐ Heavy ☐ Medium ☐ Light

Type of Street Surface _____ Side of Street: ☐ Sunny ☐ Shady

Type of Soil _____ Resistivity _____ ohms/cm

Electrolysis Indicated: ☐ Yes ☐ No   Corrosion: ☐ Outside ☐ Inside

Conditions Found:

☐ Rocks ☐ Voids   Proximity to Other Utilities _____

Depth of Frost _____ in.   Depth of Snow _____ in.

Office Data for Main Break Evaluation

Weather conditions previous two weeks _____

Sudden change in air temp.:   ☐ Yes ☐ No ☐ Temp. __°F Rise __°F Fall __°F

Water Temp. Sudden Change: ☐ Yes ☐ No ☐ Temp. __°F Rise __°F Fall __°F

Spec. of Main _____ Class or Thickness _____ Laying Length _____ ft

Date Laid _____ Operating Pressure_____

Previous Break psi Reported _____

Initial Installation Data:

Trench Preparation:

☐ Native Material _____   ☐ Sand Bedding ☐ Gravel Bedding

Backfill:

☐ Native Material   Describe_____ ☐ Bank Run Sand & Gravel

☐ Gravel ☐ Sand ☐ Crushed Rock ☐ Other _____

Settlement:

☐ Natural ☐ Water ☐ Compactors ☐ Vibrators ☐ Other_____

Additional Data for Local Utility

Location of Break _____ Map No. _____

Reported by _____

Damage to Paving and/or Private Property _____

Repair Made (Materials, Labor, Equipment) _____

Repair Difficulties (if any) _____

Installing Contractor_____

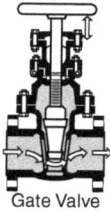
Gate Valve

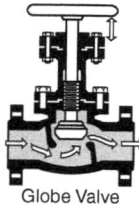
Globe Valve

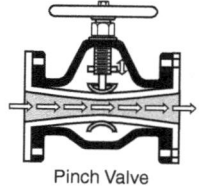
Pinch Valve

Diaphragm Valve

Needle Valve

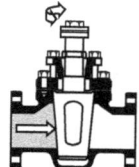
Plug Valve

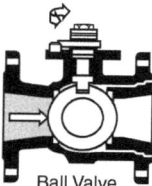
Ball Valve

Butterfly Valve

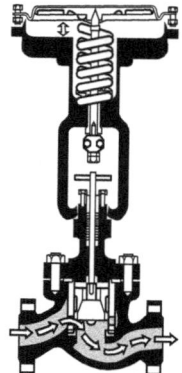
Control Valve

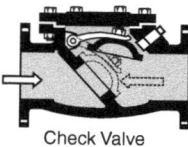
Check Valve

Relief Valve

Courtesy of the Valve Manufacturers Association of America, Washington, D.C.

Types of Water Utility Valves

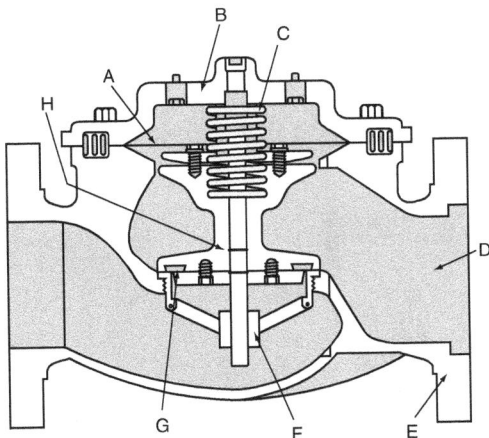

A. **Diaphragm**—Separates upper chamber operating pressure from low chamber line pressure. Buna-N diaphragm standard; Viton available if required; all nylon reinforced for high strength and long life.

B. **Bonnet**—Four tapped ports for pilot piping. Center port for valve position indicator or valve-actuated switches. Primed and painted like body.

C. **Valve Spring**—Stainless-steel spring aids in closing the valve.

D. **Diaphragm Assembly**—The only moving part of the Model 65 valve. Ductile-iron spool, seat retainer, diaphragm plate. Guided top and bottom by bronze or Teflon bushings.

E. **Body**—Globe pattern 1¼–12 in.: 250-lb iron, 150- & 300-lb steel, 150-lb aluminum. Screwed ends 1¼–3 in. globe and angle. Iron and steel bodies epoxy primed inside and out with baked enamel exterior. Four tapped ports for pilot piping.

F. **Seat Ring**—Bronze or stainless-steel ring is replaceable and provides a lower guide for the stainless-steel valve stem.

G. **Valve Seat**—Buna-N or Viton compensates for wear on seating surface and maintains a drip-tight seal over extended service life.

H. **O-Ring**—Creates a static seal. No packing glands required, therefore breakaway friction is eliminated and valve will operate even at extremely low pressures.

NOTE: Basic valve can be used as a pressure, relief, altitude-control, or pressure-reducing valve depending on the type of "brains" equipped to the valve.

Basic Valve

Valve Installation and Operation

Because they are operated frequently, many of the valves located in treatment plants and pumping stations are power operated. Distribution system valves are usually operated infrequently, and so they are manually operated.

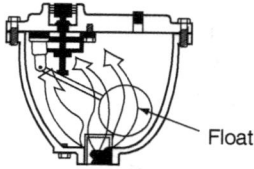

Float

1. During the filling of the line, air entering the valve body will be exhausted to atmosphere. When the air is expelled and water enters the valve, the float will rise and cause the orifices to be closed.

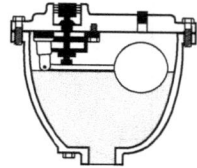

2. The large and small orifices of the air-and-vacuum valve are normally held closed by the buoyant force of the float.

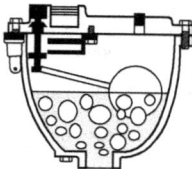

3. While the line is working under pressure, small amounts of trapped or entrained air are exhausted to atmosphere through the small orifice.

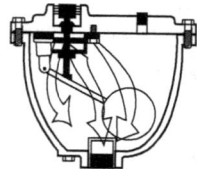

4. Air is permitted to enter the valve and replace the water while the line is being emptied.

Courtesy of GA Industries, Inc.

Air-and-Vacuum Release Valves

Approximate Number of Turns to Operate Valves

Most Double Disk Valves Used in Water Systems

Valve Size, in.	Number of Turns	Valve Size, in.	Number of Turns
3	7½	12	38½
4	14½	14	46
6	20½	16	53
8	27	18	59
10	33½	20	65

Most Metal-Seated Sewerage Valves

Valve Size, in.	Number of Turns	Valve Size, in.	Number of Turns
3	11	12	38
4	14	14	46
6	20	16	53
8	27	18	57
10	33	20	65

NOTE: Most large gate valves are furnished with geared operators so the number of turns required to operate the valve is several times the number shown. For specific numbers, contact the valve manufacturer. This information does not apply to butterfly valves.

Potential Cross-Connections

Situations in which some of the conditions for a cross-connection exist, but require something else to be done to complete the connection, are called *potential cross-connections*. In these situations, the end of the hose must be immersed in liquid for there to be a cross-connection. Although the likelihood is remote that there will be a vacuum on the water system just at the time the hose is submerged in liquid, the tank truck or sink could hold a toxic substance, and consequences of backsiphonage could be very serious.

One common cross-connection is a chemical dispenser connected to a garden hose. If a vacuum should occur while the unit is in use, the chemical solution would be sucked back into the house plumbing.

Many potential locations for cross-connections exist in factories, restaurants, canneries, mortuaries, and hospitals. Any place where a water fill line is below the rim of a container, a cross-connection can exist. A summary of common cross-connections and potential hazards is presented in the table below.

Some Cross-Connections and Potential Hazards

Connected System	Hazard Level
Access hole flush	High
Agricultural pesticide mixing tanks	High
Aspirators	High
Boilers	High
Chlorinators	High
Cooling towers	High
Flush valve toilets	High
Laboratory glassware or washing equipment	High
Plating vats	High
Sewage pumps	High
Sinks	High
Sprinkler systems	High
Sterilizers	High
Car washes	Moderate to high
Photographic developers	Moderate to high
Pump primers	Moderate to high
Baptismal founts	Moderate
Dishwashers	Moderate
Swimming pools	Moderate
Watering troughs	Moderate
Auxiliary water supplies	Low to high
Garden hoses (sill cocks)	Low to high
Irrigation systems	Low to high
Solar energy systems	Low to high
Water systems	Low to high
Commercial food processors	Low to moderate

Backflow Control Devices

When a cross-connection situation is identified, one of two actions must be taken. Either the cross-connection must be removed, or some means must be installed to protect the public water supply from possible contamination.

Air Gaps

The least expensive and most positive method of protecting against backflow is to install an air gap. There are no moving parts to maintain or break, and surveillance is necessary only to ensure that it is not altered. The only requirement for an air gap between the supply outlet and the maximum water surface of a nonpotable substance is that it must be at least twice the internal diameter of the supply pipe, but no less than 1 in. (25 mm) in any situation.

Typical uses of an air gap are for supplying water to tank trucks, to a nonpotable supply, or to a surge tank in a factory.

Reduced-Pressure-Zone Devices

A device that can be used in every cross-connection situation is the reduced-pressure-zone (RPZ) backflow preventer. It consists of two spring-loaded check valves with a pressure-regulated relief valve located between them. As illustrated in the figure on the following page, if there is a potential backsiphonage situation, both check valves will close, and the space between them is opened to atmospheric pressure. If there is backpressure in excess of the water main pressure, both check valves will close, and if there is any leakage in the second valve, it will be allowed to escape through the center relief valve.

An RPZ device is much safer than one or two check valves because there is always the potential of a check valve leaking. Even though RPZ devices are designed to be dependable, they are mechanical devices that must be tested and maintained regularly in accordance with the manufacturer's recommendations. They must be installed in locations where the relief port cannot be submerged, and where they are protected from freezing and vandalism.

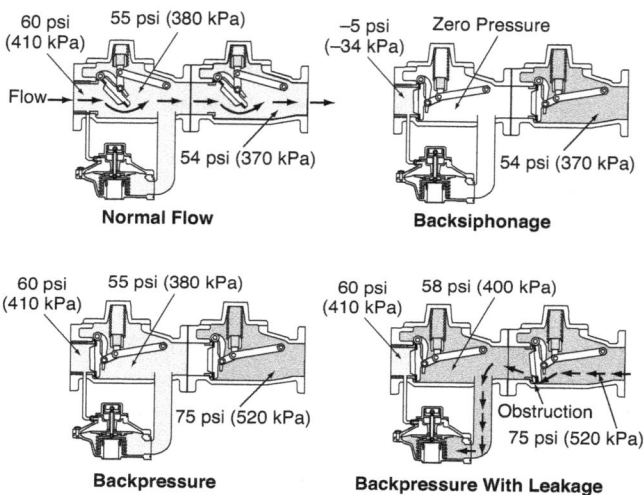

60 psi (410 kPa)	55 psi (380 kPa)		−5 psi (−34 kPa)	Zero Pressure
Flow→				
	54 psi (370 kPa)			54 psi (370 kPa)
Normal Flow			**Backsiphonage**	

60 psi (410 kPa)	55 psi (380 kPa)		60 psi (410 kPa)	58 psi (400 kPa)
	75 psi (520 kPa)			Obstruction
				75 psi (520 kPa)
Backpressure			**Backpressure With Leakage**	

Courtesy of Cla-Val Company, Backflow Preventer Division.

Valve Position and Flow Direction in a Reduced Pressure Zone Device

Double Check Valves

A double-check-valve (DCV) backflow preventer is designed similarly to an RPZ device except there is no relief valve between the two check valves. The protection is not as positive as an RPZ device because of the possibility of the check valves leaking, so they are not recommended for use in situations where a health hazard may result from valve failure. Local and state officials should be contacted for approval before a DCV is installed for cross-connection protection in a potable water line.

Some water utilities install check valves on some or all customer water services. This is particularly desirable for customers with operable private wells because of the potential of well water being forced backward into the utility's system if there is some piping change by the customer. Several manufacturers have developed DCV assemblies for this use.

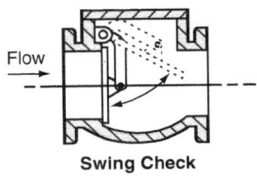

Swing Check

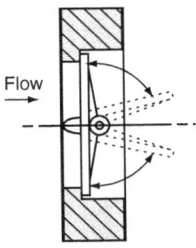

Double Disk Check

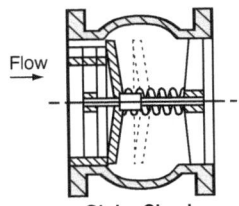

Globe Check

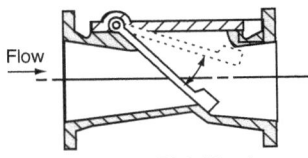

Slanting Disk Check

NOTE: All valves are in the closed position. The dashed lines show the open position

Source: Office of Water Programs, California State University, Sacramento Foundation, in Small Water System Operation and Maintenance. *For additional information, visit <www.owp.csus.edu> or call 916-278-6142.*

Types of Check Valves

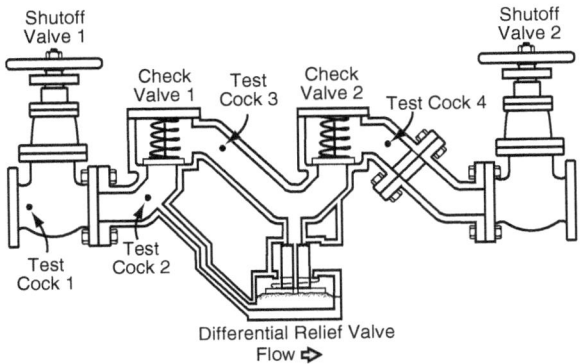

Shutoff Valve 1

Shutoff Valve 2

Check Valve 1

Test Cock 3

Check Valve 2

Test Cock 4

Test Cock 1

Test Cock 2

Differential Relief Valve
Flow ⇨

Reduced Pressure Zone (Backflow-Prevention) Device

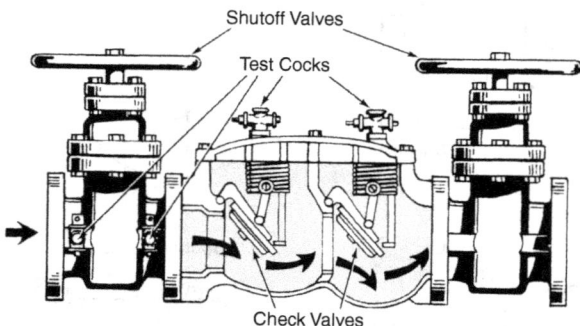

Double Check Valve Assembly

Complete Isolation

The most positive method of preventing connection between piping systems from two different sources is complete separation. When piping systems from two sources are located in the same building, they can be identified by signs and color coding. The need for monitoring continues, however, to ensure that the systems are not inadvertently connected.

Someone who does not realize the potential consequences may install a temporary connection between two systems using a spool piece or a swing connection. Such connections are not recommended for use regardless of the degree of risk involved, and they should be completely removed to eliminate any possibility of a cross-connection.

Air Inlet Port

Check Valve

Normal Flow

Air Inlet Port

Check Valve

Backsiphonage

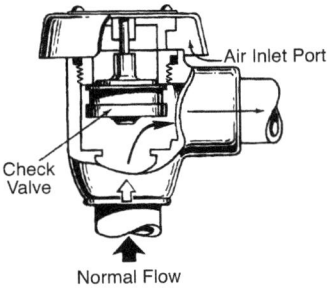

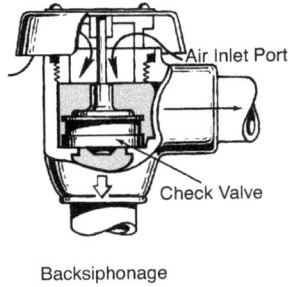

Atmospheric Vacuum Breaker

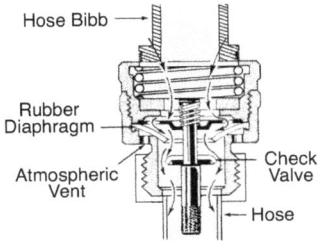

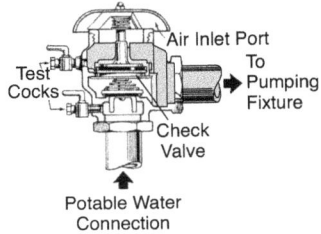

Hose Bibb

Rubber Diaphragm

Atmospheric Vent

Check Valve

Hose

Hose-Bibb Type of Atmospheric Vacuum Breaker

Air Inlet Port

Test Cocks

To Pumping Fixture

Check Valve

Potable Water Connection

Pressure Vacuum Breaker

TYPES OF HYDRANTS

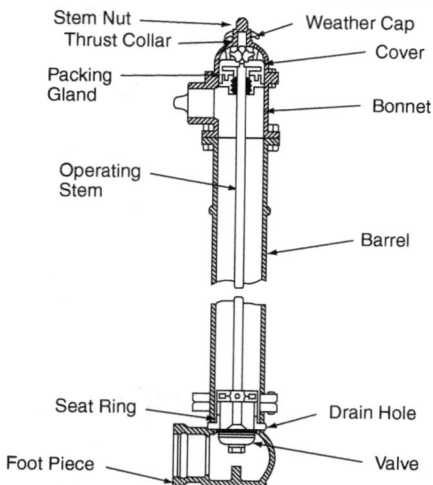

Stem Nut — Weather Cap
Thrust Collar — Cover
Packing Gland
Bonnet
Operating Stem
Barrel
Seat Ring — Drain Hole
Foot Piece — Valve

Dry Barrel Hydrant

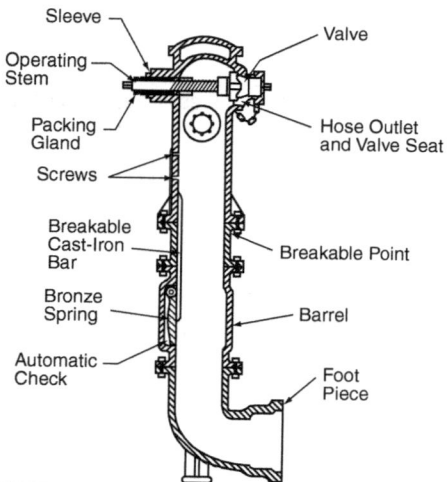

Sleeve — Valve
Operating Stem
Packing Gland — Hose Outlet and Valve Seat
Screws
Breakable Cast-Iron Bar — Breakable Point
Bronze Spring — Barrel
Automatic Check — Foot Piece

Wet Barrel Hydrant

SERVICE CONNECTIONS

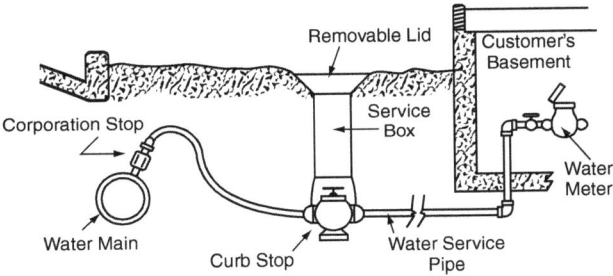

Typical Service Connection

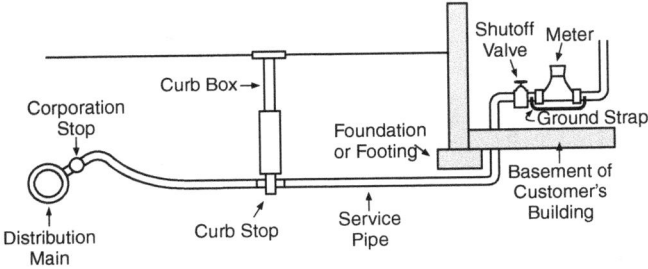

Small Service Connection With the Meter Located in a Basement

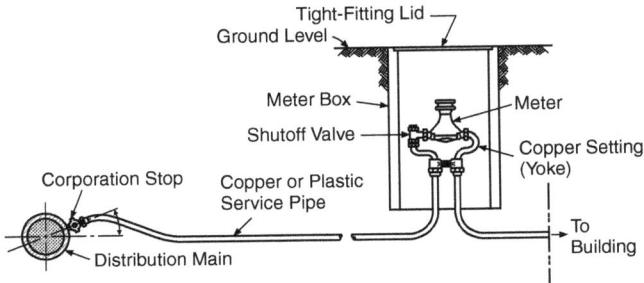

Small Service Connection With Shallow Meter Box

Comparison of Computed Water Flow Friction Losses for Service Line,* psi/ft

Flow Rate, gpm	¾ in.			1 in.			1¼ in.	
	Copper Tube	Plastic* Iron Pipe ID	Plastic Copper Tube OD	Copper Tube	Plastic* Iron Pipe ID	Plastic Copper Tube OD	Copper Tube	Plastic* Iron Pipe ID
1	0.003	0.002	0.005					
2	0.008	0.005	0.013					
3	0.016	0.010	0.026					
4	0.025	0.015	0.041					
5	0.037	0.024	0.065	0.009	0.007	0.022		
6	0.048	0.031	0.081	0.011	0.008	0.029		
7	0.063	0.041	0.108	0.018	0.011	0.038		
8	0.075	0.049	0.128	0.021	0.013	0.045		
9	0.095	0.062	0.160	0.026	0.017	0.057		
10	0.121	0.078	0.203	0.035	0.021	0.072		
11	0.166	0.108	0.274	0.057	0.040	0.104		
12	0.192	0.128	0.317	0.069	0.048	0.126		
13	0.206	0.134	0.338	0.071	0.050	0.130		

Table continued on next page

Comparison of Computed Water Flow Friction Losses for Service Line,* *psi/ft* (continued)

Flow Rate, gpm	¾ in.			1 in.			1¼ in.	
	Copper Tube	Plastic* Iron Pipe ID	Plastic Copper Tube OD	Copper Tube	Plastic* Iron Pipe ID	Plastic Copper Tube OD	Copper Tube	Plastic* Iron Pipe ID
14	0.228	0.149	0.377	0.086	0.059	0.156		
15	0.263	0.169	0.428	0.107	0.076	0.200		
20	0.433	0.286	0.734	0.120	0.085	0.225	0.042	0.030
25	0.656	0.429	1.118	0.226	0.136	0.343	0.059	0.046
30	0.920	0.599		0.242	0.188	0.450	0.083	0.060
35		0.809		0.370	0.254	0.611	0.108	0.076
40				0.411	0.317	0.763	0.135	0.091
45				0.517	0.397	0.924	0.156	0.109
50				0.623	0.482	1.100	0.200	0.131
55				0.749	0.583		0.234	0.161
60				0.892	0.685		0.278	0.185

Note: Metric conversions: gpm × 0.2268 = m³/hr, in. × 25.4 = mm, psi/ft × 22.62 = kPa/m.
*Plastic per ANSI/AWWA Standard C901.

Distribution

Computed Pressure Losses for Service Components, *psi*

Flow Rate,	Corporation Stop		Curb Stop			Globe Valve		Gate or Ball Valve		Meter		
gpm	¾ in.*	1 in.†	¾ in.‡	1 in.§	¾ in.**	1 in.††	¾ in.‡	1 in.§§	⅝ in.	¾ in.	1 in.
1	0.015	0.004	0.011	0.003	0.060	0.017					
2	0.050	0.014	0.034	0.008	0.195	0.058	0.001				
3	0.103	0.029	0.071	0.017	0.301	0.09	0.002				
4	0.170	0.03	0.117	0.028	0.664	0.20	0.003		0.39		
5	0.246	0.07	0.170	0.04	0.962	0.29	0.005	0.002	0.62		
6	0.340	0.11	0.234	0.05	1.33	0.43	0.007	0.002	0.89	0.39	
7	0.457	0.14	0.315	0.07	1.79	0.57	0.009	0.004	1.22	0.53	
8	0.557	0.17	0.384	0.08	2.18	0.71	0.011	0.004	1.60	0.71	0.25
9	0.703	0.22	0.485	0.10	2.75	0.85	0.013	0.005	2.02	0.89	0.32
10	0.879	0.23	0.606	0.13	3.44	1.0	0.016	0.007	2.50	1.10	0.39
11	1.20	0.28	0.72	0.15	4.08	1.2	0.020	0.009	3.00	1.35	0.48
12	1.31	0.32	0.83	0.18	4.72	1.4	0.024	0.011	3.60	1.60	0.57
13	1.40	0.37	0.94	0.21	5.36	1.6	0.028	0.013	4.20	1.88	0.67
14	1.54	0.42	1.05	0.24	6.00	1.8	0.032	0.015	4.90	2.18	0.77
15	1.70	0.47	1.17	0.27	6.64	2.0	0.036	0.017	5.60	2.50	0.89
16	2.11	0.53	1.31	0.30	7.46	2.3	0.040	0.019	6.40	2.85	1.02

Table continued on next page

Computed Pressure Losses for Service Components, *psi* (continued)

Flow Rate,	Corporation Stop		Curb Stop		Globe Valve		Gate or Ball Valve			Meter		
gpm	¾ in.*	1 in.†	¾ in.‡	1 in.§	¾ in.**	1 in.††	¾ in.‡‡	1 in.§§	⅝ in.	¾ in.	1 in.
17	2.25	0.59	1.45	0.34	8.28	2.6	0.044	0.021	7.20	3.22	1.16
18	2.36	0.65	1.60	0.38	9.10	2.9	0.049	0.023	8.10	3.60	1.30
19	2.54	0.72	1.75	0.42	9.93	3.2	0.054	0.025	9.00	4.00	1.45
20	2.75	0.80	1.90	0.46	10.76	3.5	0.059	0.028	10.00	4.45	1.60
25		1.2		0.67		5.1		0.047		7.10	2.50
30		1.6		0.92		7.0		0.062		10.00	3.60
35		2.1		1.2		9.0		0.077			4.80
40		2.7		1.5		11.6		0.086			6.40
45		3.3		1.9		14.6		0.093			8.10
50		4.0		2.3		17.6		0.130			10.00

Note: Metric conversions: gpm × 0.2268 = m³/hr, in. × 25.4 = mm, psi/ft × 6.89476 = kPa.

*Based on equivalent loss of 5.86 ft of ¾-in. copper tubing.
†Based on equivalent loss of 6.67 ft of 1-in. copper tubing.
‡Based on equivalent loss of 4.04 ft of ¾-in. copper tubing.
§Based on equivalent loss of 3.85 ft of 1-in. copper tubing.
**Based on equivalent loss of 22.90 ft of ¾-in. copper tubing.
††Based on equivalent loss of 29.1 ft of 1-in. copper tubing.
‡‡Based on equivalent loss of 0.14 ft of ¾-in. copper tubing.
§§Based on equivalent loss of 0.21 ft of 1-in. copper tubing.

Distribution

Discharge From Fixtures and Faucets at Various Operating Pressures

Type of Fixture or Faucet	5 psi	Operating Pressure, 30 psi	90 psi
		Discharge, *gpm*	
¾-in. compression sink faucet:			
Wide open	8.1	20.0	33.4
Half open	7.6	19.0	32.9
One-fourth open	7.0	17.4	29.9
½-in. compression sink faucet, wide open	6.0	14.8	24.5
½-in. ground key sink faucet, wide open	9.5	23.4	36.4
¾-in. ground key sink faucet, wide open	13.8	31.7	51.0
¾-in. compression sink faucet, wide open	9.0	22.1	36.0
½-in. self-closing compression faucet, wide open	2.6	6.8	11.7
⅜-in. ground key sink faucet, wide open	6.8	16.7	27.7
⅜-in. compression sink faucet, wide open	3.2	8.2	14.1
½-in. compression sink faucet, wide open	4.8	12.3	21.3
½-in. compression laundry tray faucet, open	6.3	17.3	25.3
Compression wash basin, wide open	5.0	11.9	21.3
1-in. ground key sink faucet, wide open	30.7	78.9	118.8

Table continued on next page

Discharge From Fixtures and Faucets at Various Operating Pressures (continued)

Type of Fixture or Faucet	Operating Pressure,		
	5 psi	30 psi	90 psi
		Discharge, *gpm*	
1-in. compression sink faucet, wide open	12.7	39.9	64.8
Combination compression laundry faucet:			
Both outlets open	9.6	22.4	38.6
Either hot or cold, wide open	6.1	14.4	24.8
Combination compression bath tub:			
Both hot and cold open, no nozzle	8.0	20.4	34.4
Both hot and cold open, with nozzle	5.9	14.3	24.8
Either hot or cold only, open, no nozzle	4.3	11.1	19.9
Either hot or cold only, open, with nozzle	3.8	9.2	16.1
Combination compression sink faucet with swinging nozzle:			
Hot and cold open	4.6	12.2	21.4
Either hot or cold open	3.2	8.4	14.8
Water closets:			
Tank type	2.9	8.0	14.6
Flush valves	9.7	30.0	45.7

Note: Metric conversions: gpm × 0.2268 = m³/hr, in. × 25.4 = mm, psi × 6.89476 = kPa.

Distribution

263

Thawing Frozen Services

The best way of preventing frozen services is to insist that all service piping be installed to below the recommended maximum frost depth for that part of the country. The most common cause of a shallow service is that the installing contractor did not comply with local requirements. The best policy is to insist on an inspection of the installation by a city or water utility inspector before the trench is backfilled. Some utilities go so far as to make the contractor reexcavate the trench if it is backfilled before inspection, to make sure the proper depth is maintained.

Continuous snow cover typically allows relatively little frost penetration in areas where the snow has not been disturbed. But at the same time, during a very cold winter, there can be several feet of frost in the ground under streets and driveways. The most common services to freeze are those that run under roadways.

A copper, lead, and galvanized-iron service can usually be thawed by running an electrical current through it. A portable source of direct current, such as a welding unit, is connected to the main and the service at the building, and will usually generate enough heat in the line to release the ice blockage. Electrical thawing can be dangerous and should only be performed by someone with experience.

One of the problems that can arise in electrical thawing is a point of poor conductivity in the service line connections. Another potential problem is if the service is in contact with another conductor, such as a gas pipe, which will divert the current and could cause it to enter adjacent buildings. The current may also damage O-rings or gaskets in the service fittings. Thawing is a service performed by the utility or a contractor on the customer's property, so a waiver should be signed by the property owner, and there must be confirmation by the insurer that the person doing the work has adequate liability insurance in effect to cover any possible consequences of the work.

Hot-water thawing is becoming more common because it is less dangerous and can be used for plastic pipe. A plastic tube carrying hot water is fed into the service line and pushed against the ice blockage. The same process can also be used with a steam generator, but it should first be determined if the extreme heat might damage plastic pipe.

If only a meter or small section of service is frozen in a meter pit, it is best thawed using a hair dryer or heat gun. A propane torch should be used for thawing with extreme caution because of the possibility of igniting explosive gases in the pit and the potential of damaging fittings by overheating them. If the line is heated too quickly, there is also the danger of generating steam, which may have no place to escape, and could rupture the meter or piping.

After a frozen service pipe has been thawed, the only way to prevent it from freezing again is to open a faucet to allow a small, continuous flow until the ground has thawed. In that this could be several weeks or months, some water utilities will remove the meter or make an allowance for the unusual water use when billing the customer.

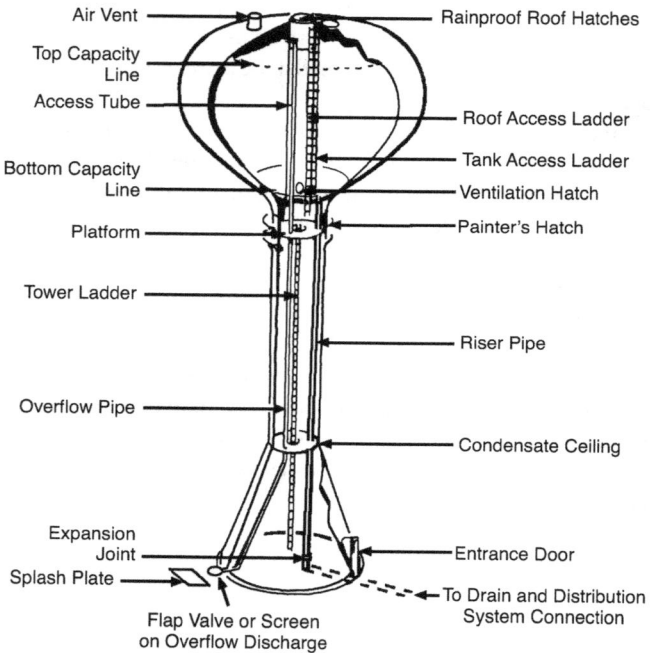

Air Vent

Top Capacity Line

Access Tube

Bottom Capacity Line

Platform

Tower Ladder

Overflow Pipe

Expansion Joint

Splash Plate

Flap Valve or Screen on Overflow Discharge

Rainproof Roof Hatches

Roof Access Ladder

Tank Access Ladder

Ventilation Hatch

Painter's Hatch

Riser Pipe

Condensate Ceiling

Entrance Door

To Drain and Distribution System Connection

Courtesy of CB&I.

Principal Accessories for an Elevated Storage Tank

Capacities of Vertical Tanks

Tank Diameter, *in.*	Gallons per Foot Depth
12	5.86
18	13.20
24	23.42
30	36.6
36	52.6
42	71.6
48	93.6
54	119.0
60	146.0
72	211.0

Capacities of Horizontal Tanks*

Ratio of Water Depth to Total Depth	Percent of Total Volume
0.1	5.22
0.2	14.22
0.3	26.2
0.4	37.4
0.5	50.0
0.6	62.6
0.7	73.8
0.8	85.8
0.9	94.8
1.0	100.0

*These figures assume flat tank ends. The actual volume of tanks with dished ends will be somewhat greater.

Distribution

Capacities of Cylindrical Tanks

Inside Diameter, in.	Gallons per Inch Depth	Inside Diameter, in.	Gallons per Inch Depth
12	0.49	5	12.24
13	0.57	6	17.63
14	0.67	7	23.99
15	0.76	8	31.33
16	0.87	9	39.66
17	0.98	10	48.96
18	1.10	11	59.24
19	1.23	12	70.50
20	1.36	13	82.74
21	1.50	14	95.96
22	1.65	15	110.16
23	1.80	16	125.34
24	1.96	17	141.49
25	2.12	18	158.63
26	2.30	19	176.75
27	2.48	20	195.84
28	2.67	21	215.91
29	2.86	22	236.97
30	3.06	23	259.00
31	3.27	24	282.01
32	3.48	25	306.00
33	3.70	26	330.97
34	3.93	27	356.92
35	4.16	28	383.85
36	4.41	29	411.75
37	4.65	30	440.64
38	4.91	31	470.50
39	5.17	32	501.35
40	5.44	33	533.17
41	5.72	34	565.98

Table continued on next page

Capacities of Cylindrical Tanks (continued)

Inside Diameter, in.	Gallons per Inch Depth	Inside Diameter, in.	Gallons per Inch Depth
42	6.00	35	599.76
43	6.29	36	634.52
44	6.58	37	670.26
45	6.88	38	706.98
46	7.19	39	744.68
47	7.51	40	783.36
48	7.83	41	823.02
49	8.16	42	863.65
50	8.50	43	905.27
51	8.84	44	947.86
52	9.19	45	991.44
53	9.55	46	1,035.99
54	9.91	47	1,081.52
55	10.28	48	1,128.04
56	10.66	49	1,175.53
57	11.05	50	1,224.00
58	11.44		
59	11.84		
60	12.24		

Wells

There are between 10,000,000 and 20,000,000 water wells scattered throughout the United States. Most are situated in valleys or river-bottom land, although many are located in hilly and mountainous regions. They range from shallow hand-dug wells to carefully designed, large production wells.

recovery time   The time it takes after pumping has stopped for the water level to return to the static water level.

residual drawdown   A lowered water level, below the original static level that remains after pumping has been stopped for a period of time.

specific capacity   The well yield per unit of drawdown, or

$$\text{specific capacity} = \frac{\text{well yield}}{\text{drawdown}}$$

well yield   The rate of water withdrawal that a well can supply over a long period of time. In other words, it is the recharge rate that the aquifer can continuously sustain to the well. The yield of small wells is usually expressed in gallons or liters per minute. For large wells, it may be expressed in cubic feet or cubic meters per second.

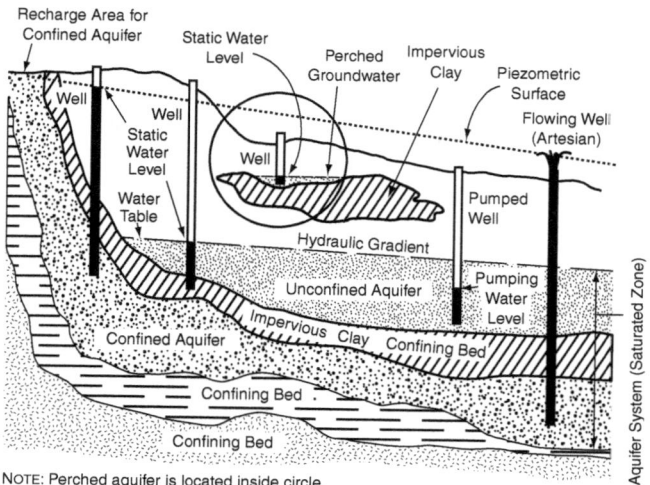

NOTE: Perched aquifer is located inside circle.

Adapted from Water Wells and Pumps: Their Designs, Construction, Operation and Maintenance, *Division of Agricultural Science, University of California, Davis.*

Groundwater and Wells

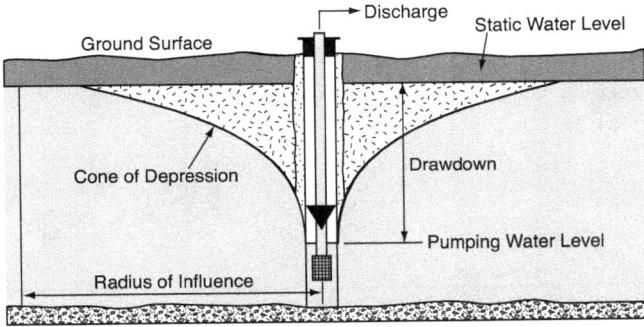

Hydraulic Characteristics of a Well

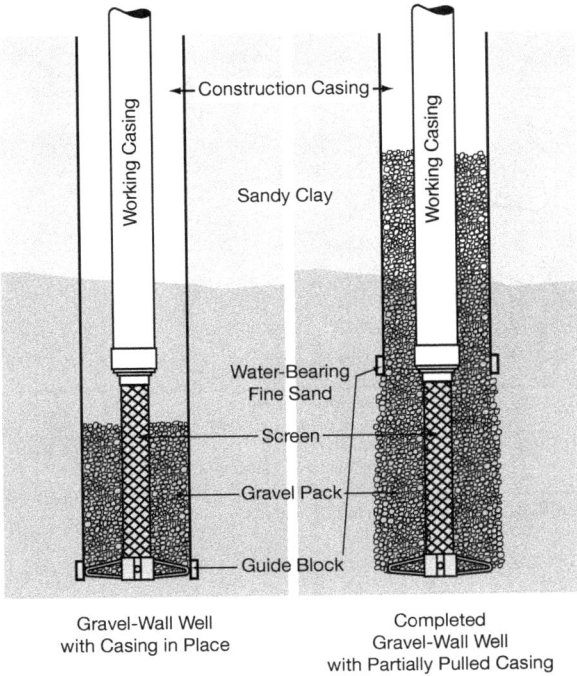

Gravel-Wall Well
with Casing in Place

Completed
Gravel-Wall Well
with Partially Pulled Casing

Gravel-Wall Well Construction

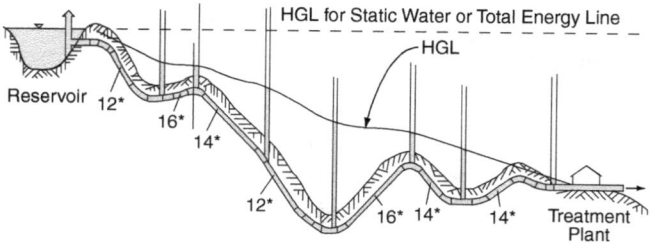

Hydraulic Grade Line of Water Transmission Line

HGL for Static Water or Total Energy Line

HGL

Reservoir

12*

16*

14*

12*

16* 14*

14*

Treatment
Plant

*Pipe Diameter in Inches

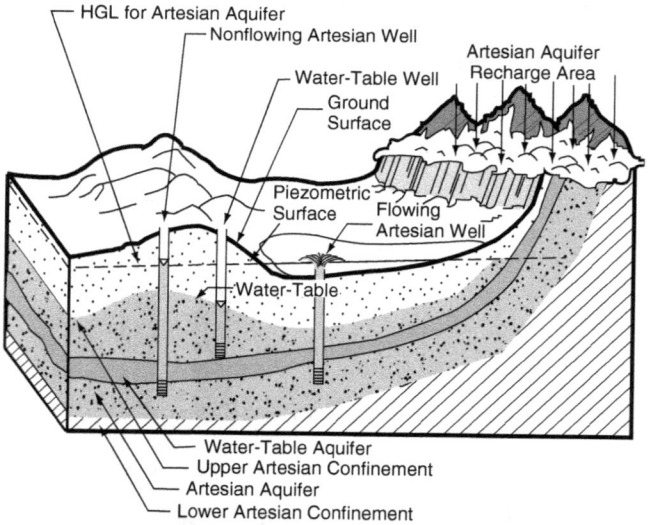

Hydraulic Grade Line and Artesian Wells

HGL for Artesian Aquifer

Nonflowing Artesian Well

Water-Table Well

Ground
Surface

Artesian Aquifer
Recharge Area

Piezometric
Surface

Flowing
Artesian Well

Water-Table

Water-Table Aquifer

Upper Artesian Confinement

Artesian Aquifer

Lower Artesian Confinement

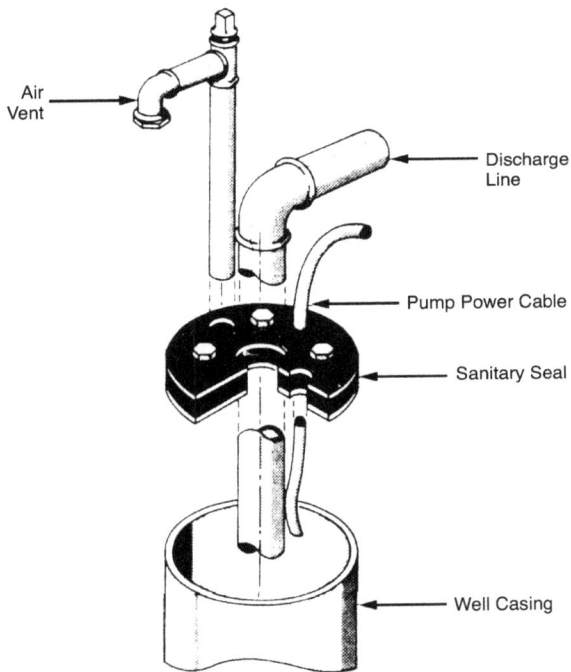

Components of a Sanitary Seal

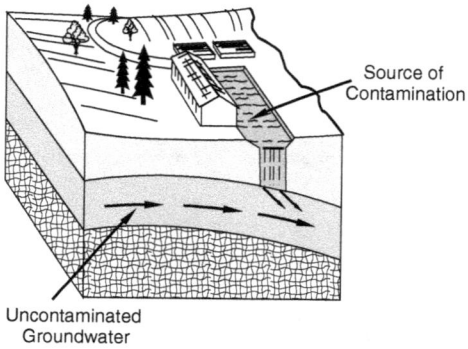

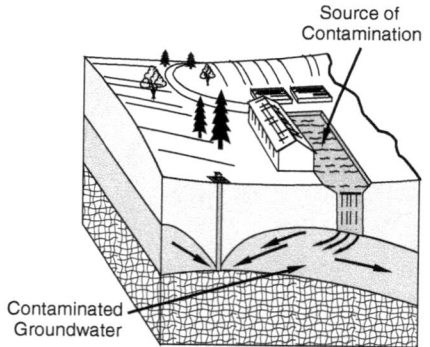

Reversal of Flow in an Aquifer Resulting From Well Drawdown

Approximate Amount of Water in a Well

Diameter of Casing or Hole, *in.*	Gallons per Foot of Depth	Cubic Feet per Foot of Depth	Liters per Meter of Depth
1	0.041	0.0055	0.509
1½	0.092	0.0123	1.142
2	0.163	0.0218	2.204
2½	0.255	0.0341	3.167
3	0.367	0.0491	4.558
3½	0.500	0.0668	6.209
4	0.653	0.0873	8.110
4½	0.826	0.1104	10.28
5	1.020	0.1364	12.67
5½	1.234	0.1650	15.33
6	1.469	0.1963	18.24
7	2.000	0.2673	24.84
8	2.611	0.3491	32.43
9	3.305	0.4418	41.04
10	4.080	0.5454	50.67
11	4.937	0.6600	61.31
12	5.875	0.7854	72.96
14	8.000	1.069	99.33
16	10.44	1.396	129.65
18	13.22	1.767	164.18
20	16.32	2.182	202.68
22	19.75	2.640	245.28
24	23.50	3.142	291.85
26	27.58	3.687	342.52
28	32.00	4.276	397.41
30	36.72	4.909	456.02
32	41.78	5.585	518.87
34	47.16	6.305	585.68
36	52.88	7.069	656.72

Wells

Amounts of Chlorine-Containing Compounds Used For Well Disinfection*

Casing Diameter		Casing Volume†		65% Available Chlorine, as Calcium Hypochlorite (e.g., HTH, perchloron, etc.) Dry Weight		25% Available Chlorine as Calcium Hypochlorite (Chloride of Lime) Dry Weight		5.25% Available Chlorine as Sodium Hypochlorite (e.g., Purex, Chlorox, etc.) Liquid Volume	
in.	mm	gal	m³	oz	g	oz	g	oz	L
2	51	16.3	0.06	0.4	11.4	1.0	28.4	4	0.12
4	102	65.3	0.25	1.4	39.60	4	113.4	18	0.6
6	152	147	0.56	4	113.4	8	226	40	1.2
8	203	261	0.99	6	170.2	14	396	68	2.0
10	254	408	1.5	8	226	22	624	112	3.4
12	305	588	2.2	12	340	32	908	160	4.8
16	406	1,045	4.0	22	624	56	1,588	256	7.6
20	508	1,632	6.2	34	964	86	2,438	428	12.8
24	610	2,350	8.9	48	1,360	126	3,572	596	17.4

* Amounts indicated are necessary to form a chlorine concentration of 100 mg/L in 100 ft (30.5 m) of water-filled well casing. Adjust the indicated amounts as a direct multiple based on the actual amount of water-filled casing.

† Based on 100-ft (30.5-m) length of water filled casing.

Pumps

Two basic categories of pumps are used in water supply operations: velocity pumps and positive-displacement pumps. Velocity pumps, which include centrifugal and vertical turbine pumps, are used for most distribution system applications. Positive-displacement pumps are most commonly used in water treatment plants for chemical metering.

KEY FORMULAS FOR PUMPS

Pumping Rate

volume, gal = gpm × time, min

$$\text{rate, gpm} = \frac{\text{tank volume, gal}}{\text{time, min}}$$

$$\text{time, min} = \frac{\text{tank volume, gal}}{\text{fill rate, pgm}}$$

Pump Size

$$\text{water horsepower} = \frac{\text{gpm} \times \text{total head, ft}}{3,960}$$

$$\text{brake horsepower} = \frac{\text{gpm} \times \text{total head, ft}}{3,960 \times \% \text{ efficiency}}$$

$$\frac{\% \text{ overall efficiency}}{\text{pump/motor}} = \text{motor, \% efficiency} \times \text{pump, \% efficiency}$$

Pumping Cost

$$\text{cost, \$} = \text{bhp} \times 0.746 \text{ kW/hp} \times \text{operating hr} \times \frac{\text{¢/kW·hr}}{100}$$

Wells

drawdown, ft = pumping level, ft − static level, ft

$$\text{specific capacity, gpm/ft} = \frac{\text{well yield, gpm}}{\text{drawdown, ft}}$$

Head and Pressure (for water at 60°F)

$$\text{head in feet} = \frac{\text{head in psi} \times 2.31}{\text{specific gravity}}$$

$$\text{head in psi} = \frac{\text{head in feet} \times \text{specific gravity}}{2.31}$$

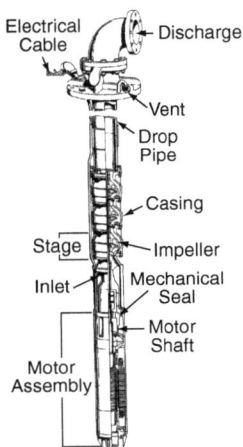

Submersible Pump

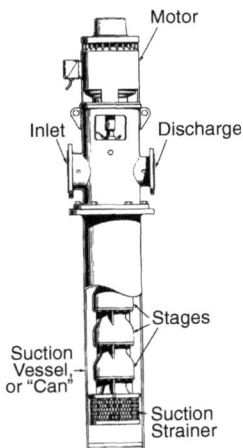

Turbine Booster Pump

Pumping Power

$$
\text{horsepower} \quad
\begin{array}{rcl}
\times\ 550 & = & \text{ft-lb/sec} \\
\times\ 33{,}000 & = & \text{ft-lb/min} \\
\times\ 2{,}546 & = & \text{Btu/hr} \\
\times\ 745.7 & = & \text{W} \\
\times\ 0.7457 & = & \text{kW} \\
\times\ 1.014 & = & \text{metric hp}
\end{array}
$$

$$
\begin{array}{rcl}
\text{brake horsepower} \\ \text{(bhp)}
\end{array}
= \frac{\text{gpm} \times H \times \text{specific gravity}}{3{,}960 \times \text{efficiency}}
$$

$$
= \frac{\text{bph} \times H \times \text{specific gravity}}{5{,}657 \times \text{efficiency}}
$$

$$
\text{bhp} = \frac{\text{gpm} \times \text{psi}}{1{,}714 \times \text{efficiency}}
$$

$$
= \frac{\text{bph} \times \text{psi}}{2{,}449 \times \text{efficiency}}
$$

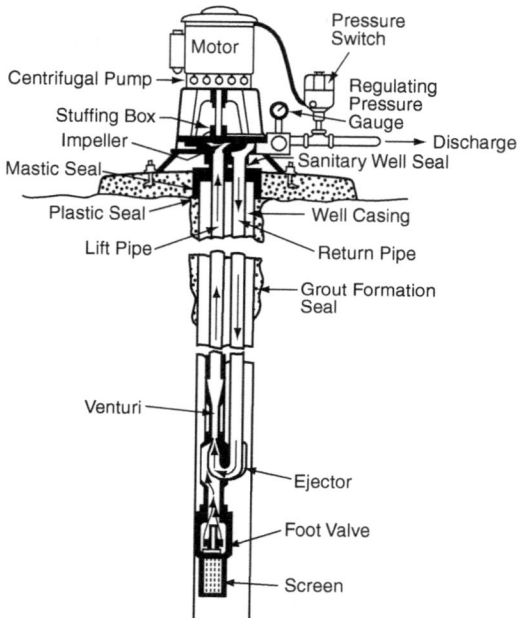

Jet Pump

Where:

gpm = US gallons per minute delivered
(1 gal = 8.338 lb at 60°F)

bph = barrels (42 gal) per hour delivered = 0.7 gpm

H = total head, in feet of liquid—differential

psi = pounds per square inch—differential

efficiency = expressed as a decimal

NOTE: To obtain the hydraulic horsepower from the above expressions, assume a pump efficiency of 100%.

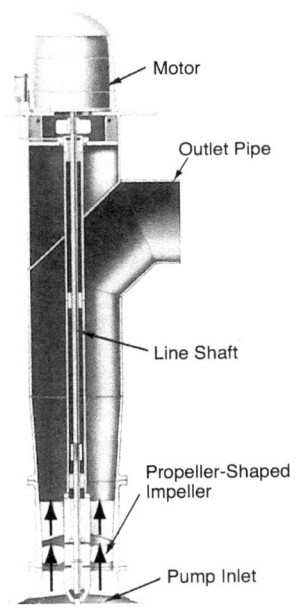

Courtesy of Ingersoll-Dresser
Pump Company.

Axial-Flow Pump

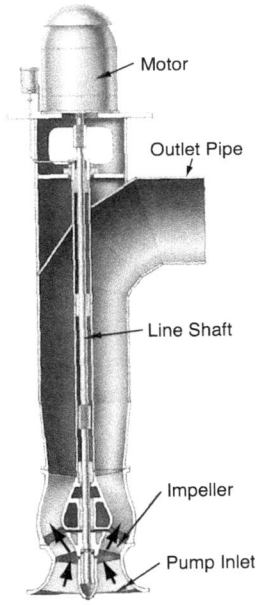

Courtesy of Ingersoll-Dresser
Pump Company.

Mixed-Flow Pump

$$\text{electrical hp input to motor} = \frac{\text{pump bhp}}{\text{motor efficiency}}$$

$$\text{kW input to motor} = \frac{\text{pump bhp} \times 0.7457}{\text{motor efficiency}}$$

Torque

$$\text{torque, lb-ft} = \frac{\text{bhp} \times 5,250}{\text{rpm}}$$

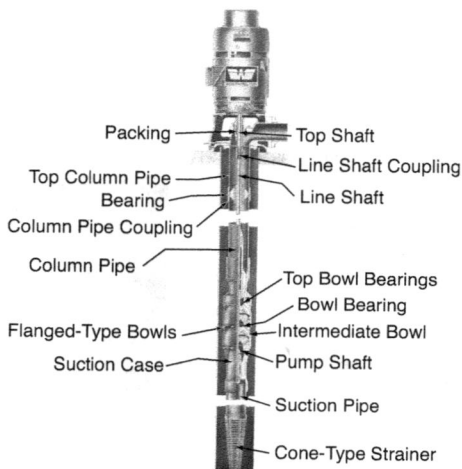

Packing — Top Shaft
Top Column Pipe — Line Shaft Coupling
Bearing — Line Shaft
Column Pipe Coupling —
Column Pipe — Top Bowl Bearings
Bowl Bearing
Flanged-Type Bowls — Intermediate Bowl
Suction Case — Pump Shaft
Suction Pipe
Cone-Type Strainer

Courtesy of Ingersoll-Dresser Pump Company.

Deep-Well Pump

Specific Speed

$$\mathcal{N}_s = \frac{\text{rpm}\sqrt{\text{gpm}}}{H^{3/4}}$$

Where:

$\mathcal{N}_s$ = impeller-specific speed
rpm = speed
gpm = design capacity at best efficiency point
H = head per stage, in feet, at best efficiency point

$$S = \frac{\text{rpm}\sqrt{\text{gpm}}}{\text{NPSHR}^{3/4}}$$

Where:

S = suction-specific speed
rpm = speed

gpm = design capacity at best efficiency point for single-suction first-stage impellers, or one-half design capacity for double-suction impellers

NPSHR = net positive suction head rate

Affinity Laws

At constant impeller diameter—variable speed:

$$\frac{\text{rpm}_1}{\text{rpm}_2} = \frac{\text{gpm}_1}{\text{gpm}_2} = \frac{\sqrt{H_1}}{\sqrt{H_2}}$$

Where:

H_1 = head, in feet, original

H_2 = head, in feet, new

At constant speed—variable impeller diameter:

$$\frac{D_1}{D_2} = \frac{\text{gpm}_1}{\text{gpm}_2} = \frac{\sqrt{H_1}}{\sqrt{H_2}}$$

Where:

D = diameter, in feet

H_1 = head, in feet, original

H_2 = head, in feet, new

Horsepower and Efficiency Calculations

$$\text{water horsepower} = \frac{\text{flow rate, gpm} \times \text{total head, ft}}{3,960}$$

1 hp = 746 W power

1 hp = 0.746 kW power

% efficiency = hp output/hp supplied × 100

% motor efficiency = bhp/mhp × 100

% pump efficiency = whp/bhp × 100

% overall efficiency = whp/mhp × 100

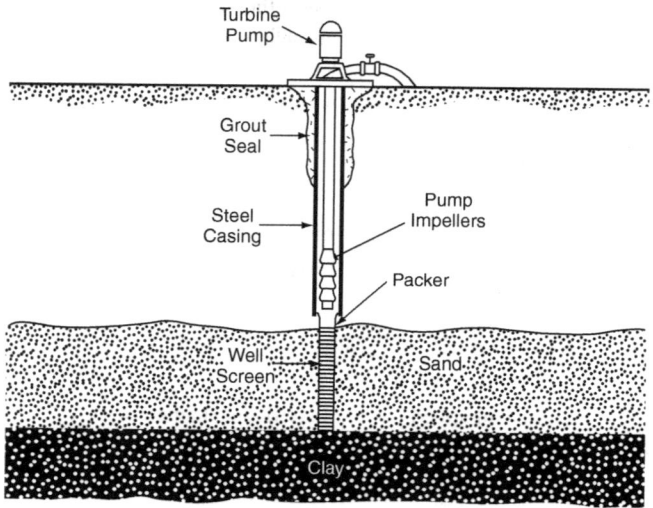

Turbine Pump

Grout Seal

Steel Casing

Pump Impellers

Packer

Well Screen

Sand

Clay

Exposed Well Screen

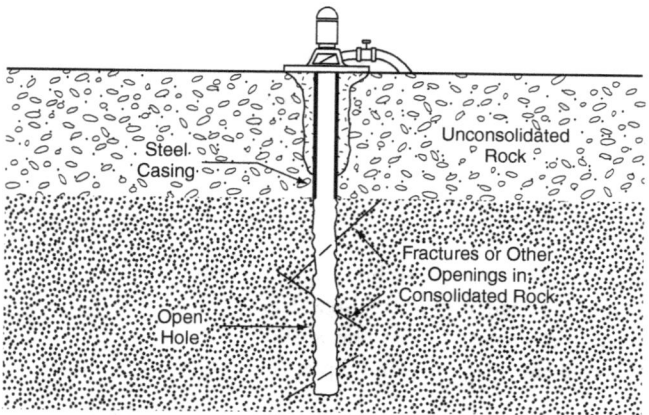

Steel Casing

Unconsolidated Rock

Fractures or Other Openings in Consolidated Rock

Open Hole

Casing Seated at Top of Rock Layer With an Open Hole Underneath

ELECTRICAL MEASUREMENTS

A simple explanation of electrical measurements can be made by comparing the behavior of electricity to the behavior of water.

- Volts (potential) can be compared to the pressure in a water pipe (psi).
- Amperage (current) can be compared to quantity of flow in a pipe (gpm).
- Resistance (ohms) can be likened to the friction loss in a pipe.

Frequently Used Formulas

volts = amperes × resistance
watts = volts × amperes
watts = amperes^2 × resistance

Single-Phase AC Motor

$$\text{horsepower (output)} = \frac{\text{volts} \times \text{amps} \times \text{efficiency \%} \times \text{power factor}}{746}$$

$$\text{kilowatts} = \frac{\text{volts} \times \text{amps} \times \text{power factor}}{1,000}$$

Two-Phase AC Motor

$$\text{kilowatts} = \frac{\text{volts} \times \text{amps} \times \text{power factor}}{1,000}$$

Three-Phase AC Motor

$$\text{horsepower (output)} = \frac{1.73 \times \text{volts} \times \text{amps} \times \text{efficiency \%} \times \text{power factor}}{746}$$

$$\text{kilowatts} = \frac{1.73 \times \text{amps} \times \text{power factor} \times \text{volts}}{1,000}$$

North American Standard System Voltages, in volts

Minimum Tolerable	Minimum Favorable	Nominal System	Maximum Favorable	Maximum Tolerable	Type (Phase) of System
107	110	120	125	127	1
200	210	240	240	250	3
214/428	220/440	240/480	250/500	254/508	1
244/422	250/434	265/460	227/480	288/500	3
400	420	480	480	500	3
2,100	2,200	2,400	2,450	2,540	3
3,630	3,810	4,160	4,240	4,400	3
6,040	6,320	6,900	7,050	7,300	3
12,100	12,600	13,200	13,800	14,300	3
12,600	13,000	14,400	14,500	15,000	3
30,000		34,500		38,000	3
60,000		69,000		72,500	3
100,000		115,000		121,000	3
120,000		138,000		145,000	3
140,000		161,000		169,000	3

North American Standard Nominal Voltages

Nominal System	Generator Rated	Transformer Secondary	Switchgear Rated	Capacitor Rated	Motor Rated	Starter Rated	Ballast Rated
Single-Phase Systems							
120	120	120	120	—	115	115	118
120/240	120/240	120/240	240	230	230	230	236
208/120	208/120	208/240	240	230	115	115	118
Three-Phase Systems							
240	240	240	240	230	240	220	236
480/277	480/277	480/277	480	460	460	440	460
480	480/277	480/277	480	480	460	440	460
2,400	2,400/1,388	2,400	2,400	2,400	2,300	2,300	—
4,160	4,160/2,400	4,160/2,400	4,160	4,160	4,000	4,000	—
6,900	6,900/3,980	6,900/3,980	7,200	6,640	6,600	6,600	—
7,200	6,900/3,980	7,200/4,160	13,800	7,200	7,200	7,200	—
12,000	12,500/7,210	12,000/6,920	13,800	12,470	11,000	11,000	—
13,200	13,800/7,970	13,800/7,610	13,800	13,200	13,200	13,200	—
14,400	14,000/8,320	13,800/7,970	14,400	14,400	13,200	13,200	—

Pumps

Approximate Full Load Current and Fuse Size Required by AC Motors*

hp	115 V Amperes	Ordinary Fuse	Time Delay Fuse	230 V, Single-Phase Amperes	Ordinary Fuse	Time Delay Fuse
⅙	4.4	15	8	2.2		
¼	5.8	20	10	2.9		
⅓	7.2	25	12	3.6	16	6
½	9.8	30	15	4.9	25	8
¾	13.8	45	20	6.9	25	12
1	16	50	25	8	25	15
1½	20	60	30	10	30	15
2	24	80	35	12	40	20
3				17	60	25
5				28	90	40
7½				40	125	60
10				50	150	80

*Assumes motors running at usual speeds, with normal torque characteristics.

Three-Phase Induction Motors

hp	220 V Amperes	Ordinary Fuse	Time Delay Fuse	460 V Amperes	Ordinary Fuse	Time Delay Fuse
½	2	15	4	1	15	2
¾	2.8	15	4	1.4	15	3
1	3.6	15	6	1.8	15	3
1½	5.2	15	8	2.6	15	4
2	6.8	25	10	3.4	15	5
3	9.6	30	15	4.8	15	8
5	15.2	50	25	7.6	25	15
7½	22	75	35	11	35	20
10	28	90	40	14	45	20
15	42	125	60	21	70	30
20	54	175	80	27	90	40
25	68	225	100	34	110	50
30	80	250	125	40	125	60
40	104	350	150	52	175	80
50	130	400	200	65	200	100
60	154	500	250	77	250	125
75	192	600	300	96	300	150
100	248	—	400	124	400	200
125	312	—	450	156	500	250
150	360			180	600	300
200	480			240	—	400

Standard Classification of NEMA Enclosures for Nonhazardous Locations*

Type	Intended Use
1	Intended for indoor use, primarily to provide a degree of protection from persons or equipment contacting the electrical components.
2	Intended for indoor use, to provide some protection against limited amounts of falling water and dirt.
3	Intended for outdoor use, primarily to provide a degree of protection against windblown dust, rain, and sleet, and ice on the enclosure.
3R	Intended for outdoor use, primarily to provide a degree of protection against falling rain and sleet, and undamaged by the formation of ice on the enclosure.
4	Intended for indoor or outdoor use, primarily to provide a degree of protection against windblown dust and rain, splashing water, and hose-directed water; undamaged by the formation of ice on the enclosure.
4X	Intended for indoor or outdoor use, primarily to provide a degree of protection against corrosion, windblown dust and rain, splashing water, and hose-directed water; undamaged by the formation of ice on the enclosure.
6	Intended for use indoors or outdoors where occasional submersion is encountered.
12	Intended for indoor use, primarily to provide a degree of protection against dust, galling dirt, and dripping noncorrosive liquids.
13	Intended for indoor use, primarily to provide a degree of protection against dust, spraying of water, oil, and noncorrosive coolant.

*These descriptions are in summary form only and are not complete representations of the National Electric Manufacturers Association (NEMA) standards for enclosures.

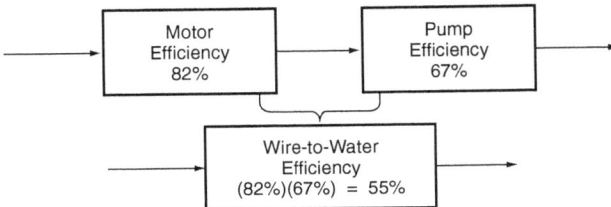

```
┌─────────────┐        ┌─────────────┐
│   Motor     │        │    Pump     │
│  Efficiency │───────▶│  Efficiency │────────▶
│    82%      │        │    67%      │
└─────────────┘        └─────────────┘
         │                    │
         └────────┬───────────┘
          ┌───────────────────┐
          │   Wire-to-Water   │
          │     Efficiency    │────────▶
          │  (82%)(67%) = 55% │
          └───────────────────┘
```

Wire-to-Water Efficiency

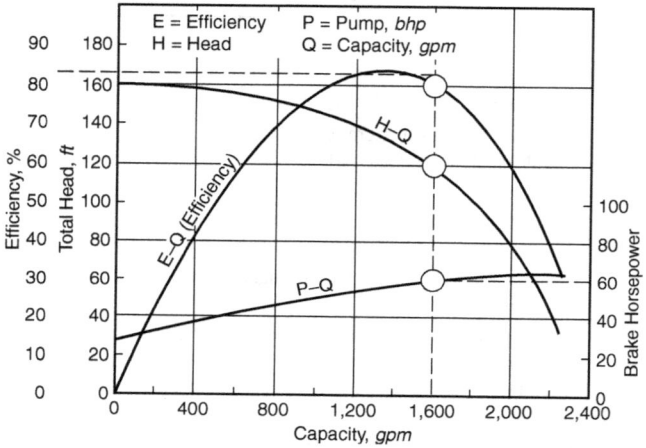

Example Pump Performance Curve

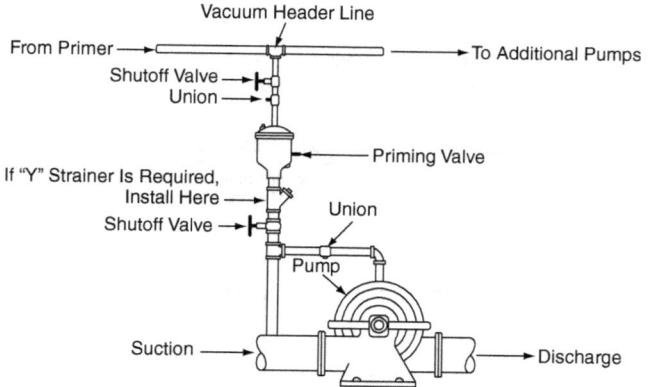

NOTE: Use pipe sizes recommended by the priming valve manufacturer.

Vacuum Priming System

Pump Troubleshooting Guide

Symptom	Probable Cause	Corrective Action
Pump will not start.	Circuit breaker or overload relay tripped, motor cold.	Reset breaker or reset manual overload relay.
	Fuses burned out.	Check for cause and correct. Replace fuses.
	No power to switch box.	Confirm with voltmeter by checking incoming power source. Notify power company.
	Motor is hot and overload relay has tripped.	Allow motor to cool. Check supply voltage. If low, notify power company. If normal, reset overload relay, start motor, check amperage. If above normal, call electrician.
	Loose or broken wire or short.	Tighten wiring terminals. Replace any broken wires. Check for shorts and correct.
	Low line voltage.	Check incoming power; use voltmeter. If low, notify power company.
	Defective motor.	Meg* out motor. If bad, replace.
	Defective pressure switch.	With contact points closed, check for voltage through switch. If no voltage, replace switch; if low voltage, clean contact points; if full voltage, proceed to next item.
	Line to pressure switch is plugged or valve in line has been accidentally shut off.	Open valve if closed. Clean or replace line.

Table continued on next page

Pump Troubleshooting Guide (continued)

Symptom	Probable Cause	Corrective Action
	Pump control valve malfunctioning.	Check limit switch for proper travel and contact. Adjust or replace as required.
	Defective time delay relay or pump start timer.	Check for voltage through relay or timer; replace as necessary. Check for loose linkage.
	Float switch or transducer malfunctioning.	If pump is activated by float switch or pressure transducer on storage tank, check for incoming signal. If no signal, check out switch or transducer with voltmeter. If okay, look for broken cable between storage tank and pump station.
Pump will not shut off.	Defective pressure switch.	Points in switch stuck or mechanical linkage broken; replace switch.
	Line to pressure switch is plugged or valve in line has been accidentally shut off.	Open valve if closed. Clean or replace plugged line.
	Cut-off pressure setting too high.	Adjust setting.
	Pump control valve malfunctioning.	Check limit switch for proper travel and contact. Adjust or replace as required.
	Float switch or transducer malfunctioning.	Defective incoming signal; check and replace components as required. Check cable.
	Defective timer in pump stop mode.	Check for voltage through pump stop timer; replace if defective.

Table continued on next page

294

Pump Troubleshooting Guide (continued)

Symptom	Probable Cause	Corrective Action
Pump starts too frequently.	Pressure switch cut-in and cut-off settings too close.	Adjust settings. Maintain minimum 20 psi (138 kPa or 1.4 kg/cm^2) differential.
	Water-logged tank.	Add air to tank. Check air charging system and air release valve. Also check tank and connections for air leaks.
	Leaking foot valve.	Check for backflow into well. If excessive or if pump shaft is turning backward, correct problem as soon as possible.
	Time delay relay or pump start/stop timers are malfunctioning.	Check relay or timers for proper operation. Replace defective components.
Fuses blow, circuit breaker or overload relays trip when pump is in operation.	Switch box or control not properly vented, in full sunshine, or in dead air location. Overload relay may be tripping because of external heat.	Provide adequate ventilation (may require small fan) and shelter from sun. Paint box or panel with heat-reflective paint, preferably white.
	Incorrect voltage.	Check incoming power source. If not within prescribed limits, notify power company.
	Overload relays tripped.	Check motor running amperage. Verify that thermal relay components are correctly sized to operating conditions. Repeated tripping will weaken units. Replace if necessary.

Table continued on next page

Pumps

Pump Troubleshooting Guide (continued)

Symptom	Probable Cause	Corrective Action
	Motor overloaded and running very hot.	Modern motors are designed to run hot. If the hand can be held on the motor for 10 seconds without extreme discomfort, the temperature is not damaging. Motor current should not exceed nameplate rating. Fifteen percent overload reduces motor life by 50%.
Pump will not deliver normal amount of water.	Pump breaking suction.	Check water level to be certain water is above pump bowls when operating. If not, lower bowls.
	Pump impellers improperly adjusted.	Check adjustment and lower impellers (qualified personnel only).
	Rotation incorrect.	Check rotation.
	Impellers worn.	If well pumps sand, impellers could be excessively worn, reducing amount of water pump can deliver. Evaluate and recondition pump bowls if required.
	Pump control valve malfunctioning.	Check limit switch for proper travel and contact. Adjust or replace as required.
	Impeller or bowls partially plugged.	Wash down pump by forcing water back through discharge pipe. Evaluate sand production from well.
	Drawdown more than anticipated.	Check pumping water level. Reduce production from pump or lower bowls.

Table continued on next page

Pump Troubleshooting Guide (continued)

Symptom	Probable Cause	Corrective Action
	Pump motor speed too slow.	Check speed and compare with performance curves. Also check lift and discharge pressure for power requirements.
Pump takes too much power.	Impellers not properly adjusted.	Refer to manufacturer's bulletin for adjustment of open or closed impellers.
	Well is pumping sand.	Check water being pumped for presence of sand. Restrict discharge until water is clear. Care should be taken not to shut down pump if it is pumping very much sand.
	Crooked well, pump shaft binding.	Reshim between pump base and pump head to center shaft in motor quill. Never shim between pump head and motor.
	Worm bearings or bent shaft.	Check and replace as necessary.
Excessive operating noise.	Motor bearings worn.	Replace as necessary.
	Bent line shaft or head shaft.	Check and replace.
	Line shaft bearings not receiving oil.	Make sure there is oil in the oil reservoir and that the oiler solenoid is opening. Check sight gauge drip rate. Adjust drip feed oiler for 5 drops/min plus 1 drop/min for each 40 ft (12 m) of column.

Source: Office of Water Programs, California State University, Sacramento Foundation, in Small Water System Operation and Maintenance. For additional information, visit <www.owp.csus.edu> or call 916-278-6142.

* Meg is a procedure used for checking the insulation resistance on motors, feeders, buss bar systems, grounds, and branch circuit wiring.

Pumps

PUMP AND MOTOR MAINTENANCE CHECKLIST _____

Refer to the manufacturer's operations and maintenance recommendations for specific guidance. These suggestions are general in nature. The type of equipment that is in operation drives how and when maintenance takes place. Water quality and equipment history play a predominant role in scheduling maintenance. Above all, safety is the main concern when performing any duty on equipment. Electrical, mechanical, and confined-space safety practices must be a part of any preventive maintenance checklist.

Daily (or during routine visits when pump is in operation)

1. Visually observe pump and motor operation.
2. Read the amperage, voltage, flows, run hours, and other information from motor control center.
3. Inspect mechanical seals.
4. Check operating temperature.
5. Check warning indicator lights.
6. Check oil levels.
7. Note any unusual vibration.

Weekly

1. Test per-square-inch levels of the relief valve system; these should be set just above the normal operating pressure of the system.
2. Inspect stuffing box and note the amount of leakage and adjust or lubricate packing gland as necessary. A leakage rate of 20 to 60 drops of seal water per minute is normal for a properly adjusted gland; inadequate or excessive leakage are signs of trouble. Do not overtighten packing gland bolts. Clean drain line if necessary.
3. Check valve lubricant levels.
4. Test the priming system and perform preventive maintenance as necessary.
5. Inspect motor for indications of overload or electrical failure. Check for burnt insulation, melted solder, or discoloration around terminals and wires.

6. Check for and remove any obstructions in or around the impeller, screens, or intake, as appropriate. (Be sure to shut off the pump.)
7. Test transfer valve, if applicable.

Monthly

1. Check bearing temperatures with a thermometer.
2. Clean strainers on system piping including strainers on automatic control valves.
3. Perform dry vacuum test.
4. Check oil level in pump gearbox; add oil as necessary.
5. Inspect gaskets.
6. Check motor ventilation screens and clean or replace as necessary.
7. Check pressure gauge reliability.
8. Check foundation bolts.
9. Clean pump control sensors (may be required weekly, depending on water quality).
10. Check drive flange bolts, if applicable, and tighten as necessary.

Semiannually

1. Perform pump and motor performance test. Check at least three performance test points and plot on the pump's performance curve. Compare this data to the design specifications. Capacity and efficiency determine the degree of pump maintenance necessary.
2. Check pump–motor shaft alignment.
3. Calibrate gauges as necessary.
4. Record vibration levels using vibration level test equipment.
5. Note condition of pump casing, base, and foundation and of pipe supports and bracing, and correct any deficiencies.
6. Calibrate meters, level sensors, controls, and recording devices as necessary.
7. Inspect and clean check valves, pump control valves, wear rings, and individual drain lines.
8. Inspect intake or screen; replace or clean as necessary.

9. Inspect condition of the impeller, pump shaft, and shaft sleeve; replace as necessary.
10. Lubricate power transfer cylinder, power shift cylinder, shift control valve, and transfer valve mechanism (on two-stage pumps).

Annually

1. Analyze changes in daily data readings.
2. Determine pumping capacity.
3. Determine pumping efficiency.
4. Check all other pumping performance levels, including engine speed and pump pressure.

Pressure, Flows, and Meters

Maintaining proper water pressure and flow in a system is an important component of successful water delivery. The operator needs to understand fundamental pressure and flow requirements, along with basic formulas for computing water pressure and flows. In addition, the operator should know how to compute head, pressure, and flow equivalents in various units of measure; how to size, install, and maintain the different types of meters; and how to deal with undesirable phenomena such as water hammer.

KEY FORMULAS FOR PRESSURE

Hydraulic (Water Column Height) Pressure

$$\text{psi} = \frac{\text{head, ft}}{2.31 \text{ ft/psi}}$$

$$\text{psi} = \text{head, ft} \times 0.433 \text{ psi/ft}$$

or

$$\text{head, ft} = \text{psi} \times 2.31 \text{ ft/psi}$$

$$\text{head, ft} = \frac{\text{psi}}{0.433 \text{ psi/ft}}$$

Pounds of Force on the Face of a Valve

$$\text{force, lb} = \text{area} \times \text{pressure}$$

or

$$\text{force, lb} = 0.785 \times \text{diameter, ft}^2 \times 144 \text{ in.}^2/\text{ft}^2 \times \text{psi}$$

Bottom Force and Buoyancy

Tank Bottom Forces

Rectangular Basins

$$\text{force, lb} = \text{length, ft} \times \text{width, ft} \times \text{height, ft} \times 62.4 \text{ lb/ft}^3$$

Right Cylinders

$$\text{force, lb} = 0.785 \times \text{diameter, ft}^2 \times \text{height, ft} \times 62.4 \text{ lb/ft}^3$$

Pounds per Square Foot on a Tank Bottom

Rectangular Basins

$$\text{force, lb/ft}^2 = \frac{\text{length, ft} \times \text{width, ft} \times \text{height, ft} \times 62.4 \text{ lb/ft}^2}{\text{bottom area, ft}^2}$$

Right Cylinders

$$\text{force, lb/ft}^2 = \frac{0.785 \times \text{diameter, ft}^2 \times \text{height, ft} \times 62.4 \text{ lb/ft}^2}{\text{bottom area, ft}^2}$$

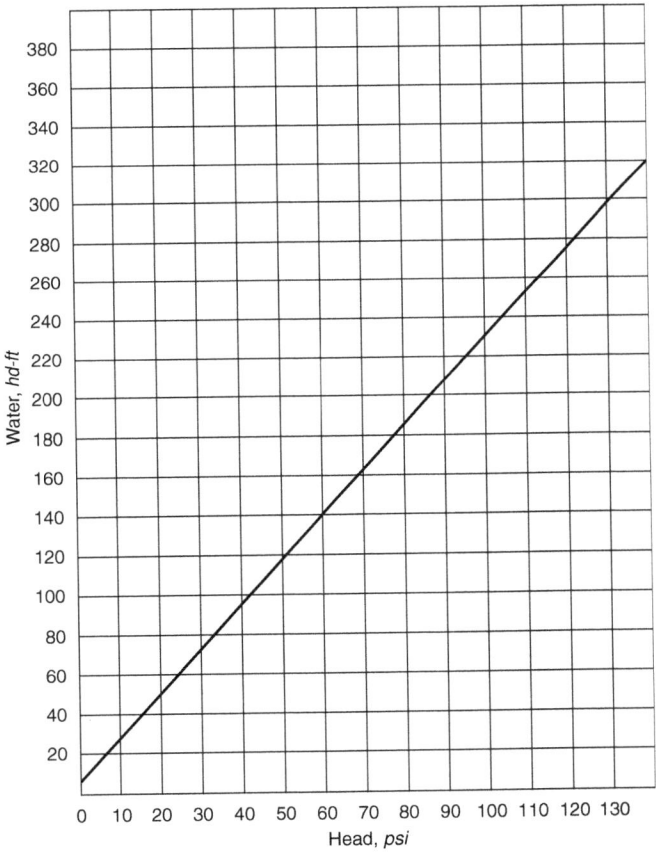

Conversion Chart for Feet of Water/Pounds per Square Inch

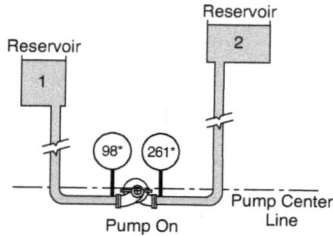

*Pressure, in psi.

Convert pressure head to elevation head, in ft.

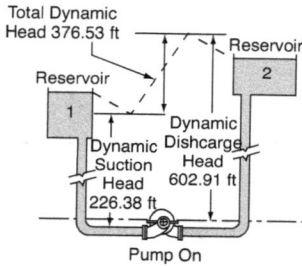

Schematic for Total Dynamic Head

The total dynamic head is the *difference* between the head on the discharge side of the pump and the head on the suction side of the pump:

$$\text{total dynamic head} = 602.91 \text{ ft} - 226.38 \text{ ft}$$
$$= 376.53 \text{ ft}$$

Conversions for Head and Pressure Equivalents

Convert from \ Convert to	lb/in.²	lb/ft²	atm	kg/cm²	kg/m²	in. water*	ft water*	in. mercury†	mm mercury†	bars
lb/in.²	1	144	0.068046	0.070307	703.070	27.7276	2.3106	2.03602	51.7150	0.06895
lb/ft²	0.0069444	1	0.000473	0.000488	4.88241	0.1926	0.01605	0.014139	0.35913	0.000479
atm	14.696	2116.22	1	1.0332	10332.27	407.484	33.9570	29.921	760	1.01325
kg/cm²	14.2233	2048.155	0.96784	1	10000	394.38	32.8650	28.959	735.559	0.98067
kg/m²	0.001422	0.204768	0.0000968	0.0001	1	0.03944	0.003287	0.002896	0.073556	0.000098
in. water*	0.036092	5.1972	0.002454	0.00253	25.375	1	0.08333	0.073430	1.8651	0.00249
ft water*	0.432781	62.3205	0.029449	0.03043	304.275	12	1	0.88115	22.3813	0.029839
in. mercury†	0.491154	70.7262	0.033421	0.03453	345.316	13.6185	1.1349	1	25.40005	0.033864
mm mercury†	0.0193368	2.78450	0.0013158	0.0013595	13.59509	0.53616	0.044680	0.03937	1	0.001333
bars	14.5038	2088.55	0.98692	1.01972	10197.2	402.156	33.5130	29.5300	750.062	1

* Water at 68°F (20°C).
† Mercury at 32°F (0°C).
Example: 15 lb/ft² × 4.88241 = 73.236 kg/m².

Summary of Pressure Requirements

Requirement	Value, psi	(kPa)	Location
Minimum pressure	35	(241)	All points within distribution system
	20	(140)	All ground level points
Desired maximum	100	(690)	All points within distribution system
Fire flow minimum	20	(140)	All points within distribution system
Ideal range	50–75	(345–417)	Residences
	35–60	(241–414)	All points within distribution system

KEY CONVERSIONS FOR FLOWS

Conversion of US customary flow units can be easily made using the block diagram below. When moving from a smaller to a larger block, multiply by the factor shown on the connecting line. When moving from a larger to a smaller block, divide.

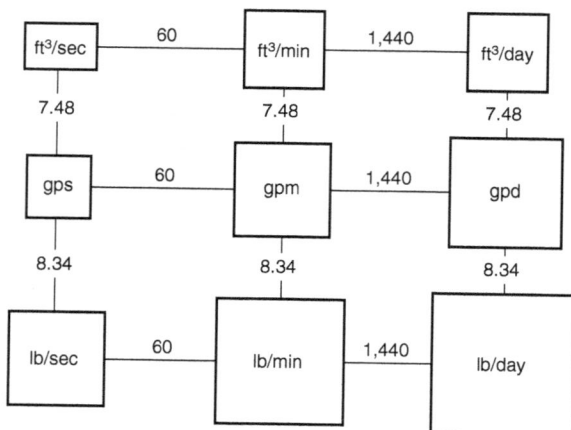

$$\text{flow, gpm} = \text{flow, cfs} \times 448.8 \text{ gpm/cfs}$$

$$\text{flow, cfs} = \frac{\text{flow, gpm}}{448.8 \text{ gpm/cfs}}$$

$$\text{pipe diameter, in.} = \frac{\text{area, ft}^2}{0.785} \times 12 \text{ in./ft}$$

$$\text{actual leakage, gpd/mi./in.} = \frac{\text{leak rate, gpd}}{\text{length, mi.} \times \text{diameter, in.}}$$

NOTE: minimum flushing velocity: 2.5 fps

maximum pipe velocity: 5.0 fps

key conversions: 1.55 cfs/mgd; 448.8 gpm/cfs

KEY FORMULAS FOR FLOWS AND METERS _____

Velocity

$$\text{flow, cfs} = \text{area, ft} \times \text{velocity, fps}$$

$$\frac{\text{gpm}}{448.8 \text{ gpm/cfs}} = 0.785 \times \text{diameter, ft}^2 \times \frac{\text{distance, ft}}{\text{time, sec}}$$

$$\text{velocity, fps} = \frac{\text{flow, cfs}}{\text{area, ft}^2}$$

$$\text{area, ft}^2 = \frac{\text{flow, cfs}}{\text{velocity, fps}}$$

Head Loss Resulting From Friction

Darcy–Weisbach Formula

$$h_L = f(L/D)(V^2/2g)$$

Where (in any consistent set of units):

h_L = head loss

f = friction factor, dimensionless

L = length of pipe

D = diameter of the pipe

V = average velocity

g = gravity constant

Hazen–Williams Formula

$$h_f = k_1 \frac{LQ^{1.85}}{C^{1.85}D^{4.87}}$$

Where:

h_f = head loss, in ft

k_1 = 4.72, in units of seconds$^{1.85}$ per feet$^{0.68}$

L = pipe length, in ft

Q = flow rate, in cfs

C = Hazen–Williams roughness coefficient

D = pipe diameter, in ft

The value of C ranges from 60 for corrugated steel to 150 for clean, new asbestos–cement pipe.

Manning Formula

$$v = \frac{1.486}{n} R^{\frac{2}{3}} S^{\frac{1}{2}}$$

Where:

v = flow velocity, in fps

n = Manning coefficient of channel roughness

R = hydraulic radius, in ft

S = channel slope (for uniform flow) or the energy slope (for nonuniform flow), dimensionless

The energy slope is calculated as $-dH/dx$, where H is the total energy, which is expressed as

$$H = z + y + \frac{v^2}{2g}$$

Where (in any consistent set of units):

x = elevation head

y = water depth

v = velocity

g = gravitational constant

x = distance between any two points

Approximate Flow Through Venturi Tube

$$Q = 19.05 d_1^2 \sqrt{H} \sqrt{\dfrac{1}{1 - \left(\dfrac{d_1}{d_2}\right)^4}}$$

for any Venturi tube.

$$Q = 19.17 \, d_1^2 \sqrt{h}$$

for a Venturi tube in which $d_1 = \frac{1}{3} d_2$.

Where:

Q = flow, in gpm

d_1 = diameter of Venturi throat, in in.

H = difference in head between upstream end and throat, in ft

d_2 = diameter of main pipe, in in.

These formulas are suitable for any liquid with viscosities similar to water. The values given here are for water. A value of 32.174 ft/sec^2 was used for the acceleration of gravity and a value of 7.48 gal/ft^3 was used in computing the constants.

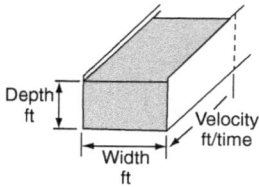

General Case, Open Channel

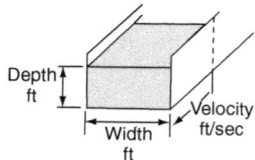

$$\underbrace{Q}_{\substack{\text{(Flow Rate)}\\ \text{ft}^3/\text{time}}} = \underbrace{A}_{\substack{\text{(Width)(Depth)}\\ \text{ft} \quad \text{ft}}} \quad \underbrace{V}_{\substack{\text{(Velocity)}\\ \text{ft/time}}}$$

Cubic-Feet-per-Second Flow

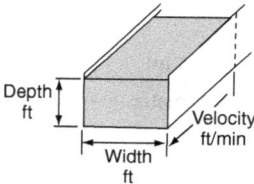

$$\underbrace{Q}_{\substack{\text{(Flow Rate)}\\ \text{ft}^3/\text{sec}}} = \underbrace{A}_{\substack{\text{(Width)(Depth)}\\ \text{ft} \quad \text{ft}}} \quad \underbrace{V}_{\substack{\text{(Velocity)}\\ \text{ft/sec}}}$$

Cubic-Feet-per-Minute Flow

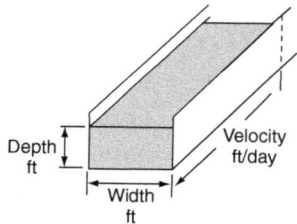

$$\underbrace{Q}_{\substack{\text{(Flow Rate)}\\ \text{ft}^3/\text{min}}} = \underbrace{A}_{\substack{\text{(Width)(Depth)}\\ \text{ft} \quad \text{ft}}} \quad \underbrace{V}_{\substack{\text{(Velocity)}\\ \text{ft/min}}}$$

Cubic-Feet-per-Day Flow

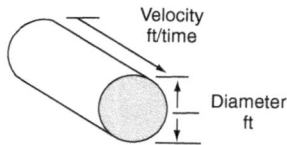

$$\underbrace{Q}_{\substack{\text{(Flow Rate)}\\ \text{ft}^3/\text{day}}} = \underbrace{A}_{\substack{\text{(Width)(Depth)}\\ \text{ft} \quad \text{ft}}} \quad \underbrace{V}_{\substack{\text{(Velocity)}\\ \text{ft/day}}}$$

$$\underbrace{Q}_{\substack{\text{(Flow Rate)}\\ \text{ft}^3/\text{time}}} = \underbrace{A}_{\substack{(0.785)(\text{Diameter})^2\\ \text{ft}}} \quad \underbrace{V}_{\substack{\text{(Velocity)}\\ \text{ft/time}}}$$

General Case, Circular Pipe Flowing Full

The $Q = AV$ Equation As It Pertains to Flow in an Open Channel

310

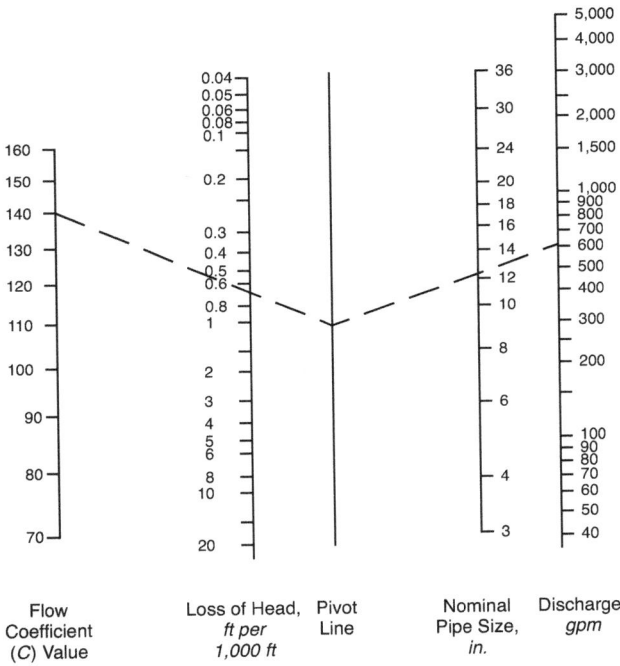

Flow Coefficient (C) Value	Loss of Head, ft per 1,000 ft	Pivot Line	Nominal Pipe Size, in.	Discharge, gpm

Draw a line between two known values on the same side of the pivot line and extend it so that it touches the pivot line. Draw a line between the point on the pivot line and the other known value on the other side of the pivot line. Read the unknown value where the second line intersects the graph.

Flow of Water in Ductile-Iron Pipe

WEIRS

$$\text{weir overflow rate} = \frac{\text{flow, gpd}}{\text{weir length, ft}}$$

Discharge From a V-Notch Weir With End Contractions*

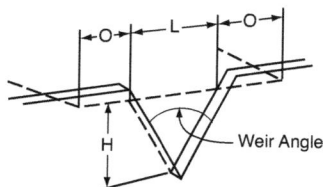
Weir Angle

		Discharge Over Weir, *gpm*				
Head (H),		**Weir Angle, *degrees***				
in.	**10th of foot**	**22.5**	**30**	**45**	**60**	**90**
1	.083	0.4	0.5	0.8	1.2	2.0
1¼	.104	0.8	1.0	1.6	2.2	3.9
1½	.125	1.2	1.7	2.6	3.5	6.1
1¾	.146	1.8	2.4	3.8	5.2	9.1
2	.167	2.6	3.4	5.3	7.3	12.7
2¼	.188	3.4	4.6	7.1	9.8	17.1
2½	.208	4.4	5.9	9.1	12.7	22.0
2¾	.229	5.6	7.5	11.6	16.1	27.9
3	.250	7.0	9.4	14.4	20.1	34.8
3¼	.271	8.7	11.4	17.9	24.9	43.1
3½	.292	10.3	13.8	21.3	29.6	51.3
3¾	.313	12.3	15.4	25.3	35.2	61.0
4	.333	14.4	19.2	29.6	41.1	71.2
4¼	.354	16.7	22.3	34.5	47.8	83.0
4½	.375	19.3	25.8	39.8	55.3	95.8
4¾	.396	22.1	29.5	45.6	63.3	109.9
5	.417	25.2	33.6	51.8	71.9	124.8
5¼	.437	28.3	37.8	58.4	81.1	140.6
5½	.458	31.9	42.5	65.6	91.1	158.0
5¾	.479	35.6	47.4	73.3	101.7	176.4
6	.500	39.7	53.0	81.8	113.6	196.9

* The distance (0) on either side of the weir must be at least ¾ L.

Discharge From a Rectangular Weir With End Contractions*

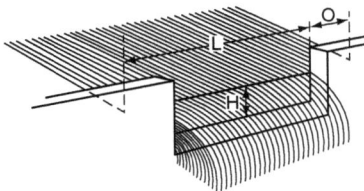

Head (H),		Discharge Over Weir, *gpm*			Additional gpm for Each Foot Over 5 ft
		Length (L) of Weir, *ft*			
in.	*10th of foot*	1	3	5	
1	.083	35.4	107.5	179.8	36.05
1¼	.104	49.5	150.4	250.4	50.4
1½	.125	64.9	197	329.5	66.2
1¾	.146	81	240	415	83.5
2	.167	98.5	302	506	102
2¼	.188	117	361	605	122
2½	.208	136.2	422	706	143
2¾	.229	157	485	815	165
3	.250	177.8	552	926	187
3¼	.271	199.8	624	1,047	211
3½	.292	222	695	1,167	236
3¾	.313	245	769	1,292	261
4	.333	269	846	1,424	288
4¼	.354	293.6	925	1,559	316
4½	.375	318	1,006	1,696	345
4¾	.936	344	1,091	1,835	374
5	.417	370	1,175	1,985	405
5¼	.437	395.5	1,262	2,130	434
5½	.458	421.6	1,352	2,282	465
5¾	.479	449	1,442	2,440	495
6	.500	476.5	1,535	2,600	528

* The distance (0) on either side of the weir must be at least 3 H.

Flow Equivalents

Cubic Feet per Second							Gallons per 24 Hours						
ft³/sec	to	gpm	to	gal/24 hr	to	m³/hr	gal/24 hr	to	gpm	to	ft³/sec	to	m³/hr
0.2		90		129,263		20.39	100,000		69		0.15		15.77
0.4		180		258,526		40.78	125,000		87		0.19		19.71
0.6		269		387,789		61.17	200,000		139		0.31		31.54
0.8		359		517,052		81.56	400,000		278		0.62		63.08
1.0		449		646,315		102.0	500,000		347		0.77		78.85
1.2		539		775,578		122.3	600,000		417		0.93		94.62
1.4		628		904,841		142.7	700,000		486		1.08		110.4
1.6		718		1,034,104		163.1	800,000		556		1.24		126.2
1.8		808		1,163,367		183.5	900,000		625		1.39		141.9
2.0		898		1,292,630		203.9	1,000,000		694		1.55		157.7
2.2		987		1,421,893		224.3	2,000,000		1,389		3.09		315.4
2.4		1,077		1,551,156		244.7	3,000,000		2,083		4.64		473.1
2.6		1,167		1,680,420		265.1	4,000,000		2,778		6.19		630.8
2.8		1,257		1,809,683		285.5	5,000,000		3,472		7.74		788.5
3.0		1,346		1,938,946		305.9	6,000,000		4,167		9.28		946.2
3.2		1,436		2,068,209		326.2	7,000,000		4,861		10.83		1,104
3.4		1,526		2,197,472		346.6	8,000,000		5,556		12.38		1,262
3.6		1,616		2,326,735		367.0	9,000,000		6,250		13.92		1,419

Table continued on next page

Flow Equivalents (continued)

		Cubic Feet per Second						Gallons per 24 Hours				
ft³/sec	to	gpm	to	gal/24 hr	to	m³/hr	gal/24 hr	to	gpm	to	ft³/sec	to	m³/hr
3.8		1,705		2,455,998		387.4	10,000,000		6,944		15.47		1,577
4.0		1,795		2,585,261		407.8	12,000,000		8,333		18.56		1,892
4.2		1,885		2,714,524		428.2	12,500,000		8,680		19.34		1,971
4.4		1,975		2,843,787		448.6	14,000,000		9,722		21.65		2,208
4.6		2,068		2,973,050		469.0	15,000,000		10,417		23.20		2,366
4.8		2,154		3,102,313		489.4	16,000,000		11,111		24.75		2,523
5.0		2,244		3,231,576		509.8	18,000,000		12,500		26.85		2,839
10.0		4,488		6,463,152		1,020	20,000,000		13,889		30.94		3,154
20.0		8,987		12,926,304		2,039	25,000,000		17,361		38.68		3,943
30.0		13,464		19,389,456		3,059	30,000,000		20,833		46.41		4,731
40.0		17,952		25,852,261		4,078	40,000,000		27,778		61.88		6,308
50.0		22,440		32,315,760		5,098	50,000,000		34,722		77.35		7,885
60.0		26,928		38,778,912		6,117	60,000,000		41,667		92.82		9,462
70.0		31,416		45,242,064		7,137	70,000,000		48,611		108.29		11,039
75.0		33,660		48,473,640		7,646	75,000,000		52,083		116.04		11,828
80.0		35,904		51,705,216		8,156	80,000,000		55,556		123.76		12,616
90.0		40,392		58,168,368		9,176	90,000,000		62,500		139.23		14,193
100.0		44,880		64,631,520		10,195	100,000,000		69,444		154.72		15,770

Table continued on next page

Pressure, Flows, and Meters

Flow Equivalents (continued)

		Cubic Feet per Second					Gallons per 24 Hours				
ft³/sec	to	gpm	to	gal/24 hr	to	m³/hr	gal/24 hr	to	gpm	to	ft³/sec	to	m³/hr

ft³/sec	to	gpm	to	gal/24 hr	to	m³/hr	gal/24 hr	to	gpm	to	ft³/sec	to	m³/hr
101.0		45,329		65,277,835		10,297	125,000,000		86,805		193.40		19,713
102.0		45,778		65,924,150		10,399	150,000,000		104,167		232.08		23,665
103.0		46,226		66,570,466		10,501	175,000,000		121,528		270.76		27,598
104.0		46,675		67,216,781		10,603	200,000,000		138,889		309.44		31,540
105.0		47,124		67,863,096		10,705	225,000,000		156,250		348.12		35,483
106.0		47,572		68,509,411		10,807	250,000,000		173,611		386.80		39,425
107.0		48,022		69,155,726		10,909	300,000,000		208,333		464.16		47,310
108.0		48,470		69,802,042		11,011	400,000,000		277,778		618.88		63,080
109.0		48,919		70,448,357		11,113	500,000,000		347,220		773.60		78,850
110.0		49,368		71,094,672		11,215	600,000,000		416,664		928.32		94,620
120.0		53,856		77,557,824		12,234	700,000,000		486,108		1,083.04		110,390
125.0		56,100		80,798,400		12,744	750,000,000		520,328		1,160.40		118,275
130.0		58,344		84,020,976		13,254	800,000,000		555,552		1,237.76		126,160
140.0		62,832		90,484,128		14,273	900,000,000		624,996		1,392.48		141,930
150.0		67,320		96,947,230		15,293	1,000,000,000		694,440		1,547.20		157,700

NOTE: gpm and gal/24 hr given to nearest whole number. The value 7.48 gal = 1 ft³ is used in calculating the values in this table.

Flow Data—Nozzles, Theoretical Discharge of Nozzles in US Gallons per Minute (1/16–1 3/8-in. diameters)

Head,*		Velocity of Discharge,	Diameter of Nozzle, in.												
psi	ft	ft/sec	1/16	1/8	3/16	1/4	3/8	1/2	5/8	3/4	7/8	1	1 1/8	1 1/4	1 3/8
10	23.1	38.6	0.37	1.48	3.32	5.91	13.3	23.6	36.9	53.1	72.4	94.5	120	148	179
15	34.6	47.25	0.45	1.81	4.06	7.24	16.3	28.9	45.2	65.0	88.5	116.0	147	181	219
20	46.2	54.55	0.52	2.09	4.69	8.35	18.8	33.4	52.2	75.1	102.0	134.0	169	209	253
25	57.7	61.0	0.58	2.34	5.25	9.34	21.0	37.3	58.3	84.0	114.0	149.0	189	234	283
30	69.3	66.85	0.64	2.56	5.75	10.2	23.0	40.9	63.9	92.0	125.0	164.0	207	256	309
35	80.8	72.2	0.69	2.77	6.21	11.1	24.8	44.2	69.0	99.5	135.0	177.0	224	277	334
40	92.4	77.2	0.74	2.96	6.64	11.8	26.6	47.3	73.8	106.0	145.0	188.0	239	296	357
45	103.9	81.8	0.78	3.13	7.03	12.5	28.2	50.1	78.2	113.0	153.0	200.0	253	313	379
50	115.5	86.25	0.83	3.30	7.41	13.2	29.7	52.8	82.5	119.0	162.0	211.0	267	330	399
55	127.0	90.5	0.87	3.46	7.77	13.8	31.1	55.3	86.4	125.0	169.0	221.0	280	346	418
60	138.6	94.5	0.90	3.62	8.12	14.5	32.5	57.8	90.4	130.0	177.0	231.0	293	362	438
65	150.1	98.3	0.94	3.77	8.45	15.1	33.8	60.2	94.0	136.0	184.0	241.0	305	376	455
70	161.7	102.1	0.98	3.91	8.78	15.7	35.2	62.5	97.7	141.0	191.0	250.0	317	391	473
75	173.2	105.7	1.01	4.05	9.08	16.2	36.4	64.7	101.0	146.0	198.0	259.0	327	404	489
80	184.8	109.1	1.05	4.18	9.39	16.7	37.6	66.8	104.0	150.0	205.0	267.0	338	418	505
85	196.3	112.5	1.08	4.31	9.67	17.3	38.8	68.9	108.0	155.0	211.0	276.0	349	431	521

Table continued on next page

Pressure, Flows, and Meters

Flow Data—Nozzles, Theoretical Discharge of Nozzles in US Gallons per Minute (1/16–1 3/8-in. diameters) (continued)

Head,*		Velocity of Discharge,	Diameter of Nozzle, in.												
psi	ft	ft/sec	1/16	1/8	3/16	1/4	3/8	1/2	5/8	3/4	7/8	1	1 1/8	1 1/4	1 3/8
90	207.9	115.8	1.11	4.43	9.95	17.7	39.9	70.8	111.0	160.0	217	284.0	359	443	536
95	219.4	119.0	1.14	4.56	10.2	18.2	41.0	72.8	114.0	164.0	223.0	292.0	369	456	551
100	230.9	122.0	1.17	4.67	10.5	18.8	42.1	74.7	117.0	168.0	229.0	299.0	378	467	565
105	242.4	125.0	1.20	4.79	10.8	19.2	43.1	76.5	120.0	172.0	234.0	306.0	388	479	579
110	254.0	128.0	1.23	4.90	11.0	19.6	44.1	78.4	122.0	176.0	240.0	314.0	397	490	593
115	265.5	130.9	1.25	5.01	11.2	20.0	45.1	80.1	125.0	180.0	245.0	320.0	406	501	606
120	277.1	133.7	1.28	5.12	11.5	20.5	46.0	81.8	128.0	184.0	251.0	327.0	414	512	619
130	300.2	139.1	1.33	5.33	12.0	21.3	48.0	85.2	133.0	192.0	261.0	341.0	432	533	645
140	323.3	144.3	1.38	5.53	12.4	22.1	49.8	88.4	138.0	199.0	271.0	354.0	448	553	668
150	346.4	149.5	1.43	5.72	12.9	22.9	51.6	91.5	143.0	206.0	280.0	366.0	463	572	692
175	404.1	161.4	1.55	6.18	13.9	24.7	55.6	98.8	154.0	222.0	302.0	395.0	500	618	747
200	461.9	172.6	1.65	6.61	14.8	26.4	59.5	106.0	165.0	238.0	323.0	423.0	535	660	799
250	577.4	193.0	1.85	7.39	16.6	29.6	66.5	118.0	185.0	266.0	362.0	473.0	598	739	894
300	692.8	211.2	2.02	8.08	18.2	32.4	72.8	129.0	202.0	291.0	396.0	517.0	655	808	977

NOTE: The actual quantity discharged by a nozzle will be less than shown in this table. A well-tapered smooth nozzle may be assumed to give 97% to 99% of the values in the table.
* Where there is both an upstream and downstream pressure, the head is a differential head.

Flow Data—Nozzles, Theoretical Discharge of Nozzles in US Gallons per Minute (1½–6-in. diameters)

Head*		Velocity of Discharge,	Diameter of Nozzle, in.												
psi	ft	ft/sec	1½	1¾	2	2¼	2½	2¾	3	3½	4	4½	5	5½	6
10	23.1	38.6	213	289	378	479	591	714	851	1,158	1,510	1,915	2,365	2,855	3,405
15	34.6	47.25	260	354	463	585	723	874	1,041	1,418	1,850	2,345	2,890	3,490	4,165
20	46.2	54.55	301	409	535	676	835	1,009	1,203	1,638	2,135	2,710	3,340	4,040	4,810
25	57.7	61.0	336	458	598	756	934	1,128	1,345	1,830	2,385	3,025	3,730	4,510	5,380
30	69.3	66.85	368	501	655	828	1,023	1,236	1,473	2,005	2,615	3,315	4,090	4,940	5,895
35	80.8	72.2	398	541	708	895	1,106	1,335	1,591	2,168	2,825	3,580	4,415	5,340	6,370
40	92.4	77.2	425	578	756	957	1,182	1,428	1,701	2,315	3,020	3,830	4,725	5,610	6,810
45	103.9	81.8	451	613	801	1,015	1,252	1,512	1,802	2,455	3,200	4,055	5,000	6,050	7,120
50	115.5	86.25	475	647	845	1,070	1,320	1,595	1,900	2,590	3,375	4,275	5,280	6,380	7,600
55	127.0	90.4	498	678	886	1,121	1,385	1,671	1,991	2,710	3,540	4,480	5,530	6,690	7,970
60	138.6	94.5	521	708	926	1,172	1,447	1,748	2,085	2,835	3,700	4,685	5,790	6,980	8,330
65	150.1	98.3	542	737	964	1,220	1,506	1,819	2,165	2,950	3,850	4,875	6,020	7,270	8,670
70	161.7	102.1	563	765	1,001	1,267	1,565	1,888	2,250	3,065	4,000	5,060	6,250	7,560	9,000
75	173.2	105.7	582	792	1,037	1,310	1,619	1,955	2,330	3,170	4,135	5,240	6,475	7,820	9,320
80	184.8	109.1	602	818	1,070	1,354	1,672	2,020	2,405	3,280	4,270	5,410	6,690	8,080	9,630
85	196.3	112.5	620	844	1,103	1,395	1,723	2,080	2,480	3,375	4,400	5,575	6,890	8,320	9,920

Table continued on next page

Pressure, Flows, and Meters

Flow Data—Nozzles, Theoretical Discharge of Nozzles in US Gallons per Minute (1½–6-in. diameters) (continued)

Head,*		Velocity of Discharge,	Diameter of Nozzle, *in.*												
psi	ft	ft/sec	1½	1¾	2	2¼	2½	2¾	3	3½	4	4½	5	5½	6
90	207.9	115.8	638	868	1,136	1,436	1,773	2,140	2,550	3,475	4,530	5,740	7,090	8,560	10,210
95	219.4	119.0	656	892	1,168	1,476	1,824	2,200	2,625	3,570	4,655	5,900	7,290	8,800	10,500
100	230.9	122.0	672	915	1,196	1,512	1,870	2,255	2,690	3,660	4,775	6,050	7,470	9,030	10,770
105	242.4	125.0	689	937	1,226	1,550	1,916	2,312	2,755	3,750	4,890	6,200	7,650	9,260	11,020
110	254.0	128.0	705	960	1,255	1,588	1,961	2,366	2,820	3,840	5,010	6,340	7,840	9,470	11,300
115	265.5	130.9	720	980	1,282	1,621	2,005	2,420	2,885	3,930	5,120	6,490	8,010	9,680	11,550
120	277.1	133.7	736	1,002	1,310	1,659	2,050	2,470	2,945	4,015	5,225	6,630	8,180	9,900	11,800
130	300.2	139.1	767	1,043	1,365	1,726	2,132	2,575	3,070	4,175	5,450	6,900	8,530	10,300	12,290
140	323.3	144.3	795	1,082	1,415	1,790	2,212	2,650	3,180	4,330	5,650	7,160	8,850	10,690	12,730
150	346.4	149.5	824	1,120	1,466	1,853	2,290	2,760	3,295	4,485	5,850	7,410	9,150	11,070	13,200
175	404.1	161.4	890	1,210	1,582	2,000	2,473	2,985	3,560	4,840	6,310	8,000	9,890	11,940	14,250
200	461.9	172.6	950	1,294	1,691	2,140	2,645	3,190	3,800	5,175	6,760	8,550	10,580	12,770	15,220
250	577.4	193.0	1,063	1,447	1,891	2,392	2,955	3,570	4,250	5,795	7,550	9,570	11,820	14,290	17,020
300	692.8	211.2	1,163	1,582	2,070	2,615	3,235	3,900	4,650	6,330	8,260	10,480	12,940	15,620	18,610

Note: The actual quantity discharged by a nozzle will be less than shown in this table. A well-tapered smooth nozzle may be assumed to give 97% to 99% of the values in the table.
* Where there is both an upstream and downstream pressure, the head is a differential head.

Rates of Flow for Certain Plumbing, Household, and Farm Fixtures

	Flow Pressure*		Flow Rate	
Location	psi	(kPa)	gpm	(L/min)
Ordinary basin faucet	8	(55)	2.0	(7.5)
Self-closing basin faucet	8	(55)	2.5	(9.5)
Sink faucet, ⅜ in. (10 mm)	8	(55)	4.5	(17.0)
Sink faucet, ½ in. (13 mm)	8	(55)	4.5	(17.0)
Bathtub faucet	8	(55)	6.0	(23.0)
Laundry tub faucet, ½ in. (13 mm)	8	(55)	5.0	(19.0)
Faucets per Energy Policy Act of 1992	—	—	≤2.5	
Shower	8	(55)	5.0	(19.0)
Showers per Energy Policy Act of 1992	—	—	≤2.5	
Ball-cock for toilet	8	(55)	3.0	(11.0)
Flush valve for toilet	15	(103)	15.0–40.0	(57.0–151.0)†
Toilets gal or L per flush per Energy Policy Act of 1992	—	—	≤1.6	(≤6.1)
Flushometer valve for urinal	15	(103)	15.0	(57.0)
Garden hose, 50 ft (15 m) (¾-in. [13-mm] sill cock)	30	(207)	5.0	(19.0)
Garden hose, 50 ft (15 m) (⅝-in. [16-mm] outlet)	15	(103)	3.33	(13.0)
Drinking fountain	15	(103)	0.75	(3.0)
Fire hose, 1½ in. (6 mm) (½-in. [13-mm] nozzle)	30	(207)	40.0	(151.0)

* Flow pressure is the pressure in the supply near the faucet or water outlet while the faucet or water outlet is wide open and flowing. Flow pressure is measured in pounds per square inch (kilopascals).

† Wide range because designs and types of toilet flush valves vary.

Pressure, Flows, and Meters

321

METERS FOR FLOW MEASUREMENTS

Positive-displacement meters use a nutating disk or piston and are quite accurate. They are used in a wide range of flows but should not be used for extended periods of time at high flow rates. Usually used for residential installation.

Single/multijet meters use a multivaned rotor. They are used on residential installations.

Turbine meters are used for higher volumes of flow. They are not as accurate for low flows. They can handle turbid waters.

Compound meters are really two meters combined into one unit. A smaller positive-displacement meter measures lower flows, and a valve switches the higher flow rates to a larger turbine meter. They are often used in master meter installations.

See AWWA Manual M6, *Water Meters—Selection, Installation, Testing, & Maintenance*. See ANSI/AWWA standards C700 series for meter standards.

Required Accuracy Limits for Compliance With Guidelines

Meter Type (all sizes)	Accuracy Limits as Found by Testing, %	
	Normal Test Flow Rates	Minimum Test Flow Rates
Displacement	96–102	80–102
Multijet	96–102	80–104
Propeller and turbine	96–103	Not applicable
Compound and fire service	95–104	Not applicable

Type and Sizing (ANSI/AWWA Standard for Minimum Flow Range) Guidelines

I. Residential Meters

- Same line provides fire service (hydrants, sprinklers): Proceed to section III.

- Line does not provide fire service: Use Disc or Piston Meter chart. (Typical flow should be ½ of maximum rated capacity. Select size from chart below.)

Disc or Piston Meter

Size, *in.*	Flow Range, *gpm*
⅝	¼–20
¾	½–30
1	¾–50
1½	1½–100
2	2–160*

* For higher flow ranges, proceed to section II.

II. Commercial/Industrial Meters

- Same line provides fire service (hydrants, sprinklers): Proceed to section III.

- Line does not provide fire service: Continue this section.

Typical Flow

Always High: Use Turbo (Turbine) Meter chart. (Can be used at maximum rated capacity, peak to 125% rating. Select size from chart below.)

Turbo Meter

Size, *in.*	Flow Range, *gpm*
1½	4–160
2	4–160
3	8–350
4	15–630
6	30–1,400
8	50–2,400
10	75–3,800
12	120–5,000

Turbo Meter (continued)

Size, *in.*	Flow Range, *gpm*
16	130–13,200 ±
24	200–19,800 ±

Always Low: Use Disc or Piston Meter chart (section I).

High and Low: Use Compound Meter chart. (Typical flow should be ½ of maximum rated capacity; for complete accountability and to minimize excessive switching wear, ensure that crossover range is not within pipe's normal flow range. Select size from chart below.)

Compound Meter

Size, *in.*	Flow Range, *gpm*
2	¼–160
3	½–320
4	¾–500
6	1½–1,000

III. Master Meters

- Line does not provide fire service: Select meter from section II.

- Same line provides fire service (hydrants, sprinklers): Continue this section.

Typical Flow

Always High (or fire service only): Use Fire Service Meter chart. (Turbo meter with Underwriters Laboratories/Factory Mutual Engineering Corporation [UL/FM] strainer. Select size from chart below.)

Fire Service Meter

Size, *in.*	Flow Range, *gpm*
3	4–350
4	10–700
6	20–1,600
8	30–2,800
10	35–4,400

High and Low: Use Fire Service Assembly chart. (A compound meter with UL/FM strainer, providing domestic and fire service; low flow varies with size of bypass line; for complete accountability and to minimize excessive switching wear, ensure that crossover range is not within pipe's normal flow range. Select size from chart below.)

Fire Service Assembly

Size, *in.*	Flow Range, *gpm*
4	4–700
6	5–1,600
8	8–2,800
10	8–4,400

IV. Special Use Meters

Mag Meter: Use when sand, debris, or bi-directional flow exist. (Select size from chart below.)

Mag Meter (No ANSI/AWWA standard)

Size, *in.*	Flow Range, *gpm*
4	4–1,300
6	9–2,900
8	16–5,200
10	25–8,000
12	35–11,600

Hydrant Meter: To measure fire hydrant water use; use with RPZ or DCV backflow-prevention device.

AWWA Meter Standards

Meter, in.	Minimum Flow Rate, gpm	Low Normal Flow Rate, gpm	Change-over Range (Compound Meters)	High Normal Flow Rate, gpm	Maximum Flow Rate, gpm	Head Loss at Maximum Flow Rate, psi
Positive Displacement						
½	0.25	1	NA	7.5	15	15
⅝	0.25	1		10	20	15
¾	0.5	2		15	30	15
1	0.75	3		25	50	15
1½	1.5	5	NA	50	100	15
2	2	8		80	160	15
Multijet						
⅝	0.25	1	NA	10	20	15
¾	0.5	2		15	30	15
1	0.75	3		25	50	15
1½	1.5	5		50	100	15
2	2.0	8		80	160	15
Turbine Class 1						
¾	1.5	NA	NA	20	30	15
1	2			35	50	15
1½	3			65	100	15

*NA = not applicable.

Table continued on next page

AWWA Meter Standards (continued)

Meter, in.	Minimum Flow Rate, gpm	Low Normal Flow Rate, gpm	Change-over Range (Compound Meters)	High Normal Flow Rate, gpm	Maximum Flow Rate, gpm	Head Loss at Maximum Flow Rate, psi
2	4			100	160	15
3	6			220	350	15
4	8			420	630	15
6	15			865	1,300	15
Turbine Class 2						
1½	4	NA	NA	80	120	7
2	4			100	160	7
3	8			240	350	7
4	15			420	630	7
6	30			920	1,400	7
8	50			1,600	2,400	7
10	75			2,500	3,800	7
12	120			3,300	5,000	7
14	150			5,200	7,500	7
16	200			6,500	10,000	7
18	250			8,500	12,500	7
20	300			10,000	15,000	7

* NA = not applicable.

Table continued on next page

Pressure, Flows, and Meters

AWWA Meter Standards (continued)

Meter, in.	Minimum Flow Rate, gpm	Low Normal Flow Rate, gpm	Change-over Range (Compound Meters)	High Normal Flow Rate, gpm	Maximum Flow Rate, gpm	Head Loss at Maximum Flow Rate, psi
Compound						
2	0.25	2	20	80	160	20
3	0.5	4	23	160	320	20
4	0.75	6	28	250	500	20
6	1.5	10	32	500	1,000	20
8	2	16	50	800	1,600	20
Singlejet						
1½	0.5	1.5	NA	50	100	15
2	0.5	2.0		80	160	15
3	0.5	2.5		160	320	15
4	0.75	3.0		250	500	15
6	1.5	4.0		500	1,000	15

* NA = not applicable.

328

Characteristics of Displacement-Type Meters

Meter Size		Safe Maximum Operating Capacity		Maximum Pressure Loss at Safe Maximum Operating Capacity		Recommended Maximum Rate for Continuous Operations		Minimum Test Flow		Normal Test Flow Limits	
in.	(mm)	gpm	(m³/hr)	psi	(kPa)	gpm	(m³/hr)	gpm	(m³/hr)	gpm	(m³/hr)
½	(13)	15	(3.4)	15	(103)	7.5	(1.7)	¼	(0.06)	1–15	(0.2–3.4)
½ × ¾	(13 × 19)	15	(3.4)	15	(103)	7.5	(1.7)	¼	(0.06)	1–15	(0.2–3.4)
⅝	(16)	20	(4.5)	15	(103)	10	(2.3)	¼	(0.06)	1–20	(0.2–4.5)
⅝ × ¾	(16 × 19)	20	(4.5)	15	(103)	10	(2.3)	¼	(0.06)	1–20	(0.2–4.5)
¾	(19)	30	(6.8)	15	(103)	15	(3.4)	½	(0.11)	2–30	(0.5–6.8)
1	(25)	50	(11.4)	15	(103)	25	(5.7)	¾	(0.17)	3–50	(0.7–11.4)
1½	(38)	100	(22.7)	15	(103)	50	(11.3)	1½	(0.34)	5–100	(1.1–22.7)
2	(51)	160	(36.3)	15	(103)	80	(18.2)	2	(0.45)	8–160	(1.8–36.3)

Pressure, Flows, and Meters

WATER METERS INSTALLATION CHECKLIST _____

Is meter installed horizontally, with isolation valves installed before and after the meter?	☐ Yes	☐ No	☐ NA
Are min. 5-diameter straight pipe (no bends, valves, reducers, etc.) installed before, and 3–5-diameter after, the meter?	☐ Yes	☐ No	☐ NA
Is a strainer installed before a turbo or compound meter?	☐ Yes	☐ No	☐ NA
Is a UL/FM strainer installed before any meter providing fire service?	☐ Yes	☐ No	☐ NA
Is a bypass (possibly with meter) installed for use during shutdowns or repairs?	☐ Yes	☐ No	☐ NA
Is meter installed before any RPZ or DCV device?	☐ Yes	☐ No	☐ NA
Is a test plug (tee or 2-in. corp. with iron pipe thread) installed after the meter (and before valve) for meter testing?	☐ Yes	☐ No	☐ NA
If within a pit or insulated enclosure (Hot Box), is remote reading device installed?	☐ Yes	☐ No	☐ NA
If within a pit, is opening large enough to allow eventual removal of meter/valves/strainer?	☐ Yes	☐ No	☐ NA
If within a pit, is opening positioned so meter/valves/strainer are within overhead lifting capabilities?	☐ Yes	☐ No	☐ NA
For optimal accountability and performance, is appropriate smallest-sized meter selected (see Type and Sizing Guidelines, p. 325)?	☐ Yes	☐ No	☐ NA

Courtesy of Ralph Harstad and Ronald Rudio of the New York State Department of Health; Cindy Kranslwer of Badger Meter, Inc.; Dan Reed of Sensus Metering Systems; and Terry Wilson of T. Wilson and Associates.

Volume of Various Diameter Pipe

Actual Inside Diameter, in.	Gallons in One Lineal Foot	Actual Inside Diameter, in.	Gallons in One Lineal Foot	Actual Inside Diameter, in.	Gallons in One Lineal Foot
0.125	0.001	3.375	0.465	6.625	1.791
0.250	0.003	3.500	0.500	6.750	1.859
0.375	0.006	3.625	0.536	6.875	1.928
0.500	0.010	3.750	0.574	7.000	1.999
0.625	0.016	3.875	0.613	7.125	2.071
0.750	0.023	4.000	0.653	7.250	2.145
0.875	0.031	4.125	0.694	7.375	2.219
1.000	0.041	4.250	0.737	7.500	2.295
1.125	0.052	4.375	0.781	7.625	2.372
1.250	0.064	4.500	0.826	7.750	2.451
1.375	0.077	4.625	0.873	7.875	2.530
1.500	0.092	4.750	0.921	8.000	2.611
1.625	0.108	4.875	0.970	8.125	2.693
1.750	0.125	5.000	1.020	8.250	2.777
1.875	0.143	5.125	1.072	8.375	2.862
2.000	0.163	5.250	1.125	8.500	2.948
2.125	0.184	5.375	1.179	8.625	3.035
2.250	0.207	5.500	1.234	8.750	3.124
2.375	0.230	5.625	1.291	8.875	3.214
2.500	0.255	5.750	1.349	9.000	3.305
2.625	0.281	5.875	1.408	9.125	3.397
2.750	0.309	6.000	1.469	9.250	3.491
2.875	0.337	6.125	1.531	9.375	3.586
3.000	0.367	6.250	1.594	9.500	3.682
3.125	0.398	6.375	1.658	9.625	3.780
3.250	0.431	6.500	1.724	9.750	3.879

Table continued on next page

Pressure, Flows, and Meters

Volume of Various Diameter Pipe (continued)

Actual Inside Diameter, in.	Gallons in One Lineal Foot	Actual Inside Diameter, in.	Gallons in One Lineal Foot	Actual Inside Diameter, in.	Gallons in One Lineal Foot
9.875	3.979	11.625	5.514	13.375	7.299
10.000	4.080	11.750	5.633	13.500	7.436
10.125	4.183	11.875	5.753	13.625	7.574
10.250	4.287	12.000	5.875	13.750	7.714
10.375	4.392	12.125	5.998	13.875	7.855
10.500	4.498	12.250	6.123	14.000	7.997
10.625	4.606	12.375	6.248	14.125	8.140
10.750	4.715	12.500	6.375	14.250	8.285
10.875	4.825	12.625	6.503	14.375	8.431
11.000	4.937	12.750	6.633	14.500	8.578
11.125	5.050	12.875	6.763	14.625	8.727
11.250	5.164	13.000	6.895	14.750	8.877
11.375	5.279	13.125	7.028	14.875	9.028
11.500	5.396	13.250	7.163	15.000	9.180

Friction Loss of Water, in Feet per 100-ft Length of Pipe, Based on Hazen–Williams Formula for C = 100

gpm	½-in. Pipe		¾-in. Pipe		1-in. Pipe		1¼-in. Pipe		1½-in. Pipe		2-in. Pipe		2½-in. Pipe		3-in. Pipe		4-in. Pipe		5-in. Pipe		6-in. Pipe	
	Vel., ft/sec	Loss, ft	Vel., ft/sec	Loss, ft	Vel., ft/sec	Loss, ft	Vel., ft/sec	Loss, ft	Vel., ft/sec	Loss, ft	Vel., ft/sec	Loss, ft	Vel., ft/sec	Loss, ft	Vel., ft/sec	Loss, ft	Vel., ft/sec	Loss, ft	Vel., ft/sec	Loss, ft	Vel., ft/sec	Loss, ft
2	2.10	7.4	1.20	1.9																		
4	4.21	27.0	2.41	7.0	1.49	2.14	.86	.57	.63	.26												
6	6.31	57.0	3.61	14.7	2.23	4.55	1.29	1.20	.94	.56	.61	.20										
8	8.42	98.0	4.81	25.0	2.98	7.8	1.72	2.03	1.26	.95	.82	.33	.52	.11								
10	10.52	147.0	6.02	38.0	3.72	11.7	2.14	3.05	1.57	1.43	1.02	.50	.65	.17	.45	.07						
12			7.22	53.0	4.46	16.4	2.57	4.3	1.89	2.01	1.23	.79	.78	.23	.54	.10						
15			9.02	80.0	5.60	25.0	3.21	6.5	2.36	3.00	1.53	1.08	.98	.36	.68	.15						
18			10.84	108.2	6.69	35.0	3.86	9.1	2.83	4.24	1.84	1.49	1.18	.50	.82	.21						
20			12.03	136.0	7.44	42.0	4.29	11.1	3.15	5.20	2.04	1.82	1.31	.61	.91	.25	.51	.06				
25					9.30	64.0	5.36	16.6	3.80	7.30	2.55	2.73	1.63	.92	1.13	.38	.64	.09				
30					11.15	89.0	6.43	23.0	4.72	11.0	3.06	3.84	1.96	1.29	1.36	.54	.77	.13	.49	.04		
35					13.02	119.0	7.51	31.2	5.51	14.7	3.57	5.10	2.29	1.72	1.59	.71	.89	.17	.57	.06		
40					14.88	152.0	8.58	40.0	6.30	18.8	4.08	6.6	2.61	2.20	1.82	.91	1.02	.22	.65	.08		
45							9.65	50.0	7.08	23.2	4.60	8.2	2.94	2.80	2.04	1.15	1.15	.28	.73	.09		
50							10.72	60.0	7.87	28.4	5.11	9.9	3.27	3.32	2.27	1.38	1.28	.34	.82	.11	.57	.04
55							11.78	72.0	8.66	34.0	5.62	11.8	3.59	4.01	2.45	1.58	1.41	.41	.90	.14	.62	.05
60							12.87	85.0	9.44	39.6	6.13	13.9	3.92	4.65	2.72	1.92	1.53	.47	.98	.16	.68	.06
65							13.92	99.7	10.23	45.9	6.64	16.1	4.24	5.4	2.89	2.16	1.66	.53	1.06	.19	.74	.076
70							15.01	113.0	11.02	53.0	7.15	18.4	4.58	6.2	3.18	2.57	1.79	.63	1.14	.21	.80	.08
75							16.06	129.0	11.80	60.0	7.66	20.9	4.91	7.1	3.33	3.00	1.91	.73	1.22	.24	.85	.10
80							17.16	145.0	12.59	68.0	8.17	23.7	5.23	7.9	3.63	3.28	2.04	.81	1.31	.27	.91	.11
85							18.21	163.8	13.38	75.0	8.68	26.5	5.56	8.1	3.78	3.54	2.17	.91	1.39	.31	.96	.12

Table continued on next page

Pressure, Flows, and Meters

Friction Loss of Water, in Feet per 100-ft Length of Pipe, Based on Hazen–Williams Formula for C = 100 (continued)

gpm	½-in. Pipe Vel, ft/sec	½-in. Pipe Loss, ft	¾-in. Pipe Vel, ft/sec	¾-in. Pipe Loss, ft	1-in. Pipe Vel, ft/sec	1-in. Pipe Loss, ft	1¼-in. Pipe Vel, ft/sec	1¼-in. Pipe Loss, ft	1½-in. Pipe Vel, ft/sec	1½-in. Pipe Loss, ft	2-in. Pipe Vel, ft/sec	2-in. Pipe Loss, ft	2½-in. Pipe Vel, ft/sec	2½-in. Pipe Loss, ft	3-in. Pipe Vel, ft/sec	3-in. Pipe Loss, ft	4-in. Pipe Vel, ft/sec	4-in. Pipe Loss, ft	5-in. Pipe Vel, ft/sec	5-in. Pipe Loss, ft	6-in. Pipe Vel, ft/sec	6-in. Pipe Loss, ft
90							19.30	180.0	14.71	84.0	9.19	29.4	5.88	9.8	4.09	4.08	2.30	1.00	1.47	.34	1.02	.14
95									14.95	93.0	9.70	32.6	6.21	10.8	4.22	4.33	2.42	1.12	1.55	.38	1.08	.15
100									15.74	102.0	10.21	35.8	6.54	12.0	4.54	4.96	2.55	1.22	1.63	.41	1.13	.17
110									17.31	122.0	11.23	42.9	7.18	14.5	5.00	6.0	2.81	1.46	1.79	.49	1.25	.21
120									18.89	143.0	12.25	50.0	7.84	16.8	5.45	7.0	3.06	1.17	1.96	.58	1.36	.24
130	8-in. Pipe								20.46	166.0	13.28	58.0	8.48	18.7	5.91	8.1	3.31	1.97	2.12	.67	1.47	.27
140	.90	.08							22.04	190.0	14.30	67.0	9.15	22.3	6.35	9.2	3.57	2.28	2.29	.76	1.59	.32
150	.96	.09									15.32	76.0	9.81	25.5	6.82	10.5	3.82	2.62	2.45	.88	1.70	.36
160	1.02	.10									16.34	86.0	10.46	29.0	7.26	11.8	4.08	2.91	2.61	.98	1.82	.40
170	1.08	.11									17.36	96.0	11.11	34.1	7.71	13.3	4.33	3.26	2.77	1.08	1.92	.45
180	1.15	.13									18.38	107.0	11.76	35.7	8.17	14.0	4.60	3.61	2.94	1.22	2.04	.50
190	1.21	.14									19.40	118.0	12.42	39.6	8.63	15.5	4.84	4.01	3.10	1.35	2.16	.55
200	1.28	.15	10-in. Pipe								20.42	129.0	13.07	43.1	9.08	17.8	5.11	4.4	3.27	1.48	2.27	.62
220	1.40	.18	.90	.06							22.47	154.0	14.38	52.0	9.99	21.3	5.62	5.2	3.59	1.77	2.50	.73
240	1.53	.22	.98	.07							24.51	182.0	15.69	61.0	10.89	25.1	6.13	6.2	3.92	2.08	2.72	.87
260	1.66	.25	1.06	.08							26.55	211.0	16.99	70.0	11.80	29.1	6.64	7.2	4.25	2.41	2.95	1.00
280	1.79	.28	1.15	.09									18.30	81.0	12.71	33.4	7.15	8.2	4.58	2.77	3.18	1.14
300	1.91	.32	1.22	.11									19.61	92.0	13.62	38.0	7.66	9.3	4.90	3.14	3.40	1.32
320	2.05	.37	1.31	.12									20.92	103.0	14.52	42.8	8.17	10.5	5.23	3.54	3.64	1.47
340	2.18	.41	1.39	.14									22.22	116.0	15.43	47.9	8.68	11.7	5.54	3.97	3.84	1.62
360	2.30	.45	1.47	.15	12-in. Pipe								23.53	128.0	16.34	53.0	9.19	13.1	5.87	4.41	4.08	1.83
380	2.43	.50	1.55	.17	1.08	.069							24.84	142.0	17.25	59.0	9.69	14.0	6.19	4.86	4.31	2.00

Friction Loss of Water, in Feet per 100-ft Length of Pipe, Based on Hazen–Williams Formula for C = 100 (continued)

Table continued on next page

Embedded sub-pipe sizes appear in otherwise-blank columns: **14-in. Pipe** data begins at 500 gpm in the 1¼-in. column; **16-in. Pipe** at 850 gpm in the 1½-in. column; **20-in. Pipe** at 1,000 gpm in the 2-in. column; **24-in. Pipe** at 1,800 gpm in the 2½-in. column. (Vel = ft/sec, Loss = ft.)

gpm	½-in Vel	½-in Loss	¾-in Vel	¾-in Loss	1-in Vel	1-in Loss	1¼-in Vel	1¼-in Loss	1½-in Vel	1½-in Loss	2-in Vel	2-in Loss	2½-in Vel	2½-in Loss	3-in Vel	3-in Loss	4-in Vel	4-in Loss	5-in Vel	5-in Loss	6-in Vel	6-in Loss
400	2.60	.54	1.63	.19	1.14	.075							26.14	156.0	18.16	65.0	10.21	16.0	6.54	5.4	4.55	2.20
450	2.92	.68	1.84	.23	1.28	.095									20.40	78.0	11.49	19.8	7.35	6.7	5.11	2.74
500	3.19	.82	2.04	.28	1.42	.113	1.04	.06							22.70	98.0	12.77	24.0	8.17	8.1	5.68	3.30
550	3.52	.97	2.24	.33	1.56	.135	1.15	.07							24.96	117.0	14.04	28.7	8.99	9.6	6.25	3.96
600	3.84	1.14	2.45	.39	1.70	.159	1.25	.08							27.23	137.0	15.32	33.7	9.80	11.3	6.81	4.65
650	4.16	1.34	2.65	.45	1.84	.19	1.37	.09									16.59	39.0	10.62	13.2	7.38	5.40
700	4.46	1.54	2.86	.52	1.99	.22	1.46	.10									17.87	44.9	11.44	15.1	7.95	6.21
750	4.80	1.74	3.06	.59	2.13	.24	1.58	.11									19.15	51.0	12.26	17.2	8.50	7.12
800	5.10	1.90	3.26	.66	2.27	.27	1.67	.13									20.42	57.0	13.07	19.4	9.08	7.96
850	5.48	2.20	3.47	.75	2.41	.31	1.79	.14	1.36	.08							21.70	64.0	13.89	21.7	9.65	8.95
900	5.75	2.46	3.67	.83	2.56	.34	1.88	.16	1.44	.084							22.98	71.0	14.71	24.0	10.20	10.11
950	6.06	2.87	3.88	.91	2.70	.38	2.00	.18	1.52	.095									15.52	26.7	10.77	11.20
1,000	6.38	2.97	4.08	1.03	2.84	.41	2.10	.19	1.60	.10	1.02	.04							16.34	29.2	11.34	12.04
1,100	7.03	3.52	4.49	1.19	3.13	.49	2.31	.23	1.76	.12	1.12	.04							17.97	34.9	12.48	14.55
1,200	7.66	4.17	4.90	1.40	3.41	.58	2.52	.27	1.92	.14	1.23	.05							19.61	40.9	13.61	17.10
1,300	8.30	4.85	5.31	1.62	3.69	.67	2.71	.32	2.08	.17	1.33	.06									14.72	18.4
1,400	8.95	5.50	5.71	1.87	3.98	.78	2.92	.36	2.24	.19	1.43	.064									15.90	22.60
1,500	9.58	6.24	6.12	2.13	4.26	.89	3.15	.41	2.39	.21	1.53	.07									17.02	25.60
1,600	10.21	7.00	6.53	2.39	4.55	.98	3.34	.47	2.56	.24	1.63	.08									18.10	26.9
1,800	11.50	8.78	7.35	2.95	5.11	1.21	3.75	.58	2.87	.30	1.84	.10	1.28	.04								
2,000	12.78	10.71	8.16	3.59	5.68	1.49	4.17	.71	3.19	.37	2.04	.12	1.42	.05								
2,200	14.05	12.78	8.98	4.24	6.25	1.81	4.59	.84	3.51	.44	2.25	.15	1.56	.06								

Pressure, Flows, and Meters

Friction Loss of Water, in Feet per 100-ft Length of Pipe, Based on Hazen–Williams Formula for $C = 100$ (continued)

Table continued on next page

30-in. Pipe

gpm	½-in. Pipe Vel., ft/sec	Loss, ft	¾-in. Pipe Vel., ft/sec	Loss, ft	1-in. Pipe Vel., ft/sec	Loss, ft	1¼-in. Pipe Vel., ft/sec	Loss, ft	1½-in. Pipe Vel., ft/sec	Loss, ft	2-in. Pipe Vel., ft/sec	Loss, ft	2½-in. Pipe Vel., ft/sec	Loss, ft	3-in. Pipe Vel., ft/sec	Loss, ft	4-in. Pipe Vel., ft/sec	Loss, ft	5-in. Pipe Vel., ft/sec	Loss, ft	6-in. Pipe Vel., ft/sec	Loss, ft
2,400	15.32	14.2	9.80	5.04	6.81	2.08	5.00	.99	3.83	.52	2.45	.17	1.70	.07	1.09	.02						
2,600			10.61	5.81	7.38	2.43	5.47	1.17	4.15	.60	2.66	.20	1.84	.08	1.16	.027						
2,800			11.41	6.70	7.95	2.75	5.84	1.32	4.47	.68	2.86	.23	1.98	.09	1.27	.03						
3,000			12.24	7.62	8.52	3.15	6.01	1.49	4.79	.78	3.08	.27	2.13	.10	1.37	.037						
3,200			13.05	7.8	9.10	3.51	6.68	1.67	5.12	.88	3.27	.30	2.26	.12	1.46	.041						
3,500			14.30	10.08	9.95	4.16	7.30	1.97	5.59	1.04	3.59	.35	2.49	.14	1.56	.047						
3,800			15.51	13.4	10.80	4.90	7.98	2.36	6.07	1.20	3.88	.41	2.69	.17	1.73	.05						
4,200					11.92	5.88	8.76	2.77	6.70	1.44	4.29	.49	2.99	.20	1.91	.07						
4,500					12.78	6.90	9.45	3.22	7.18	1.64	4.60	.56	3.20	.22	2.04	.08						
5,000					14.20	8.40	10.50	3.92	8.01	2.03	5.13	.68	3.54	.27	2.26	.09						
5,500							11.55	4.65	8.78	2.39	5.64	.82	3.90	.33	2.50	.11						
6,000							12.60	5.50	9.58	2.79	6.13	.94	4.25	.38	2.73	.13						
6,500							13.65	6.45	10.39	3.32	6.64	1.10	4.61	.45	2.96	.15						
7,000							14.60	7.08	11.18	3.70	7.15	1.25	4.97	.52	3.18	.17						
8,000									12.78	4.74	8.17	1.61	5.68	.66	3.64	.23						
9,000									14.37	5.90	9.20	2.01	6.35	.81	4.08	.28						
10,000									15.96	7.19	10.20	2.44	7.07	.98	4.54	.33						
12,000											12.25	3.41	8.50	1.40	5.46	.48						
14,000											14.30	4.54	9.95	1.87	6.37	.63						
16,000													11.38	2.40	7.28	.81						
18,000													12.76	2.97	8.18	1.02						
20,000													14.20	3.60	9.10	1.23						

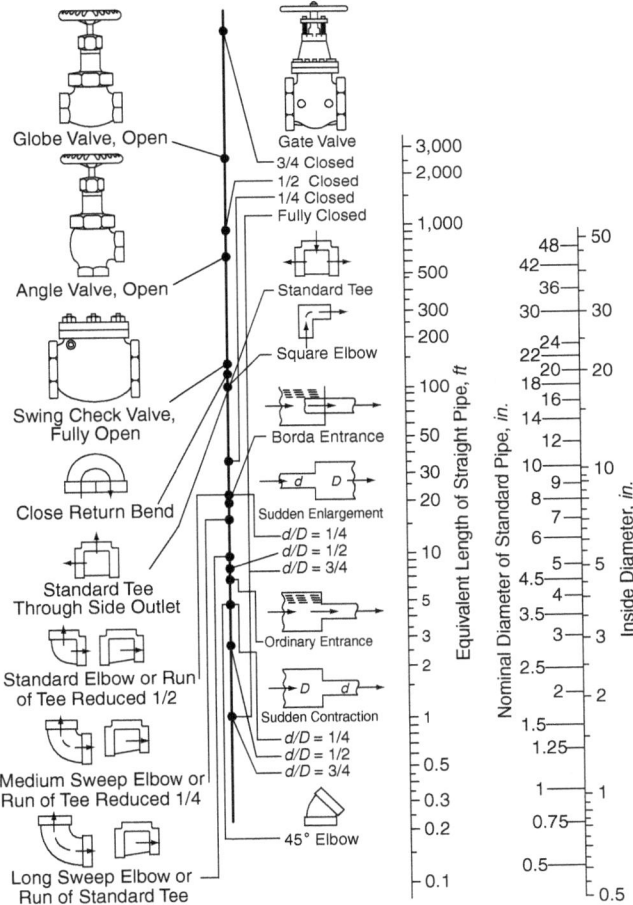

NOTE: Put one end of the ruler on indicated point on valve line and the other end of the ruler on the correct diameter line and read the equivalent length.

Resistance of Valves and Fittings to Flow of Fluids

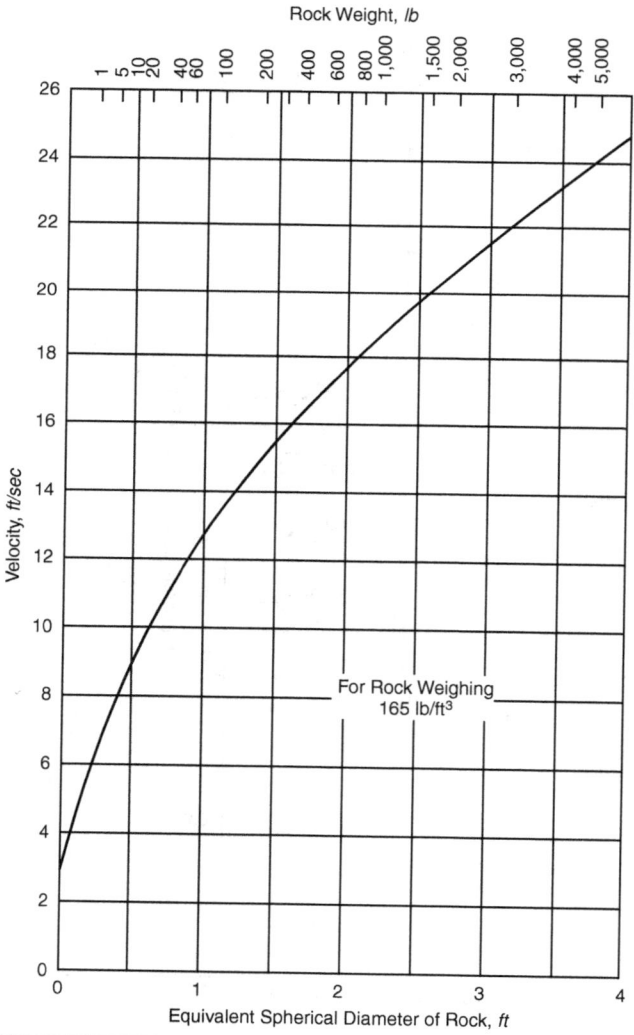

Rock Weight, *lb*

Velocity, *ft/sec*

For Rock Weighing
165 lb/ft³

Equivalent Spherical Diameter of Rock, *ft*

Recommended Riprap Sizes as a Function of Water Velocity

WATER HAMMER

When a valve is closed in a pipe through which water is flowing, the water before the valve is retarded, and a dynamic pressure is produced. If the valve is closed quickly, this dynamic pressure may be very great, leading to "water hammer" or "water ram." In many cases, this phenomenon causes bursting of the pipe, or the failure of valves or fittings.

This danger can be avoided by proper cushioning of the line with air chambers, or by installing relief valves. It may also be avoided by using slow-closing gate valves.

When a valve is closed quickly, the increase in pressure in the pipe resulting from water hammer may be calculated as follows (Merriman's Hydraulics): Shock pressure in pounds per square inch equals 63 times the velocity of the water in the pipe in feet per second.

The time that should be taken to close a valve, so that the pressure shall not exceed the normal pressure at no flow, may be found from the following formula:

$$\text{time, sec} = \frac{0.027 \times L \times V}{P - p}$$

Where:

L = length of pipe before the valve, in ft

V = velocity of flow, in ft/sec

P = pressure in the pipe, in lb/in.^{2}, when there is no flow

p = pressure in the pipe at full flow

Abbreviations
and Acronyms

*In the water industry as in other fields and
disciplines, many names, titles, programs,
organizations, legislative acts, measurements,
and activities are abbreviated to reduce the
volume of words and to simplify communications.
In this section, common abbreviations and
acronyms used in the water industry—not only
in this guide—are listed for easy reference.*

Å	angstrom
A	ampere
AACE	American Association of Cost Engineers
AAS	atomic absorption spectrophotometry
AASHTO	American Association of State Highway and Transportation Officials
ABPA	American Backflow Prevention Association
ABS	alkylbenzene sulfonate; acrilonitrile butadiene styrene
AC	alternating current
A–C	asbestos cement
ACM	asbestos-containing material
acre-ft	acre-foot
ACS	American Chemical Society
ADA	Americans with Disabilities Act
AES	atomic emission spectroscopy
AHM	acutely hazardous material
A·hr	ampere-hour
AIChE	American Institute of Chemical Engineers
AIEE	American Institute of Electrical Engineers
AMWA	Association of Metropolitan Water Agencies
ANOVA	analysis of variance
ANPRM	Advanced Notice of Proposed Rulemaking
ANSI	American National Standards Institute
AOC	assimilable organic carbon
APHA	American Public Health Association
APWA	American Public Works Association
ASCE	American Society of Civil Engineers
ASDWA	Association of State Drinking Water Administrators
ASME	American Society of Mechanical Engineers
ASSE	American Society of Safety Engineers; Association of State Sanitary Engineers
ASTM	American Society for Testing and Materials
atm	atmosphere
avdp or avoir.	avoirdupois
AWRA	American Water Resources Association
AWWA	American Water Works Association
AWWARF	AWWA Research Foundation (now Water Resarch Foundation)
BAT	best available technology
bbl	barrel

BDOM	biodegradable organic matter
Bé	Baumé
BEAC	biologically enhanced activated carbon
BeV	billion electron volts
bgd	billion gallons per day
bhp	brake horsepower
bil gal	billion gallons
BOD	biochemical oxygen demand *or* biological oxygen demand
BOM	background organic matter; biodegradable organic matter
bph	barrels per hour
bps	binary digits (bits) per second
Bq	becquerel (metric equivalent of curie)
BSA	bovine serum albumin
Btu	British thermal unit
bu	bushel
BV	bed volume
°C	degrees Celsius
C	coulomb
$C \times T$ *or* CT	disinfectant concentration $\times$ time
CAA	Clean Air Act
ccf	100 cubic feet
CCL	Contaminant Candidate List
CCR	consumer confidence report
cd	candela
CDC	Centers for Disease Control and Prevention
CERCLA	Comprehensive Environmental Response, Compensation, and Liability Act
CF	conventional filtration
CFM	cubic feet per minute
CFR	Code of Federal Regulations
cfs	cubic feet per second
cfu	colony-forming unit
CGPM	General Conference on Weights and Measures
Ci	curie
CI	cast iron
C/kg	coulombs per kilogram
cm	centimeter
CMMS	computerized maintenance management system
COD	chemical oxygen demand

Co–Pt	chloroplatinate
cpm	counts per minute
CPP	concrete pressure pipe
cps	cycles per second (1 cps = 1 Hz)
CPSC	Consumer Products Safety Commission
cpu	chloroplatinate units
CPVC	chlorinated polyvinyl chloride
CSA	Canadian Standards Association
CT	contact time
CT or $C \times T$	disinfectant concentration $\times$ time
CTS-PE	copper tubing size polyethylene
cu	color unit; cubic
CUR	activated carbon usage rate
CWA	Clean Water Act
CWS	community water system
°	degree
d	day
D	dalton
da	darcy
DAF	dissolved air flotation
dB	decibel
DBCP	dibromochloropropane
DBP	disinfection by-product
DC	direct current
DCS	distributed control system
DCV	double check valve
DCVA	double check valve assembly
D/DBP	disinfectant/disinfection by-product
DDT	dichlorodiphenyltrichloroethane
DE	diatomaceous earth (filtration)
DI	ductile iron
diam.	diameter
DIPRA	Ductile Iron Pipe Research Association
dL	deciliter
DO	dissolved oxygen
DOC	dissolved organic carbon
DOT	Department of Transportation
DPD	N,N-diethyl-p-phenylenediamine
dr	dram
DSP	disodium phosphate
DWCCL	Drinking Water Contaminant Candidate List

DWV	drain, waste, and vent (pipe)
EBCT	empty-bed contact time
EC	electrical conductivity
ED	electrodialysis *or* effective diameter
EDB	ethylene dibromide
EDR	electrodialysis reversal
EDTA	ethylenediaminetetraacetic acid
EGL	energy grade line
EIS	Environmental Impact Statement
EJC	Engineers Joint Council
ElCD	electrolytic conductivity detector
emf	electromotive force
EPA	Environmental Protection Agency (US)
EPCRA	Emergency Planning and Community Right-to-Know Act
EPDM	ethylene-propylene-diene-monomer
EPI-DMA	epichlorohydrin dimethylamine
eq/L	equivalents per liter
ES	effective size
ESA	Endangered Species Act
ESWTR	Enhanced Surface Water Treatment Rule
eV	electron volt
°F	degrees Fahrenheit
F	farad
fbm	board feet (feet board measure)
FEMA	Federal Emergency Management Agency
FIFRA	Federal Insecticide, Fungicide, and Rodenticide Act
fl oz	fluid ounce
FM	Factory Mutual Engineering Corporation
fps	foot per second
FRP	fiberglass-reinforced plastic
ft	feet
ft/hr	feet per hour
ft/min	feet per minute
ft/sec	feet per second
ft/sec/ft	feet per second per foot
ft/sec^2	feet per second squared
ft^2/sec	feet squared per second
ft^2 *or* sq ft	square foot

ft³/sec	cubic feet per second
ft³ *or* cu ft	cubic feet
ft³/hr *or* cu ft/hr	cubic feet per hour
ft³/min *or* cu ft/min	cubic feet per minute
ft³/sec *or* cu ft/sec	cubic feet per second
ft-lb	foot-pound
ftu	formazin turbidity unit
FY	fiscal year
g	gram
GAC	granular activated carbon
gal	gallon
gal/flush	gallons per flush
gal/ft²	gallons per square foot
GAO	General Accounting office
GC	gas chromatography
GC–ECD	gas chromatography–electron capture detector
GC–MS	gas chromatography–mass spectrometry
GHT	garden hose thread
GIS	geographic information system
GL	gigaliter
gpcd	gallons per capita per day
gpd	gallons per day
gpd/ft²	gallons per day per square foot
gpg	grains per gallon
gph	gallons per hour
gpm	gallons per minute
gpm/ft²	gallons per minute per square foot
gps	gallons per second
GPS	global positioning system
gpy	gallons per year
gr	grain
gsfd	gallons per square foot per day
GWUDI	groundwater under the direct influence of surface water
Gy	gray
H	henry
ha	hectare
HAA	haloacetic acid
HAA5	sum of five HAAs
HAN	haloacetonitrile

HAV	Hepatitis A virus
HDPE	high-density polyethylene
HDXLPE	high-density, cross-linked polyethylene
HF	hydrogen fluoride
HGL	hydraulic grade line
HGLE	hydraulic grade line elevation
HIV	human immunodeficiency virus
hL	hectoliter
hp	horsepower
HPC	heterotrophic plate count
hp·hr	horsepower-hour
HPLC	high-performance liquid chromatography
hr	hour
HRT	hydraulic retention time
HTH	High Test Hypochlorite
HVAC	heating, ventilating, and air conditioning
Hz	hertz
I&C	instrumentation and control
IBWA	International Bottled Water Association
ICP	inductively coupled plasma
ICR	Information Collection Rule
ID	inside diameter
IEEE	Institute of Electrical and Electronics Engineers
Imp	Imperial
in.	inch
in.-lb	inch-pound
in./min	inches per minute
in./sec	inches per second
in.^{2} *or* sq in.	square inch
in.^{3} *or* cu in.	cubic inches
IOC	inorganic contaminant
IP	iron pipe
IPS	iron pipe size
IPS-PE	iron pipe size polyethylene
IPT	iron pipe thread
IRC	International Research Center
ISA	Instrument Society of America
ISO	International Organization for Standardization
IWRA	International Water Resources Association
J	joule

K	kelvin
kB	kilobyte
kg	kilogram
kHz	kilohertz (kilocycles)
kJ	kilojoule
km	kilometer
km^2	square kilometers
kPa	kilopascal
kV	kilovolt
kVA	kilovolt-ampere
kvar	kiloreactive volt-ampere
kW	kilowatt
kW·hr	kilowatt-hour
L	liter
lb	pound
lb/day	pounds per day
lbf	pound force
lb/ft^2	pounds per square foot
lbm	pound mass
LC	liquid chromatography
L/day	liters per day
LIN	liquid nitrogen
lin ft	linear feet
LLE	liquid–liquid extraction
lm	lumen
L/min	liters per minute
LOAEL	lowest-observed-adverse-effect level
LOX	liquid oxygen
LPG	liquefied petroleum gas
LSI	Langelier saturation index
LULU	locally unacceptable land use
lx	lux
m	meter
M	molar
m^2	square meters
m^3	cubic meters
mA	milliampere
mADC	milliampere direct current
max.	maximum
MB	megabyte

MBAS	methylene blue active substances
MCL	maximum contaminant level
MCLG	maximum contaminant level goal
MDL	method detection limit
meq	milliequivalent
meq/L	milliequivalents per liter
MeV	million electron volts
MF	membrane filter; microfiltration
MFL	million fibers per liter
mg	milligram
MG	million gallons
mgd	million gallons per day
mg/L	milligrams per liter
mhp	motor horsepower
MHz	megahertz (megacycles)
μ	micron
μg	microgram
μg/L	micrograms per liter
μm	micrometer
μM	micromolar
μmhos	micromhos
μmho/cm	micromhos per centimeter
μS	microsiemens
μW	microwatt
μW-sec/cm^2	microwatt-seconds per square centimeter
mi	mile
mi^2 *or* sq mi	square miles
mil	million
mil gal	million gallons
mμ	millimicron
min	minute
min.	minimum
mJ	millijoule
MJ	megajoule
MKS	meter/kilogram/second
mL	milliliter
ML	megaliter *or* million liters
mm	millimeter
mM	millimolar
mmol	millimole
mol	mole
mol wt	molecular weight

mol/L	moles per liter
MPC	maximum permissible concentration
mph	miles per hour
MPN	most probable number
mpy	mils per year
MRDL	maximum residual disinfectant level
MRDLG	maximum residual disinfectant level goal
mS	millisiemens
MS	mass spectrometry
MSDS	material safety data sheet
m/sec/m	meters per second per meter
MSL	mean sea level
MTD	maximally tolerated dose
MTF	multiple-tube fermentation
MTZ	mass transfer zone
MUD	municipal utility district
MW	molecular weight
MWCO	molecular weight cutoff
mW-sec	megawatts per second
N	newton
NA	not applicable; not analyzed
NAS	National Academy of Science
NAWC	National Association of Water Companies
ND	not detected
NDWAC	National Drinking Water Advisory Council
NDWC	National Drinking Water Clearinghouse
NEC	National Electrical Code
NEMA	National Electrical Manufacturers Association
NEPA	National Environmental Policy Act
NEWWA	New England Water Works Association
NF	nanofiltration
NFPA	National Fire Protection Association
ng/L	nanograms per liter
NGWA	National Ground Water Association
NH	American standard fire hose coupling thread (National hose thread)
NIOSH	National Institute of Occupational Safety and Health
NIPDWR	National Interim Primary Drinking Water Regulation
nm	nanometer

NOAEL	no-observed-adverse-effect level
NOM	natural organic matter
NPDES	National Pollution Discharge Elimination System
NPDWR	National Primary Drinking Water Regulation
NPS	nominal pipe size; American standard straight pipe thread
NPSH	net positive suction head; American standard straight pipe for hose couplings (National pipe straight hose)
NPSHR	net positive suction head rate
NPSM	American standard straight pipe thread for free mechanical joints
NPT	American standard taper thread pipe (National pipe tapered)
NRWA	National Rural Water Association
NSDWR	National Secondary Drinking Water Regulation
NSFC	National Small Flows Clearinghouse
NST	American standard fire hose coupling thread (National standard thread)
NTNC	nontransient noncommunity
ntu	nephelometric turbidity unit
NWA	National Water Alliance
NWRA	National Water Resources Association

O&M	operations and maintenance
OD	outside diameter
ODM	maximum outside diameter
Ω	ohm
ORP	oxidation–reduction potential
OSHA	Occupational Safety and Health Administration
oz	ounce
ozf-in.	ounce-inch

Pa	pascal
P–A	presence–absence
PAC	powdered activated carbon
PAH	polyaromatic hydrocarbon
Pa·sec	pascal-second
PB	polybutylene
PCB	polychlorinated biphenyl
PCE	tetrachloroethylene (perchloroethylene)
pCi	picocurie

pCi/L	picocuries per liter
PCU	platinum–cobalt color unit
pE	oxidation–reduction (redox) potential
PE	polyethylene
PF	power factor
pfu	plaque-forming unit
pg	picogram
P&ID	process and instrumentation drawing
PID	proportional integral derivative control; photoionization detector
pk	peck
POE	point of entry
POTW	publicly owned treatment works
POU	point of use
ppb	parts per billion
PPE	personal protective equipment
ppm	parts per million
ppt	parts per trillion; parts per thousand
PQL	practical quantitation level
PRV	pressure-regulating valve
ps	picosecond
psi	pounds per square inch
psia	pounds per square inch absolute
psig	pounds per square inch gauge
Pt–Co	platinum–cobalt
PTFE	polytetrafluoroethylene
PVC	polyvinyl chloride
PVDF	polyvinylidene fluoride
PWD	public water district
PWL	pumping water level
PWS	public water system
QA	quality assessment
QC	quality control
qt	quart
r	roentgen
rad	radian
rad/sec	radians per second
RCRA	Resource Conservation and Recovery Act
RDL	reliable detection level
reg neg	regulatory negotiations

rem	roentgen equivalent, mammal
RMCL	recommended maximum contaminant level
RMP	risk management program
RO	reverse osmosis
rpm	revolutions per minute
rps	revolutions per second
RPZ	reduced pressure zone
RTD	resistance temperature detector
RTP	reinforced thermoset plastic
RTU	remote terminal unit
S	siemens
SARA	Superfund Amendments and Reauthorization Act
SCADA	supervisory control and data acquisition
SCBA	self-contained breathing apparatus
SCD	streaming current detector
SCFM	standard cubic feet per minute
S/cm	siemens per centimeter
SDI	sludge density index
SDR	standard dimension ratio
SDWA	Safe Drinking Water Act
sec	second
sec^{-1}	inverse seconds
SEM	scanning electron microscope
SI	Système International d'Unités (International System of Units)
SMCL	secondary maximum contaminant level
SOC	synthetic organic chemical
sp gr	specific gravity
sp ht	specific heat
SQL	Structured Query Language
sr	steradian
SSF	slow sand filtration
SUVA	specific ultraviolet absorbance
Sv	sievert
SVI	sludge volume index
SWL	static water level
SWP	State Water Plan
SWTR	Surface Water Treatment Rule
t	metric ton *or* tonne
T	tesla

TC	thermocouple
TCE	trichloroethylene (or trichloroethene)
TCLP	toxic characteristic leaching procedure
TCR	Total Coliform Rule
tcu	true color unit
TDS	total dissolved solids
TFE	tetrafluoroethylene
THM	trihalomethane
THMFP	trihalomethane formation potential
TNCWS	transient, noncommunity water system
TOC	total organic carbon
TON	threshold odor number; total organic nitrogen
TOX	total organic halogen
TPI	threads per inch
TSCA	Toxic Substances Control Act
TSP	trisodium phosphate
TSPP	tetrasodium pyrophosphate
TSS	total suspended solids
TT	treatment technique
TTHM	total trihalomethanes
TVSS	transient voltage surge suppression
uc	uniformity coefficient
UF	ultrafiltration
UFW	unaccounted-for water
UL	Underwriters Laboratories
UPS	uninterruptible power supply
URTH	unreasonable risk to health
USEPA	US Environmental Protection Agency
USPHS	US Public Health Service
UV	ultraviolet
V	volt
VA	volt-ampere
VAC	volts alternating current
VAR	volt-ampere-reactive
VDC	volts direct current
VFD	variable-frequency drive
VOC	volatile organic compound
vol.	volume
VSD	variable-speed drive

W	watt
WaterRF	Water Research Foundation
Wb	weber
WEF	Water Environment Federation
WERL	Water Engineering Research Laboratory
WFP	Water For People
WHO	World Health Organization
whp	water horsepower
WHPA	wellhead protection area
WHPP	wellhead protection program
WIDB	Water Industry Data Base
WITAF	Water Industry Technical Action Fund
WQA	Water Quality Association
WQIC	Water Quality Information Center
wt	weight
WTP	water treatment plant
WWTP	wastewater treatment plant
Xe	xenon
yd	yard
yd^2 *or* sq yd	square yards
yd^3	cubic yards
ξp *or* zp	zeta potential

Glossary

From A to Z, from absolute pressure to zone of saturation and everything in between, many terms used in the basic science—as well as the practical application of water processes and technologies—are unique to the water industry. For quick reference in the field, here is a compilation of water quantity, quality, analysis, and useage terms, along with environmental and human-health-related terms commonly used in water distribution and treatment.

absolute pressure The total pressure in a system, including both the pressure of the water and the pressure of the atmosphere (about 14.7 psi [101 kPa] at sea level).

acid Any substance that releases hydrogen ions when mixed into water.

acidic solution A solution that contains significant numbers of hydrogen ions.

acidic water Water with a pH of less than 7.0.

activated alumina The chemical compound aluminum oxide, which is used to remove fluoride and arsenic from water by adsorption.

activated carbon A highly adsorptive material used to remove organic substances from water.

activated silica A coagulant aid used to form a denser, stronger floc.

activation The process of producing a highly porous structure in carbon by exposing the carbon to high temperatures in the presence of steam.

adsorbent Any material, such as activated carbon, used to adsorb substances from water.

adsorption A physical process in which molecules adhere to a substance because of electrical charges. Used primarily to remove organic contaminants from water.

aeration The process of bringing water and air into close contact to remove or modify constituents in the water.

agar A nutrient preparation used to grow bacterial colonies in the laboratory. Agar is poured into petri dishes to form agar plates or into culture tubes to form agar slants.

agglomeration The action of microfloc particles colliding and sticking together to form larger settleable floc particles.

air binding The condition in which air has collected in the high points of distribution mains, reducing the capacity of the mains.

air gap In plumbing, the unobstructed vertical distance through the free atmosphere between (1) the lowest opening from any pipe or outlet supplying water to a tank, plumbing fixture, or other container, and (2) the overflow rim of that container.

air line A small-diameter pipe used to determine water depth in a well.

air purging A procedure to clean mains less than 4 in. (100 mm) in diameter, in which air from a compressor is mixed with the water and flushed through the main.

air release valve A small-orifice valve placed at high points of a pipeline or on a pump to automatically release small amounts of accumulated air. Also called air relief valve.

air scouring The practice of admitting air through the underdrain system to ensure complete cleaning of media during filter backwash. Normally an alternative to using a surface wash system.

air-stripping Removal of a substance from water by means of air.

air-and-vacuum relief valve A dual-function air valve that (1) permits entrance of air into a pipe being emptied, preventing a vacuum, and (2) allows air to escape in a pipe while being filled or under pressure.

alkaline water Water having a pH greater than 7.0. Also called basic water.

alluvial   Referring to a type of soil, mostly sand and gravel, deposited by flowing water.

alternating current   Electric current that flows first in one direction and then in the other. The sequence of one rise and fall in current strength in each direction is called a cycle.

altitude valve   A valve that automatically shuts off water flow when the water level in an elevated tank reaches a preset elevation, then opens again when the pressure on the system side is less than that on the tank side.

alum   The most common chemical used for coagulation (also called aluminum sulfate).

aluminum sulfate   An inorganic compound commonly used as a coagulant in water treatment. It contains waters of hydration, $Al_2(SO_4)_3 \cdot XH_2O$ (where X is a variable number). Aluminum sulfate is often called alum.

anaerobic   Characterized by the absence of air or free oxygen.

analytical balance   A sensitive balance device used to make precise weight measurements.

angle of repose   The maximum angle or slope from the horizontal that a given loose or granular material, such as sand, can maintain without caving in or sliding. Can vary considerably with changes in moisture content.

anionic polyelectrolyte   A polyelectrolyte that forms negatively charged ions when dissolved in water.

annular space   The space between the outside of a well casing and the drilled hole.

anode   The positive end (pole) of an electrolytic system.

appurtenances   Auxiliary equipment, such as valves and hydrants, attached to the distribution system to enable it to function properly.

aqueduct   A conduit, usually of considerable size, used to convey water.

aquifer   A porous, water-bearing geologic formation. Generally restricted to materials capable of yielding an appreciable supply of water. Also called groundwater aquifer.

aquifer recharge area   The land above an aquifer that contributes water to it.

arbor press   A special tool used to force a press-fitted impeller and bearings off of the pump shaft without damaging the parts.

arching   A condition that occurs when dry chemicals bridge the opening from the hopper to the dry feeder, clogging the hopper.

arterial map   A comprehensive map showing primary distribution mains 8 in. (203 mm) or larger. Generally a supplemental mapped record that is used in system analysis.

arterial-loop system   A distribution system layout involving a complete loop of arterial mains (sometimes called trunk mains or feeders) around the area being served, with branch mains projecting inward.

artesian aquifer   An aquifer in which the water is confined by both an upper and a lower impermeable layer.

artesian well   A well in which water pressure forces water up through a hole in the upper confining, or impermeable, layer of an artesian aquifer. In a flowing artesian well, the water will rise to the ground surface and flow out onto the ground.

asbestos–cement pipe   Pipe made from a mixture of asbestos fibers and cement.

aspirator   A T-shaped plumbing fixture connected to a water faucet, creating a partial vacuum for filtering operations.

atomic absorption spectrophotometer   Used to determine the concentration of metals in water and other types of samples.

atomic absorption spectrophotometric method   An analytical technique used to identify the constituents of a sample by detecting which frequencies of light the sample absorbs.

atmospheric vacuum breaker   A mechanical device consisting of a float check valve and an air-inlet port designed to prevent backsiphonage.

auxiliary tank valve   In a chlorination system, a union or yoke-type valve connected to the chlorine container or cylinder. It acts as a shutoff valve in case the container valve is defective.

average daily flow   The sum of all daily flows for a specified time period, divided by the number of daily flows added.

axial-flow pump   A pump in which a propeller-like impeller forces water out in a direction parallel to the shaft. Also called propeller pump. Compare *mixed-flow pump*, *radial-flow pump*.

backfill   (1) The operation of refilling an excavation, such as a trench, after the pipeline or other structure has been placed into the excavation. (2) The material used to fill the excavation in the process of backfilling.

backflow   A hydraulic condition, caused by a difference in pressures, in which nonpotable water or other fluids flow into a potable water system.

backpressure   A condition in which a pump, boiler, or other equipment produces a pressure greater than the water supply pressure.

backsiphonage   A condition in which the pressure in the distribution system is less than atmospheric pressure, which allows contamination to enter a water system through a cross-connection.

backwash   The reversal of flow through a filter to remove the material trapped on and between the grains of filter media.

bacterial aftergrowth   Growth of bacteria in treated water after the water reaches the distribution system.

baffle   A metal, wooden, or plastic plate installed in a flow of water to slow the water velocity and provide a uniform distribution of flow.

ball valve   A valve consisting of a ball resting in a cylindrical seat. A hole is bored through the ball to allow water to flow when the valve is open. When the ball is rotated 90°, the valve is closed.

bar screen   A series of straight steel bars welded at their ends to horizontal steel beams, forming a grid. Bar screens are placed on intakes or in waterways to remove large debris.

barrel   The body of a fire hydrant.

base   The inlet structure of a fire hydrant. An elbow-shaped piece that is usually constructed as a gray cast-iron casting. Also known as the shoe, inlet, elbow, or foot piece.

Baumé   The Baumé scale is a means of expressing the strength of a solution based on the solution's specific gravity.

bearing   An antifriction device used to support and guide pump and motor shafts.

bed life   The time it takes for a bed of adsorbent to lose its adsorptive capacity.

bedding   A select type of soil used to support a pipe or other conduit in a trench.

bell (of a pipe)   The recessed, oversized female end of a pipe into which the male, or spigot, end is inserted. Also referred to as the hub or bell end.

bell joint clamp   A ductile-iron ring with bolts and rubber gaskets designed to clamp over a bell-and-spigot joint to stop a leak.

bell-and-spigot joint   A type of joint that has an oversized bell end and a spigot end (plain end) that fits the bell and is sealed with lead or a rubber gasket.

bellows sensor   A simple, accordion-like mechanical device for sensing changes in pressure.

bench-scale study   An experimental study to evaluate the performance of unit processes performed in laboratory surroundings. See also *jar test*.

bicarbonate alkalinity   Alkalinity caused by bicarbonate ions.

biochemical oxygen demand   A measurement of the amount of oxygen used in the biochemical oxidation of organic matter over a specified time (usually 5 days) and at a specific temperature (usually 95°F [35°C]). Used to indicate the level of contamination in water or contamination potential of a waste.

biofilm   A layer of biological material that covers a surface.

blowoff valve   A valve installed in a low point or depression on a pipeline to allow drainage of the line. Also called washout valve.

body   The major part of a valve, which houses the remainder of the valve assembly.

body feed   In diatomaceous earth filters, the continuous addition of diatomaceous earth during the filtering cycle to provide a fresh filtering surface as the suspended material clogs the precoat.

bonnet   The removable top cover or closure on a hydrant's upper section.

booster disinfection   The practice of adding additional disinfectant in the distribution system.

borosilicate glass   A type of heat-resistant glass used for labware.

Bourdon tube   A semicircular tube of elliptical cross section, used to sense pressure changes.

brake horsepower   The power supplied to a pump by a motor. Compare to *water horsepower* and *motor horsepower*.

breakaway hydrant   A two-part, dry-barrel post hydrant with a coupling or other device joining the upper and lower sections.

breakpoint   The point at which the chlorine dosage has satisfied the chlorine demand.

breakpoint chlorination   The addition of chlorine to water until the chlorine demand has been satisfied and free chlorine residual is available for disinfection.

breakthrough   The point in a filtering cycle at which turbidity-causing material starts to pass through the filter.

bromine   The oxidized form of the bromide ion. In water, bromine is present as hypobromous acid and the hypobromite ion.

bronze seat ring   A machined ring, mounted in the body of a hydrant or valve, against which the moving disk of the valve closes.

brushes   Graphite connectors that rub against the spinning commutator in an electric motor or generator, connecting the rotor windings to the external circuit.

bubbler tube   A level-sensing device that forces a constant volume of air into the liquid for which the level is being measured.

buffer   A substance capable in solution of resisting a reduction in pH as acid is added.

buffering capacity   The capability of water or chemical solution to resist a change in pH.

bulk density   The weight per standard volume (usually pounds per cubic foot) of material as it would be shipped from the supplier to the treatment plant.

bushing (of a pipe fitting)   A fitting that is threaded internally and externally, which is screwed into a fitting to reduce its size.

butterfly valve   A valve whose disc rotates about a stem as it is opened and closed. When open, the disc is parallel to the pipeline and is 90° to the axis of the pipe when closed.

bypass   (1) An arrangement of pipes, conduits, gates, or valves by which the flow may be passed around an appurtenance or treatment process. (2) In cross-connection control, any pipe arrangement that passes water around a protective device, causing the device to be ineffective.

bypass valve   A small valve installed in parallel with a larger valve. Used to equalize the pressure on both sides of the disc of the larger valve before the larger valve is opened.

$C \times T$ value   The product of the residual disinfectant concentration, C, in milligrams per liter, and the corresponding disinfectant contact time, T, in minutes, or $C \times T$. Minimum $C \times T$ values are specified by the Surface Water Treatment Rule as a means of enduring adequate kill or inactivation of pathogenic microorganisms in water. Also called CT value.

C value   The Hazen–Williams roughness coefficient; a number used in the Hazen–Williams formula to determine flow capacities of pipelines.

calcium   One of the principal elements making up the earth's crust. The presence of calcium in water is a factor contributing to the formation of scale and insoluble soap curds that are a means of clearly indentifying hard water.

calcium carbonate   The principal hardness- and scale-causing compound in water.

calcium hardness   The portion of total hardness caused by calcium compounds such as calcium carbonate and calcium sulfate.

calcium hypochlorite   A chemical compound used as a bleach or disinfecting agent and a source of chlorine in water teratment. Commercial grades contain 70 percent available chlorine (99.2 percent available chlorine for the pure chemical). Calcium hypochlorite is specifically useful because it is stable as a dry powder and can be formed into pellets.

capacity   The flow rate that a pump is capable of producing.

carbon dioxide   A common gas in the atmosphere that is very soluble in water. High concentrations in water can cause the water to be corrosive. In a process known as recarbonation, carbon dioxide is added to water after the lime-softening process to lower the pH and reduce calcium carbonate scale formation.

carbonate alkalinity   Alkalinity caused by carbonate ions.

carbonate hardness   Hardness caused primarily by compounds containing carbonate, such as calcium carbonate and magnesium carbonate.

casing   (1) The enclosure surrounding a pump impeller, into which are machined the suction and discharge ports. (2) The metal pipe used to line the borehole of a well (also called well casing).

cast-iron pipe   Pipe made from pigiron, cast in a rotating, water-cooled mold.

cathode   The negative end (pole) of an electrolytic system.

cathodic protection   An electrical system for preventing corrosion to metals, particularly metallic pipe and tanks.

cation exchange   Ion exchange involving ions that have positive charges, such as calcium and sodium.

cation exchange materials   Materials that release nontroublesome ions into water in exchange for hardness-causing ions.

cationic polyelectrolyte   A polyelectrolyte that forms positively charged ions when dissolved in water.

cavitation   During cavitation, a partial vacuum forms near the pipe wall or impeller blade, causing potentially rapid pitting of the metal. Occurs when pumps are run too fast or water is forced to change direction quickly.

centrate   The water that is separated from sludge and discharged from a centrifuge.

centrifugal pump   A pump consisting of an impeller on a rotating shaft enclosed by a casing that has suction and discharge connections. The spinning impeller throws water outward at high velocity, and the casing shape converts this high velocity to a high pressure.

centrifugation   In water treatment, a method of dewatering sludge by using a mechanical device (centrifuge) that spins the sludge at a high speed.

check valve   A valve designed to open in the direction of normal flow and close with reversal of flow. An approved check valve has substantial construction and suitable materials, is positive in closing, and permits no leakage in a direction opposite to normal flow.

chelation A chemical process used to control scale formation, in which a chelating agent "captures" scale-causing ions and holds them in solution, preventing them from precipitating out and forming scale.

chemical oxidation The use of a chemical, such as chlorine or ozone, to remove or change some contaminant in water.

chemical precipitation The use of a chemical to cause some contaminant to become insoluble in water.

chemical reaction A process that occurs when atoms of certain elements are brought together and combine to form molecules, or when molecules are broken down into individual atoms.

chloramines Disinfectants produced from the mixing of chlorine and ammonia.

chlorination The process of adding chlorine to water to kill disease-causing organisms or to act as an oxidizing agent.

chlorinator Any device used to add chlorine to water.

chlorine A chemical used as a disinfectant and oxidizing agent is converted into hypochlorous acid and the hypochlorite ion; the ratio of the two substances is dependent on the pH of the solution. Chlorine is also commercially available in liquid form as a hypochlorite ion solution.

chlorine demand The quantity of chlorine consumed by reaction with substances in water.

chlorine dioxide A red-yellow gas that is very reactive and unstable. It is a strong oxidizing agent and is also used as a disinfectant. Chlorine dioxide decomposes in water to yield the chlorite ion and, to a lesser extent, the chlorate ion.

chlorine residual The concentration of chlorine remaining in solution.

circuit breaker A device that functions both as a current-overload protective device and as a switch.

clarification Any process or combination of processes that reduces the amount of suspended matter in water.

close-coupled Relating to pump assembly for which the impeller is mounted on the shaft of the motor that drives the pump. Compare *frame-mounted*.

coagulant A chemical used in water treatment for coagulation. Common examples are aluminum sulfate and ferric sulfate.

coagulant aid A chemical added during coagulation to improve the process by stimulating floc formation or by strengthening the floc so it holds together better.

coagulation The water treatment process that causes very small suspended particles to attract one another and form larger particles. Accomplished by adding a chemical, called a coagulant, that neutralizes the electrostatic charges on the particles that cause them to repel each other.

coagulation–flocculation The water treatment process that converts small particles of suspended solids into larger, more settleable clumps.

coliform group/bacteria A group of bacteria predominantly inhabiting the intestines of humans or animals, but also occasionally found elsewhere. Presence of the bacteria in water is used as an indication of fecal contamination (contamination by human or animal waste).

colloidal solid   Finely divided solid that will not settle out of water for very long periods of time unless the coagulation–flocculation process is used.

color   A physical characteristic of water. Color is most commonly tan or brown as a result of oxidized iron, but contaminants may cause other colors, such as green or blue.

color comparator   A device used for tests such as chlorine residual or pH. Concentrations of constituents are determined by visual comparison of a permanent standard (usually sealed in glass or plastic) and a water sample.

color unit   The unit of measure of the color of water, measured by comparing the color of a water sample with the color of a standard solution.

colorimeter   An instrument that measures the concentration of a constituent in a sample by measuring the intensity of color in that sample. The color is usually created by mixing a chemical reagent with the water sample according to a specific test procedure.

colorimetric method   Any analytical method that measures a constituent in water by determining the intensity of color in the water. The color is usually produced when a chemical solution specified by the particular procedure is added to the water.

combined chlorine residual   The chlorine residual produced by the reaction of chlorine with substances in the water. Because the chlorine is "combined," it is not as effective a disinfectant as free chlorine residual.

community water system   A public water system providing water to at least 15 service connections used by year-round residents or regularly serving at least 25 year-round residents.

commutator   A device that is part of the rotor of certain designs of motors and generators. The motor unit's brushes rub against the surface of the spinning commutator, allowing current to be transferred between the rotor and the external circuits.

completed test   The third major step of the multiple-tube fermentation method. Confirms that positive results from the presumptive test are the result of coliform bacteria. See also *confirmed test*; *presumptive test*.

compound meter   A water meter consisting of two single meters of different capacities and a regulating valve that automatically diverts all or part of the flow from one meter to the other. The valve senses flow rate and shifts the flow to the meter that can most accurately measure it.

compression fitting   A piping device that seals against pressure by compressing a rubber gasket.

compression-type hydrant   A hydrant that opens against the flow of water by the movement of the operating stem and in which the water pressure tends to keep the main valve closed.

concentration cell corrosion   A form of localized corrosion that can form deep pits and tubercules.

condensation   The process by which a substance changes from the gaseous form to a liquid or solid form.

cone of depression   The cone-shaped depression in the groundwater level around a well during pumping.

cone valve A valve in which the movable internal part is a cone-shaped rotating plug. The valve is opened when the plug is turned through an angle of 90°, so fluid can pass through a port machined through the plug.

confining bed A layer of material, typically consolidated rock or clay, that has very low permeability and restricts the movement of groundwater into or out of adjacent aquifers.

confirmed test The second major step of the multiple-tube fermentation method. This test confirms that positive results from the presumptive test are due to coliform bacteria. See also *completed test*; *presumptive test*.

contactor A vertical, steel cylindrical pressure vessel used to hold the activated carbon bed.

container valve The valve mounted on a chlorine container or cylinder.

contaminant Anything found in water other than hydrogen or oxygen.

contamination Any introduction into water of microorganisms, chemicals, wastes, or wastewater in a concentration that makes the water unfit for its intended use.

continuous feed method A method of disinfecting new or repaired mains in which chlorine is continuously added to the water being used to fill the pipe, maintaining a constant concentration.

conventional filtration A term that describes the treatment process used by most US surface water systems, consisting of the steps of coagulation, flocculation, sedimentation, and filtration.

corporation stop A valve for joining a service line to a street water main. Cannot be operated from the surface. Also called corporation cock.

coupling A device that connects the pump shaft to the motor shaft.

coupon In tapping, the section of the main cut out by the drilling machine.

coupon test A method of determining the rate of corrosion or scale formation by placing metal strips (coupons) of a known weight in the pipe and examining them for corrosion after a period of time.

cross-connection Any connection between a safe drinking water supply and a nonpotable water or other fluid. Also called cross contamination.

Cryptosporidium A widespread intestinal coccidian protozoan parasite about 3.5 micrometers in diameter, causing diarrhea and capable of infecting humans, birds, fish, and snakes. It is responsible for waterborne disease outbreaks.

culture tube A hollow, slender glass tube with an open top and a rounded bottom used in microbiological testing procedures such as the multiple-tube fermentation test.

curb box A cylinder placed around the curb stop and extending to the ground surface to allow access to the valve.

curb stop A shutoff valve attached to a water service line from a water main to a customer's premises, usually placed near the customer's property line. May be operated by a valve key to start or stop flow to the water supply line. Also called curb valve.

curie The activity of 1 g of radium, or 3.7×10^{10} disintegrations per second.

current meter   A device for determining flow rate by measuring the velocity of moving water. Turbine meters, propeller meters, and multijet meters are common types. Compare *positive-displacement meter*.

cut-in valve   A specially designed valve used with a sleeve that allows the valve to be placed in an existing main.

cyclone degritter   A centrifugal sand-and-grit removal device.

cyst   A resistant form of a living organism.

daily flow   The total volume of water (in gallons or liters) that passes through a plant during a 24-hour period.

dead end   A section of a water distribution system that is not connected to another section of pipe by means of a connecting loop. Such portions of a distribution system can experience lower flows than surrounding portions, which can lead to water quality problems caused by somewhat stagnant water. Examples of problems include tastes or odors, bacteriological growth, loss of chlorine residual, or any combination of these.

deionizer   A device used to remove all dissolved inorganic ions from water.

density current   A flow of water that moves through a larger body of water, such as a reservoir or sedimentation basin, and does not become mixed with the other water because of a density difference.

density stratification   The formation of layers of water in a reservoir; water is the densest at the bottom and the least dense at the surface.

design point   The mark on the head–capacity curve of a pump characteristics curve that indicates the head and capacity at which the pump is intended to operate for best efficiency in a particular installation.

destratification   The use of a method to prevent a lake or reservoir from becoming stratified. Typically consists of releasing diffused compressed air at a low point on the lake bottom.

detector-check meter   A meter that measures daily flow but allows emergency flow to bypass the meter. Consists of a weight-loaded check valve in the main line that remains closed under normal usage and a bypass around the valve containing a positive-displacement meter.

detention time   The average length of time a drop of water or a suspended particle remains in a tank or chamber. Mathematically, the volume of water in the tank divided by the flow rate through the tank.

dewatering (of reservoirs)   A physical method for controlling aquatic plants in which a water body is completely or partially drained and the plants are allowed to die.

dewatering (of sludge)   A process to remove a portion of water from sludge.

diaphragm element   A mechanical sensor used to determine liquid levels. Uses a diaphragm and an enclosed volume of air.

diaphragm-type metering pump   A pump in which a flexible rubber, plastic, or metal diaphragm is fastened at the edges in a vertical cylinder. As the diaphragm is pulled back, suction is exerted and the liquid is drawn into the pump. When it is pushed forward, the liquid is discharged.

diatomaceous earth filter   A pressure filter using a medium made from diatoms. Pumping forces the water through the diatomaceous earth.

diffuser   (1) A section of a perforated pipe or porous plates used to inject a gas, such as carbon dioxide or air, under pressure, into water. (2) A type of pump.

diffuser vanes   Vanes installed within the pump casing of diffuser centrifugal pumps to change velocity head to pressure head.

direct current   A type of electrical current, such as that produced by a battery, for which the same polarity is maintained at all times.

direct filtration   A filtration method that includes coagulation, flocculation, and filtration but excludes sedimentation. Applicable only to raw water that is relatively low in turbidity because all suspended matter must be trapped by the filters.

direct tap   The process of cutting threads directly into the wall of a pipe for insertion of a connecting valve, as opposed to installing a saddle.

directional flushing   A systematic approach to direct the flow from a clean source to the area to be flushed.

disinfectant residual   An excess of chlorine left in water after treatment. The presence of residuals indicates that an adequate amount of chlorine has been added at the treatment stage to ensure completion of all reactions with some chlorine remaining.

disinfectant/disinfection by-product   A term used in connection with state and federal regulations designed to protect public health by limiting the concentration of either disinfectants or the by-products formed by the reaction of disinfectants with other substances in the water (such as trihalomethanes).

disinfection   (1) The process of destroying or inactivating pathogenic organisms (bacteria, viruses, fungi, and protozoa) by either chemical or physical means. (2) In water treatment, the process in which water is exposed to a chemical disinfectant—chlorine, chloramines, chlorine dioxide, iodine, or ozone—for a specified time period to kill pathogenic organisms.

disinfection by-product   The new chemical compound that is formed by the reaction of disinfectants with organic compounds in water. At high concentrations, many disinfection by-products are considered a danger to human health.

displacement meter   A meter with a piston or disk that displaces water, causing a rotating assembly to register flow.

dissolved air flotation   A clarification process in which gas bubbles are generated in a basin so that they will attach to solid particles to cause them to rise to the surface.

dissolved oxygen   The oxygen dissolved in water, wastewater, or other liquid, usually expressed in milligrams per liter, parts per million, or percent of saturation.

dissolved solid   Any material that is dissolved in water and can be recovered by evaporating the water after filtering the suspended material. Also called filterable residue.

distribution main   Any pipe in the distribution system other than a service line.

distribution storage   A tank or reservoir connected with the distribution system of a water supply. Used primarily to accommodate changes in demand that occur over short periods (several hours to several days) and also to provide local storage disinfection for use during emergencies.

distribution system   The piping system, usually consisting of pipe 12 in. (305 mm) and smaller, that distributes water supply to customers.

Doppler effect   The apparent change in frequency (pitch) of sound waves resulting from the relative velocity between the source of the sound waves and the observer.

double-suction pump   A centrifugal pump in which the water enters from both sides of the impeller. Also called split-case pump.

drainage basin   An area from which surface runoff is carried away by a single drainage system. Also called catchment area or watershed drainage area.

drawdown   The amount the water level in a well drops once pumping begins. Drawdown equals static water level minus pumping water level.

drawdown method   A testing procedure that determines the characteristics of an aquifer or well.

drip leg   A small piece of pipe installed on a chlorine cylinder or container that prevents collected moisture from draining back into the container.

dry tap   A connection made to a main that is empty. Compare to *wet tap*.

dry barrel hydrant   A hydrant for which the main valve is located in the base. The barrel is pressurized with water only when the main valve is opened. When the main valve is closed, the barrel drains. This type of hydrant is especially appropriate for use in areas where freezing weather occurs.

dry top hydrant   A dry barrel hydrant in which the threaded end of the main rod and the revolving or operating nut is sealed from water in the barrel when the main valve of the hydrant is in use.

dual system   A double system of pipelines, one carrying potable water and the other carrying water of lesser quality.

dual-media filtration   A filtration method designed to operate at a higher rate by using two different types of filter media, usually sand and finely granulated anthracite.

dynamic discharge head   The difference in height measured from the pump centerline at the discharge of the pump to the point on the hydraulic grade line directly above it.

dynamic suction head   The distance from the pump centerline at the suction of the pump to the point of the hydraulic grade line directly above it. Dynamic suction head exists only when the pump is below the piezometric surface of the water at the pump suction. When the pump is above the piezometric surface, the equivalent measurement is dynamic suction lift.

dynamic suction lift   The distance from the pump centerline at the suction of the pump to the point on the hydraulic grade line directly below it. Dynamic suction lift exists only when the pump is above the piezometric surface of the water at the pump suction. When the pump is below the piezometric surface, the equivalent measurement is called *dynamic suction head*.

dynamic water system A process or system in which motion occurs, as compared to static conditions with no motion.

eddy hydrant A type of dry barrel hydrant in which the main valve closes against pressure (downward) and the barrel extends slightly below the connection to the pipe. Compare *standard compression hydrant*.

eductor A device used to mix a chemical with water. The water is forced through a constricted section of pipe (Venturi) to create a low pressure, which allows the chemical to be drawn into the stream of water.

effective height The total feet of head against which a pump must work.

efficiency The ratio of the total energy output to the total energy input, expressed as percent.

effluent Water flowing out of a structure such as a treatment plant.

ejector The portion of a chlorination system that feeds the chlorine solution into a pipe under pressure.

electrical conductivity A test that measures the ability of water to transmit electricity and indicates dissolved solids concentration.

electrode method Any analytical procedure that uses an electrode connected to a millivoltmeter to measure the concentration of constituent in water.

electrodynamic meter A device used to measure electrical power (in watts or kilowatts).

electrophotometer A photometer that uses different colored glass filters to produce wavelengths desired for analyses. Also called filter photometer.

elevation head The energy possessed per unit weight of a fluid because of its elevation above some reference point (called the reference datum). Also called position head or potential head.

empty-bed contact time The volume of the tank holding an activated carbon bed, divided by the flow rate of water. Expressed in minutes; corresponds to the detention time in a sedimentation basin.

energy grade line A line joining the elevations of the energy heads; a line drawn above the hydraulic grade line by a distance equivalent to the velocity head of the flowing water at each section along a stream, channel, or conduit. Sometimes called energy gradient line or energy line.

Enhanced Surface Water Treatment Rule A revision of the original Surface Water Treatment Rule that includes new technology and requirements to deal with newly identified problems.

epilimnion The upper, warmer layer of water in a stratified lake.

equivalent weight The weight of an element or compound that, in a given chemical reaction, has the same combining capacity as 8 g of oxygen or as 1 g of hydrogen. May vary with the reaction being considered.

Escherichia coli A bacteria of the coliform group used as a substitute for fecal coliforms in the regulations of the Total Coliform Rule.

ethylenediaminetetraacetic acid An organic chelating agent that forms very stable complexes with calcium, magnesium, and other divalent ions; is used as an analytical reagent (e.g., in hardness titration); and is in some detergents, cleaning agents, and scale preventatives.

eutrophication A process by which a lake becomes too rich in aquatic life.

evaporator A device used to increase release of chlorine gas from a container by heating the liquid chlorine.

excess-lime treatment A modification of the lime–soda ash method that uses additional lime to remove magnesium compounds.

expansion joint A fabricated pipe fitting that allows axial movement of two joined pipes.

external load Any load placed on the outside of the pipe from backfill, traffic, or other sources. Also known as superimposed load.

fecal coliform A bacteria of the coliform group indicative of fecal contamination.

feedwater Water that is added to a commercial or industrial system and subsequently used by the system, such as water that is fed to a boiler to produce steam.

ferric sulfate A chemical commonly used for coagulation.

filter agitation A method used to achieve more effective cleaning of a filter bed. Typically uses nozzles attached to a fixed or rotating pipe installed just above the filter media. Water or an air–water mixture is fed through the nozzles at high pressure to help agitate the media and break loose accumulated suspended matter. Also called auxiliary scour or surface washing.

filter backwash rate A measurement of the volume of water flowing upward (backward) through a unit of filter surface area; mathematically, the backwash flow rate divided by the total filter area.

filter loading rate A measurement of the volume of water applied to each unit of filter surface area; mathematically, the flow rate into the filter divided by the total filter area.

filter media The selected materials in a filter that form a barrier to the passage of filterable suspended solids. Filter designs include (1) loose media filters with particles lying in beds or loosely packed in column form in tank-type filters, or (2) cartridge-type filters that may contain membranes or fabric, fiber, bonded-ceramic, precoat, or cast solid-block filter media. The media used in some filters are chemically inert, such as sand, which performs only a mechanical filtration. Other filter media are mutifunctional, chemically reactive media, such as calcite, granular activated carbon, magnesia, manganese dioxide, and manganese greensand.

filter tank The concrete or steel basin that contains filter media, gravel support bed, underdrain, and wash-water troughs.

filterable residue test A test used to measure the total dissolved solids in water by first filtering out any undissolved solids and then evaporating the filtered water to dryness. The residue that remains is called filterable residue or total dissolved solids.

filtering crucible A small porcelain container with holes in the bottom, used in the total suspended solids test. Also known as Gooch crucible.

filtration A water treatment process that removes suspended matter by passing the water through a porous medium such as sand.

fire hydrant A device connected to a water main, equipped with the necessary valves and outlet nozzles for attaching a fire hose.

flaming The process of passing a flame over the end of a faucet to kill bacteria before taking a water sample for bacteriological sampling. Flaming is no longer recommended.

flange A projecting rim, edge, lip, or rib.

flanged pipe The pipe joined by bolting flanges together.

floc Collections of smaller particles (such as silt, organic matter, and microorganisms) that have come together (agglomerated) into larger, more settleable particles as a result of the coagulation–flocculation process.

flocculation A water treatment process, following coagulation, that uses gentle stirring to bring suspended particles together to form larger, more settleable clumps called floc.

floor stand A device for operating a gate valve (by hand) and indicating the extent of opening.

flotation A process for separating solids from water by using air to float the particles.

flow The general term for movement of water, commonly used to mean (imprecisely) instantaneous flow rate, average flow rate, or volume.

flow rate A measure of the volume of water moving past a given point in a given period of time. Flow rates are either instantaneous or average.

flow tube One type of primary element used in a pressure-differential meter. Measures flow velocity based on the amount of pressure drop through the tube. Similar to a Venturi tube.

flow-proportional composite A composite sample in which individual sample volumes are proportional to the flow rate at the time of sampling.

flow-proportional control A method of controlling chemical feed rates by increasing or decreasing the feed rate as the flow increases or decreases.

fluoridation The water treatment process in which a chemical is added to the water to increase the concentration of fluoride ions to an optimal level.

fluorosis The staining or pitting of the teeth resulting from excessive amounts of fluoride in the water.

fluosilicic acid A strongly acidic liquid used to fluoridate drinking water.

flush hydrant A fire hydrant with the entire barrel and head below ground elevation. The head, with operating nut and outlet nozzles, is encased in a box with a cover that is flush with the ground line. Usually a dry barrel hydrant.

foot valve A check valve placed in the bottom of the suction pipe of a pump, which opens to allow water to enter the suction pipe but closes to prevent water from passing out of it at the bottom end.

frame-mounted Relating to centrifugal pumps in which the pump shaft is connected to the motor shaft with a coupling. Compare to *close-coupled*.

free chlorine residual The residual formed once all the chlorine demand has been satisfied. The chlorine no longer combines with other constituents in the water and is "free" to kill microorganisms.

free water surface The surface of water that is in contact with the atmosphere.

friction head loss   The head lost by water flowing in a stream or conduit as the result of (1) the disturbance set up by the contact between the moving water and its containing conduit and (2) intermolecular friction.

gallons per capita per day   A measurement of the average number of gallons of water used by the average person each day in a water system; calculated by dividing the total gallons of water used each day by the total number of people using the water system.

galvanic cell   A corrosion condition created when two different metals are connected and immersed in an electrolyte such as water. Also called galvanic corrosion.

galvanic series   A listing of metals and alloys according to their corrosion potential.

gas chromatography   A technique used to measure the concentration of organic compounds in water.

gas chromatography–mass spectrophotometry   A very sophisticated analytical technique for analyzing and identifying organic compounds.

gate hydrant   A dry barrel hydrant in which the main valve is a simple gate valve, similar to one side of an ordinary rubber-faced gate valve.

gate valve   A valve in which the closing element consists of a disk that slides across an opening to stop the flow of water.

gauge pressure   The water pressure as measured by a gauge; expressed in pounds per square inch gauge.

Giardia lamblia   A protozoan that can survive in water. Causes human disease.

glass-fiber filter   Filters made of uniform glass fibers with pore sizes 0.7 to 2.7 μm. Used to filter fine particles and algae while maintaining a high flow rate.

globe valve   A valve having a round, ball-like shell and horizontal disk.

gooseneck   A flexible coupling, usually consisting of a short piece of lead on copper pipe shaped like the letter "S."

grab sample   A single water sample collected at one time from a single point.

graduated cylinder   A tall, cylindrical glass or plastic container with quantity graduation marks on the side and a pouring lip, used for measuring liquids quickly without great accuracy.

grains per gallon   A measure of the concentration of a solution. One grain per gallon equals 17.12 mg/L.

granular activated carbon   Activated carbon in a granular form, which is used in a bed, much like a conventional filter, to adsorb organic substances from water.

granular media   A material used for filtering water, consisting of grains of sand or other material.

gravel bed   The layers of gravel of specific sizes that support the filter media and help distribute the backwash water uniformly.

gravel pack   The gravel surrounding the well intake screen, artificially placed ("packed") to aid the screen in filtering out the sand of an aquifer.

gravimetric feeder   A chemical feeder that adds specific weights of dry chemical.

gravimetric procedure   Any analytical procedure that uses the weight of a constituent to determine its concentration.

grid system   A distribution system layout in which all ends of the mains are connected to eliminate dead ends.

groundwater   The subsurface water occupying the saturation zone, from which wells and springs are fed. In a strict sense, the term applies only to water below the water table. Compare to *surface water*.

Ground Water Rule   USEPA published the Ground Water Rule in the Federal Register on November 08, 2006. The purpose of the rule is to provide for increased protection against microbial pathogens in public water systems that use groundwater sources.

groundwater supply system   A water system using wells, springs, or infiltration galleries.

groundwater under the direct influence of surface water   A term used in state and federal regulations to designate groundwater sources that are considered vulnerable to contamination from surface water.

haloacetic acids   The chemicals formed as a reaction of disinfectants with contaminants in water, consisting of monochloroacetic acid, dichloroacetic acid, trichloroacetic acid, monobromoacetic acid, and dibromoacetic acid.

hardness   A characteristic of water, caused primarily by the salts of calcium and magnesium. Causes deposition of scale in boilers, damage in some industrial processes, and sometimes objectionable taste.

head   (1) A measure of the energy possessed by water at a given location in the water system, expressed in feet or meters. (2) A measure of the pressure or force exerted by water, expressed in feet or meters.

head loss   The amount of energy used by water in moving from one point to another.

helical sensor   A spiral tube used to sense pressure changes.

heterotrophic plate count   A laboratory procedure for estimating the total bacterial count in a water sample. Also called standard plate count, total plate count, or total bacterial count.

high-velocity jet   A well-screen cleaning technique using pressurized water.

hose bibb   A faucet to which a hose may be attached. Also called sill cock.

hydraulic conductivity   A measure of the ease with which water will flow through geologic formations.

hydraulic detention time   The time the water is in the system facility or system component (such as a storage tank).

hydraulic grade line   A line (hydraulic profile) indicating the piezometric level of water at all points along a conduit, open channel, or stream.

hydraulic jetting   The use of water forced through the well screen to suspend fine particles in well development.

hydraulics   The branch of science that deals with fluids at rest and in motion.

hydrochloric acid   A water-based solution of hydrogen chloride that is a strong, highly corrosive acid. Hydrochloric acid may be used as a regenerant for cation resin deionization systems operated in the hydrogren cycle. Also called muriatic acid.

hydrodynamics   The study of water in motion.

hydrogen sulfide   A toxic gas produced by the anaerobic decomposition of organic matter and by sulfate-reducing bacteria. Has a very noticeable rotten-egg odor.

hydrologic cycle   The water cycle; the movement of water to and from the surface of the earth.

hydropneumatic system   A system using an airtight tank in which air is compressed over water (separated from the air by a flexible diaphragm). The air imparts pressure to water in the tank and the attached distribution pipelines.

hydropneumatic tanks   The air-pressurized tanks used to maintain distribution system pressure.

hydrostatic pressure   The pressure exerted by water at rest (for example, in a nonflowing pipeline).

hydroxyl alkalinity   The alkalinity caused by hydroxyl ions.

hypochlorination   Chlorination using solutions of calcium hypochlorite or sodium hypochlorite.

hypochlorous acid   An acid used as a disinfectant and as a bleaching and oxidizing agent. During the disinfection of drinking water with chlorine, hypochlorous acid reacts with natural organic matter and bromide to form disinfection by-products.

hypolimnion   The lower, cooler layer of water in a stratified lake.

impeller   The rotating set of vanes that forces water through a pump.

impermeable layer   A layer that does not allow, or allows only with great difficulty, the movement of water.

impervious   Resistant to the passage of water.

impoundment   A pond, lake, tank, basin, or other space, either natural or constructed, that is used for storage, regulation, and control of water.

indicator organisms   A bacterium that does not cause disease but indicates that disease-causing bacteria may be present. See also *fecal coliform*.

indirect potable reuse   The use of water from streams that have upstream discharges of wastewater.

infiltration   The flow or movement of water through soil. Also called percolation.

infiltration gallery   A subsurface structure to receive water filtered through a streambed.

Information Collection Rule   A federal regulation requiring large water systems to collect special information to build up a database that will assist in developing new monitoring and treatment regulations.

inlet zone   The initial zone in a sedimentation basin; decreases the velocity of the incoming water and distributes it evenly across the basin.

inorganic contaminant An inorganic substance regulated by the US Environmental Protection Agency in terms of compliance monitoring for drinking water. Sometimes called an inorganic chemical.

inserting valve A shutoff valve that can be inserted by special apparatus into a pipeline while the line is in service under pressure.

instantaneous flow rate A flow rate of water measured at one particular instant, such as by a metering device, involving the cross-sectional area of the channel or pipe and the velocity of the water at that instant.

intake structure A structure or device placed in a surface water source to permit the withdrawal of water from that source.

interference fit A method of joining the pump impeller to the shaft by warming the impeller, then allowing it to cool and shrink around the shaft to provide a tight fit. Also called shrink fit.

interference substances All of the substances with which chlorine reacts before a chlorine residual can be available.

internal backflow In a pump, the leakage around the impeller from the discharge to the suction side.

internal load The load or force exerted by the water pressure on the inside of the pipe.

intersection method A method of preparing valve and hydrant maps, drawn on a very large scale, permitting valves, hydrants, and mains to be drawn to scale.

iodometric method A procedure for determining the concentration of dissolved oxygen in water. Also known as *modified Winkler method*.

ion-exchange process A process used to remove practically all hardness from water; depends on special materials known as resins. The resins trade nonhardness-causing ions (usually sodium) for the hardness-causing ions calcium and magnesium.

ion-exchange resin A beadlike material that removes ions from water, used in deionizers.

ion-exchange water softener A treatment unit that removes calcium and magnesium from water using ion-exchange resins.

iron bacteria The bacteria that use dissolved iron as an energy source. Can create serious problems in a water system because they form large, slimy masses that clog well screens, pumps, and other equipment.

isolation valve A valve installed in a pipeline to shut off flow in a portion of the pipe for the purpose of inspection or repair.

jar test A laboratory procedure for evaluating coagulation, flocculation, and sedimentation processes.

jet pump A device that pumps fluid by converting the energy of a high-pressure fluid into that of a high-velocity fluid.

lamellar plates A series of thin, parallel plates installed at 45° angles for shallow-depth sedimentation.

Langelier saturation index A numerical index that indicates whether calcium carbonate will be deposited or dissolved in a distribution system. Also used to indicate the corrosivity of water.

lantern ring A perforated ring placed around the pump shaft in the stuffing box. Water from the pump discharge is piped to the lantern ring so that it will form a liquid seal around the shaft and lubricate the packing.

lateral A smaller diameter pipe that conveys water from mains to points of use.

Leopold filter bottom A patented filter underdrain system using a series of perforated vitrified clay blocks with channels to carry the water.

lime A calcined chemical material, calcium oxide. Lime is used in lime softening and in lime–soda ash water treatment, but first it must be slaked to calcium hydroxide. Lime is also called burnt lime, calyx, fluxing lime, quicklime, or unslaked lime.

lime–soda ash method A process used to remove carbonate and noncarbonate hardness from water.

limestone contactor A treatment device consisting of a bed of limestone through which water is passed to dissolve calcium carbonate.

loading rate (1) The flow rate per unit area at which the water is passed through a filter or ion-exchange unit. (2) The maximum free chlorine, chloramine, and chlorine dioxide residual allowable in distribution system water.

magnesium One of the elements that make up the earth's crust as components of many rock-forming minerals, such as dolomite. Magnesium and calcium dissolved in water constitute hardness. The presence of magnesium in water contributes to the formation of scale and the insoluble soap curds that typify hard water.

magnesium hardness The portion of total hardness caused by magnesium compounds such as magnesium carbonate and magnesium sulfate.

magnetic flowmeter A flow-measuring device in which the movement of water induces an electrical current proportional to the rate of flow.

magnetic stirrer A device used for mixing chemical solutions in the laboratory.

main rod A rod, made of two sections, that connects the standard compression hydrant valve to the operating nut.

main valve In a dry barrel hydrant, the valve in the hydrant's base that is used to pressurize the hydrant barrel, allowing water to flow from any open outlet nozzle.

main valve assembly A standard compression-type hydrant subassembly including the lower main rod, upper valve plate, resilient hydrant valve, lower valve plate, cap nut, and bronze seat ring. Screws into a bronze subseat or directly into threads cut into the base.

manifold A pipe with several branches or fittings to allow water or gas to be discharged at several points. In aeration, manifolds are used to spray water through several nozzles.

manual solution feed   A method of feeding a chemical solution for small water systems. The chemical is dissolved in a small plastic tank, transferred to another tank, and fed to the water system by a positive-displacement pump.

maximum contaminant level   The maximum allowable concentration of a contaminant in drinking water, as established by state and/or federal regulations. Primary maximum contaminant levels are health-related and mandatory. Secondary maximum contaminant levels are related to the aesthetics of the water and are highly recommended but not required.

maximum contaminant level goal   Nonenforceable health-based goals published along with the promulgation of an maximum contaminant levels. Originally called recommended maximum contaminant levels.

measuring chamber   A chamber of known size in a positive-displacement meter; used to determine the amount of water flowing through the meter.

mechanical joint   A type of joint for ductile-iron pipe. Uses bolts, flanges, and a special gasket.

mechanical seal   A seal placed on the pump shaft to prevent water from leaking from the pump along the shaft. Also prevents air from entering the pump.

membrane filter   A filter, used for microbiological examination, made of cellulose acetate with a uniform, small pore size.

membrane filter method   A laboratory method used for coliform testing. Uses an ultrathin filter with a uniform pore size smaller than bacteria—less than 1 μm. After water is forced through the filter, the filter is incubated in a special media that promotes the growth of coliform bacteria. Bacterial colonies with a green-gold sheen indicate the presence of coliform bacteria.

membrane process   A water treatment process in which relatively pure water passes through a porous membrane while particles, molecules, or ions of unwanted matter are excluded. The membrane process used primarily for potable water treatment is reverse osmosis.

meter box   A pit-like enclosure that protects water meters installed outside of buildings and allows access for reading the meter. Also known as meter pit.

metering flume   A flow-measuring device, such as a Parshall flume, that is used to measure flow in an open channel.

metering pump   A chemical solution feed pump that adds a measured volume of solution with each stroke or rotation of the pump.

methane   A colorless, odorless, flammable gas formed by the anaerobic decomposition of organic matter. When dissolved in water, methane causes a garlic-like taste. Also called swamp gas.

methyl orange   An indicator used in the measurement of the total alkalinity of a water sample.

microfiltration   A pressure-driven membrane process that separates micrometer-diameter and submicrometer-diameter particles (down to an approximately 0.1-micrometer-diameter size) from a feed stream by using a sieving mechanism. The smallest partcle size removed is dependent on the pore size rating of the membrane.

microfloc   The initial floc formed immediately after coagulation. Composed of small clumps of solids.

microstrainer A rotating drum lined with a finely woven material such as stainless steel. Used to remove algae and small debris before they enter the treatment plant.

milk of lime The lime slurry formed when water is mixed with calcium hydroxide.

minimum-day demand The least volume per day flowing through the plant for any day of the year.

minimum-hour demand The least volume per hour flowing through a plant for any hour in the year.

minimum-month demand The least volume of water passing through the plant during a calendar month.

mixed-flow pump A pump that moves water partly by centrifugal force and partly by the lift of vanes on the liquid. The flow enters the impeller axially and leaves axially and radially. Compare to *axial-flow pump*, *radial-flow pump*.

MMO–MUG technique An approved bacteriological procedure for detecting the presence or absence of total coliforms.

modified Winkler method A modification of the standard Winkler (iodometric) method that uses an alkali-iodide-azide reagent to make the procedure less subject to interferences. See also *Winkler titration*.

molarity A measure of concentration defined as the number of moles of solute per liter of solution.

mole The quantity of a compound or element that has a weight in grams equal to the substance's molecular or atomic weight.

molecular weight The sum of the atomic weights of all the atoms in the compound. Also called formula weight.

motor horsepower The horsepower equivalent to the watts of electric power supplied to a motor. Compare to *brake horsepower* and *water horsepower*.

muffle furnace A high-temperature oven used to ignite and burn volatile solids, usually operated at temperatures near 1,112°F (600°C).

multijet meter A type of current meter in which a vertically mounted turbine wheel is spun by jets of water from several ports around the wheel.

multimedia filter A filtration method designed to operate at a high rate by utilizing three or more different types of filter media (typically silica sand, anthracite, and garnet sand).

multiple protection barriers A series of system components, each providing a barrier to contaminants entering the water supply.

multiple-tube fermentation method A laboratory method used for coliform testing. Uses a nutrient broth placed in culture tubes; gas production indicates the presence of coliform bacteria.

multiplier A number noted on the meter face, such as 10° or 100°. The reading from the meter must be multiplied by that number to provide the correct volume of water.

muriatic acid Another name for hydrochloric acid.

mutagen A substance that can change the structure of DNA, thus changing the basic blueprint for cell replication.

nanofiltration   A pressure-driven membrane separation process that generally removes substances in the nanometer size range. Its separation capability is controlled by the diffusion rate of solutes through a membrane barrier and by sieving and is dependent on the membrane type.

National Primary Drinking Water Regulation   A regulation developed under the Safe Drinking Water Act that establishes maximum contaminant levels, monitoring requirements, and reporting procedures for contaminants in drinking water that endanger human health.

negative head   A condition that can develop in a filter bed when the head loss gets too high.

negative sample   When referring to the multiple-tube fermentation or membrane filter tests, any sample that does not contain coliform bacteria. Also called absence.

nephelometric turbidimeter   An instrument that determines turbidity by measuring the amount of light scattered by turbidity in a water sample; the only instrument approved by the US Environmental Protection Agency to measure turbidity in treated drinking water. Also called nephelometer.

nephelometric turbidity unit   The unit of measure used to express the turbidity (cloudiness) in a water sample as measured by a nephelometric turbidimeter.

nomograph   A graph in which three or more scales are used to solve mathematical problems.

noncarbonate hardness   The hardness caused by the salts of calcium and magnesium.

nonionic polyelectrolyte   A polyelectrolyte that forms both positively and negatively charged ions when dissolved in water.

nonpoint source   The material entering a water body that comes from overland flow instead of out of a pipe.

nonpotable water   Water that may contain objectionable pollution, contamination, minerals, or infective agents and is considered unsafe and/or unpalatable for drinking.

nonrising-stem valve   A gate valve in which the valve stem does not move up or down as it is rotated.

nonsettleable solids   The finely divided solids, such as bacteria and fine clay particles, that will stay suspended in water for long periods of time.

nontransient noncommunity public water system   A public water system that is not a community water system and that regularly serves at least 25 of the same persons over 6 months per year.

normality   A method of expressing the concentration of a solution; the number of equivalent weights of solute per liter of solution.

nutating-disk meter   A type of positive-displacement meter that uses a hard rubber disk that wobbles (rotates) in proportion to the volume of water flowing through the meter. Also called wobble meter.

ohmmeter   An instrument for measuring the resistance of a circuit (in ohms). Usually combined with a voltmeter in test equipment.

Ohm's law An equation expressing the relationship between the potential (E) in volts, the resistance (R) in ohms, and the current (I) in amperes for electricity passing through a metallic conductor. Ohm's law is $E = I \times R$.

online turbidimeter A turbidimeter that continuously samples, monitors, and records turbidity levels in water.

operating nut A nut, usually pentagonal or square, rotated with a wrench to open or close a valve or hydrant valve. May be a single component or combined with a weather shield.

operating storage A tank supplying a given area and capable of storing water during hours of low demand, for use when demands exceed the pumps' capacity to deliver water to the district.

organobromine compound The chemical compound formed when chlorine reacts with bromine.

orifice meter A type of flowmeter consisting of a section of pipe blocked by a disk pierced with a small hole or orifice. The entire flow passes through the orifice, creating a pressure drop proportional to the flow rate.

orifice plate A type of primary element used in a pressure-differential meter, consisting of a thin plate with a precise hole through the center. Pressure drops as the water passes through the hole.

outlet nozzle A threaded bronze outlet on the upper section of a fire hydrant, providing a point of hookup for hose lines or suction hose from hydrant to pumper truck.

outlet zone The final zone in a sedimentation basin. Provides a smooth transition from the settling zone to the effluent piping.

outlet-nozzle cap The cast-iron cover that screws on to the outlet nozzle of a fire hydrant, protecting it from damage and unauthorized use.

overflow level The maximum height that water or liquid will rise in a receptacle before it flows over the overflow rim. Also known as flood level.

overflow rim The top edge of an open receptacle over which water will flow. Also known as flood rim.

overflow weir A steel or fiberglass plate designed to distribute flow evenly.

oxidation (1) The chemical reaction in which the valence of an element increases because of the loss of electrons from that element. (2) The conversion of organic substances to simpler, more stable forms by either chemical or biological means.

ozone An unstable gas that is toxic to humans and has a pungent odor. It is a more active oxidizing agent than oxygen. It is formed locally in air from lightning or in the stratosphere by ultraviolet irradiation; it inhibits penetration of ultraviolet light from the sun to the earth's surface. It also is produced in automobile engines and contributes to the formation of photochemical smog. For industrial applications, it is usually manufactured at the site of use. It serves as a strong oxidant and disinfectant in the purification of drinking water and as an oxidizing agent in several chemical processes.

ozone contactor A tank used to transfer ozone to water. A common type applies ozone under pressure through a porous stone at the bottom of the tank.

ozone generator   A device that produces ozone by passing an electrical current through air or oxygen.

packed tower   A cylindrical tank containing packing material, with water distributed at the top and airflow introduced from the bottom by a blower. Commonly called air-stripper.

packing   The rings of graphite-impregnated cotton, flax, or synthetic materials, used to control leakage along a valve stem or a pump shaft.

packing gland   A follower ring that compresses the packing in the stuffing box.

packing material   The material placed in a packed tower to provide a very large surface area over which water must pass to attain a high liquid–gas transfer.

Parshall flume   A calibrated channel for measuring the flow of liquid in an open conduit.

peak-hour demand   The greatest volume per hour flowing through a plant for any hour in the year.

pellet reactor   A conical tank, filled about halfway with calcium carbonate granules, in which softening takes place quite rapidly as water passes up through the unit.

percolation   The movement or flow of water through the pores of soil, usually downward.

permanent hardness   Another term for noncarbonate hardness, derived from the fact that the hardness-causing noncarbonate compounds do not precipitate when the water is boiled.

pH of saturation   The theoretical pH at which calcium carbonate will neither dissolve nor precipitate. Used to calculate the Langelier saturation index.

phenanthroline method   A colorimetric procedure used to determine the concentration of iron in water.

phenolphthalein indicator   A chemical color-changing indicator used in several tests, including tests for alkalinity, carbon dioxide, and pH.

photometer   An instrument used to measure the intensity of light transmitted through a sample or the degree of light absorbed by a sample.

pi   The ratio of the circumference of a circle to the diameter of that circle, approximately equal to 3.14159, or about 22/7.

piezometer   An instrument that measures pressure head in a conduit, tank, or soil by determining the location of the free water surface.

piezometric surface   The surface that coincides with the static water level in an artesian aquifer.

pig   A bullet-shaped polyurethane foam plug, often with a tough, abrasive external coating, used to clean pipelines. Forced through the pipeline by water pressure.

pilot filter   A small tube, containing the same media as treatment plant filters, through which flocculated plant water is continuously passed, with a recording turbidimeter continuously monitoring the effluent.

pilot valve   The control mechanism on an automatic altitude- or pressure-regulating valve.

pipe lateral system   A filter underdrain system using a main pipe (header) with several smaller perforated pipes (laterals) branching from it on both sides.

piston meter   A water meter of the positive-displacement type, generally used for pipeline sizes of 2 in. (50 mm) or less, in which the flow is registered by the action of an oscillating piston.

piston pump   A positive-displacement pump that uses a piston moving back and forth in a cylinder to deliver a specific volume of liquid being pumped.

pitometer   A device operating on the principle of a Pitot tube, principally used for determining velocity of flowing fluids at various points in a water distribution system.

Pitot tube   A device for measuring the velocity head of the stream as an indicator of velocity. Consists of a small tube pointed upstream, connected to a gauge on which the velocity head may be measured. Also called Pitot gauge.

plain sedimentation   The sedimentation of suspended matter without the use of chemicals or other special means.

plat   A map showing street names, mains, main sizes, numbered valves, and numbered hydrants for the plat-and-list method of setting up valve and hydrant maps.

plat-and-list method   A method of preparing valve and hydrant maps. The plat is the map position, showing mains, valves, and hydrants. The list is the text portion, which includes appropriate information for items on the plat.

platinum–cobalt method   A procedure used to determine the amount of color in water.

plug valve   A valve in which the movable element is a cylindrical or conical plug.

point source   The wastewater coming from a discharge pipe.

point-of-use treatment   A water treatment device used by a water customer to treat water at only one point, such as at a kitchen sink. Also sometimes used interchangeably with point-of-entry treatment to cover all treatment installed on customer services.

polyelectrolyte   A high molecular weight, synthetic organic compound that forms ions when dissolved in water. Also called polymer.

polystyrene resin   The most common resin used in the ion-exchange process.

porous plate   A filter underdrain system using ceramic plates supported above the bottom of the filter tank. Often used without a gravel layer so that the plates are directly beneath the filter media.

positive sample   In reference to the multiple-tube fermentation or membrane filter test, any sample that contains coliform bacteria. Also called presence.

positive-displacement meter   A meter that measures the quantity of flow by recording the number of times a known volume is filled and empties. Primarily used for low flows. Two common styles: disk type and piston type.

positive-displacement pump A pump that delivers a specific volume of liquid for each stroke of the piston or rotation of the impeller.

post hydrant A fire hydrant with an upper section that extends at least 24 in. (600 mm) above the ground.

potable Safely drinkable.

potassium permanganate A treatment chemical for iron, manganese, taste, and odors.

potential cross-connection Any arrangement of pipes, fittings, or devices that indirectly connects a potable water supply to a nonpotable source. Also known as indirect cross-connection.

potentiometric method Any laboratory procedure that measures a difference in electric potential (voltage) to indicate the concentration of a constituent in water.

pounds per square inch absolute The sum of gauge pressure and atmospheric pressure.

pounds per square inch gauge The pressure measured by a gauge and expressed in terms of pounds per square inch. See also *gauge pressure*. Compare with *pounds per square inch absolute*.

powdered activated carbon Activated carbon in a fine powder form. Added to water in a slurry form primarily for removing organic compounds that cause tastes and odors.

precipitate (1) A substance separated from a solution or suspension by a chemical reaction. (2) To form such a substance.

precoating The initial step in diatomaceous earth filtration, in which a thin coat of diatomaceous earth is applied to a support surface called a septum, providing an initial layer of media for the water to pass through.

precursor compound Any of the organic substances that react with chlorine to form trihalomethanes (and other disinfection by-products).

presedimentation A preliminary treatment process used to remove gravel, sand, and other gritty material from the raw water before it enters the main treatment plant. Usually done without using coagulating chemicals.

presedimentation impoundment A large earthen or concrete basin used for presedimentation of raw water. Also useful for storage and for reducing the impact of raw-water quality changes on water treatment processes.

presence–absence test An approved bacteriological procedure for detecting total coliforms.

pressure differential The difference in pressure between two points in a hydraulic device or system.

pressure head A measurement of the amount of energy in water resulting from water pressure.

pressure vacuum breaker A device designed to prevent backsiphonage, consisting of one or two independently operating, spring-loaded check valves and an independently operating, spring-loaded air-inlet valve.

pressure zone map A map showing zones of equal pressure. Sometimes called water gradient contour map.

pressure-differential meter   Any flow-measuring device that creates and measures a difference in pressure proportionate to the rate of flow. Examples include the Venturi meter, orifice meter, and flow nozzle.

pressure-reducing valve   A valve with a horizontal disk for automatically reducing water pressures in a main to a preset value.

pressure-relief valve   A valve that opens automatically when the water pressure reaches a preset limit to relieve the stress on a pipeline.

pressure-sand filter   A sand filter placed in a cylindrical steel pressure vessel. The water is forced through the media under pressure.

prestressed concrete   The reinforced concrete placed in compression by highly stressed, closely spaced, helically wound wire. Prestressing permits the concrete to withstand tension forces.

presumptive test   The first major step in the multiple-tube fermentation test. Step presumes (indicates) the presence of coliform bacteria on the basis of gas production in nutrient broth after incubation. See also *completed test*; *confirmed test*.

pretreatment   Any physical, chemical, or mechanical process used before the main water treatment processes. Can include screening, presedimentation, and chemical addition.

primacy   The acceptance by states of the task of enforcing the Safe Drinking Water Act. Also known as primary enforcement responsibility.

Primary Drinking Water Regulation   The regulation on drinking water quality considered essential for public health preservation.

primary element   The part of a pressure-differential meter that creates a signal proportional to the water velocity through the meter.

priming   The action of starting the flow in a pump or siphon.

propeller meter   A meter for measuring (1) flow rate by measuring the speed at which a propeller spins, and hence (2) the velocity at which the water is moving through a conduit of known cross-sectional area.

proportional meter   Any flowmeter that diverts a small portion of the main flow and measures the flow rate of that portion as an indication of the rate of the main flow. The rate of the diverted flow is proportional to the rate of the main flow.

public notification   A required notice to the public given by water systems that violate operating, monitoring, or reporting requirements.

public water system   As defined by the Safe Drinking Water Act, any system, publicly or privately owned, that serves at least 15 service connections for 60 days out of the year or serves an average of 25 people at least 60 days out of the year.

pulse-duration modulation   An analog type of telemetry-signaling protocol in which the time that a signal pulse remains on varies with the value of the parameter being measured.

pump characteristic curve   A curve or curves showing the interrelation of speed, dynamic head, capacity, brake horsepower, and efficiency of a pump.

pumper outlet nozzle A large fire-hydrant outlet, usually 4.5 in. (114.0 mm) in diameter, used to supply the suction hose for fire department pumpers. Sometimes called a steamer outlet nozzle because it was originally used to supply steam-driven fire engines.

pumping water level The water level measured when the pump is in operation.

purchased water system A water system that purchases water from another water system. Generally provides only distribution and minimal treatment.

push-on joint The joint commonly used for ductile-iron, asbestos–cement, and polyvinyl chloride piping systems. One side of the joint has a bell with a specially designed recess to accept a rubber ring gasket; the other has a beveled-end spigot.

quicklime Another name for calcium oxide, which is used in water softening and stabilization. See also *lime*.

racking A condition in which a pump is subjected to frequent start–stop operations because of pressure surges affecting the pump controller. Can also result from a malfunctioning controller.

radial flow The flow that moves across a basin from the center to the outside edges or vice versa.

radial well A very wide, relatively shallow caisson that has horizontally drilled wells with screen points at the bottom.

radial-flow pump A pump that moves water by centrifugal force, spinning the water radially outward from the center of the impeller. Compare *axial-flow pump* and *mixed-flow pump*.

rapid mixing The process of quickly mixing a chemical solution uniformly through the water.

rate-of-flow controller A control valve used to maintain a fairly constant flow through the filter.

reactivate To remove the adsorbed materials from spent activated carbon and restore the carbon's porous structure so that it can be used again.

recarbonation (1) The process of adding carbon dioxide as a final stage in the lime–soda ash softening process to convert carbonate to bicarbonates. Prevents precipitation of carbonates in the distribution system. (2) The reintroduction of carbon dioxide into the water, either during or after lime–soda ash softening, to lower the pH of the water.

recharge The addition of water to the groundwater supply from precipitation and by infiltration from surface streams, lakes, reservoirs, and snowmelt.

recovery method A procedure for aquifer evaluation that measures how quickly water levels return to normal after pumping.

rectilinear flow The uniform flow in a horizontal direction.

red water The rust-colored water resulting from the formation of ferric hydroxide from iron naturally dissolved in the water or from the action of iron bacteria.

regeneration   The process of reversing the ion-exchange softening reaction of ion-exchange materials. Hardness ions are removed from the used materials and replaced with nontroublesome ions, rendering the materials fit for reuse in the softening process.

regeneration rate   The flow rate per unit area of an ion-exchange resin at which the regeneration solution is passed through the resin.

regulatory negotiation process   A US Environmental Protection Agency process that draws on the experience of many water professionals to negotiate the various issues in preparing a new draft regulation for public comment.

reject water   The water that does not pass through a membrane, carries away the rejected matter, and must be disposed.

reservoir   (1) Any tank or basin used for the storage of water. (2) A ground-level storage tank for which the diameter is greater than the height.

residual flow control   A method of controlling the chlorine feed rate based on the residual chlorine after the chlorine feed point.

residual pressure   The pressure remaining in the mains of a water distribution system when a specified rate of flow, such as that needed for fire fighting, is being withdrawn from the system.

resilient hydrant valve   A fire-hydrant valve made of resilient materials to ensure effective shutoff.

resilient-seated gate valve   A gate valve with a disk that has a resilient material attached to it to allow leak-tight shutoff at high pressure.

resin   The synthetic, bead-like material used in the ion-exchange process.

reuse water   The wastewater treated to make it useful.

reverse osmosis   A pressure-drive process in which almost-pure water is passed through a semipermeable membrane. Water is forced through the membrane and most ions (salts) are left behind. Principally used for desalination of seawater.

riser   The vertical supply pipe to an elevated tank.

rotameter   A flow measurement device used for gases.

rotary pump   A type of positive-displacement pump consisting of elements resembling gears that rotate in a close-fitting pump case.

routine (required) sample   A sample required by the National Primary Drinking Water Regulations to be taken at regular intervals to determine compliance with the maximum contaminant levels.

rule of continuity   A physical rule that states that the flow (Q) that enters a system must also be the flow that leaves the system. Mathematically stated as $Q_1 = Q_2$ or (because $Q = AV$) $A_1 V_1 = A_2 V_2$.

rural water system   A water system that has been established to serve widely spaced homes and communities in areas having no available groundwater or water of very poor quality.

saddle   A device attached around a main to hold the corporation stop. Used with mains that have thinner walls to prevent leakage. Also called service clamp.

Safe Drinking Water Act A federal law enacted Dec. 16, 1974, which set up a cooperative program among local, state, and federal agencies to ensure safe drinking water for consumers.

safe yield The maximum dependable water supply that can be withdrawn continuously from a surface water or groundwater supply during a period of years in which the driest period or period of greatest deficiency in water supply is likely to occur.

salmonellosis A disease that affects the intestinal tract and is caused by pathogenic bacteria.

saltwater intrusion The invasion of an aquifer by salt water because of overpumping of a well.

sand barrier A layer of gravel around the curb of a dug well.

sand boil The violent washing action in a filter caused by uneven distribution of backwash water.

sand trap An enlargement of a conduit carrying raw water that allows the water velocity to slow down so that sand and other grit can settle.

saturation A stable condition of water in which the water will neither deposit scale nor cause corrosion.

saturation point The point at which a solution can no longer dissolve any more of a particular chemical. Precipitation of the chemical will occur beyond this point.

saturator A piece of equipment that feeds a sodium fluoride solution into water for fluoridation.

schmutzdecke The layer of solids and biological growth that forms on top of a slow sand filter, allowing the filter to remove turbidity effectively without chemical coagulation.

screening A pretreatment method that uses coarse screens to remove large debris from the water to prevent clogging of pipes or channels to the treatment plant.

seat The portion of a valve that the disk compresses against to achieve valve shutoff.

Secondary Drinking Water Regulation A regulation developed under the Safe Drinking Water Act that establishes maximum levels for substances affecting the taste, odor, or color (aesthetic characteristics) of drinking water.

sectional map A map that provides a detailed picture of a portion (section) of the distribution system.

sedimentation The water treatment process that involves reducing the velocity of water in basins so that the suspended material can settle out by gravity.

sedimentation basin A basin or tank in which water is retained to allow settleable matter, such as floc, to settle by gravity. Also called settling basin, settling tank, or sedimentation tank.

selective absorption A method used in gas chromatography to separate organic compounds so their concentrations can be determined.

sequestering A chemical reaction in which certain chemicals (sequestering or chelating agents) "tie up" other chemicals, particularly metal ions, so that the chemicals no longer react. Used to prevent the formation of precipitates or other compounds.

sequestering agent A chemical compound that chemically ties up (sequesters) other compounds or ions so that they cannot be involved in chemical reactions.

service connection The portion of the service line from the utility's water main to the curb stop at or adjacent to the street line or the customer's property line.

service line The pipe (and all appurtenances) that runs between the utility's water main and the customer's place of use, including fire lines.

service outlet A device used for releasing water at a dam for downstream uses.

service valve A valve, such as a corporation stop or curb stop, that is used to shut off water to individual customers.

settleability test A determination of the settleability of solids in a suspension by measuring the volume of solids settled out of a measured volume of sample in a specified interval of time, usually reported in milliliters per liter.

settling zone The zone in a sedimentation basin that provides a calm area so that the suspended matter can settle.

shaft (1) The bearing-supported rod in a pump, turned by the motor, on which the impeller is mounted. (2) The portion of a butterfly valve attached to the disk and a valve actuator. The shaft opens and closes the disk as the actuator is operated.

shaft bearing The corrosion-resistant bearing that fits around the shaft on a butterfly valve to reduce friction when the shaft turns.

shallow-depth sedimentation A modification of the traditional sedimentation process using inclined tubes or plates to reduce the distance that the settling particles must travel to be removed.

shielding A method to protect workers against cave-ins through the use of a steel box open at the top, bottom, and ends. Allows the workers to work inside the box while installing water mains.

short-circuiting A hydraulic condition in a basin in which the actual flow time of water through the basin is less than the design flow time (detention time).

shutoff A valve that can close off the source of water or a section of a distribution system in a water system.

silt stop A device placed at the outlet of water storage tanks to prevent silt or sediment from reaching the consumer.

siltation The accumulation of silt (small soil particles between 0.00016 and 0.0024 in. [0.004 and 0.061 mm] in diameter) in an impoundment.

single-suction pump A centrifugal pump in which the water enters from only one side of the impeller. Also called end-suction pump.

slaker The part of a quicklime feeder that mixes the quicklime with water to form hydrated lime (calcium hydroxide).

slaking The addition of water to quicklime to form calcium hydroxide, which can then be used in the softening or stabilization processes.

slip (1) In a pump, the percentage of water taken into the suction end that is not discharged because of clearances in the moving unit. (2) In a motor, the difference between the speed of the rotating magnetic field produced by the stator and the speed of the rotor.

sloping A method of preventing cave-ins that involves excavating the sides of the trench at an angle (the angle of repose) to make the sides stable.

slow sand filtration A filtration process that involves passing raw water through a bed of sand at low velocity, resulting in particulate removal by physical and biological mechanisms.

sludge The accumulated solids separated from water during treatment.

sludge blowdown The controlled withdrawal of sludge from a solids-contact basin to maintain the proper level of settled solids in the basin.

sludge zone The bottom zone of a sedimentation basin. Receives and stores the settled particles.

slug method A method of disinfecting new or repaired water mains in which a high dosage of chlorine is added to a portion of the water used to fill the pipe.

sluice gate A single, movable gate mounted in a frame, used in open channels or conduits to regulate flow.

sodium fluoride A dry chemical used in the fluoridation of drinking water. Commonly used in saturators.

sodium hypochlorite A substance, commonly known as liquid bleach, that is used for disinfection as an alternative to chlorine gas, especially where safety concerns over storage of the gas exist. When the hypochlorite is added to water, it hydrolyzes to form hypochlorous acid, the same active ingredient that occurs when chlorine gas is used. See also *chlorine, hypochlorous acid*.

sodium silicofluoride A dry chemical used in the fluoridation of drinking water.

softening The water treatment process that removes calcium and magnesium, the hardness-causing constituents in water.

sole-source aquifer The single water supply available in an area; also a US Environmental Protection Agency designation for water protection.

solids-contact basin A basin in which the coagulation, flocculation, and sedimentation processes are combined. Used primarily in the lime softening of water. Also called upflow clarifier or sludge-blanket clarifier.

solids-contact process A process combining coagulation, flocculation, and sedimentation in one treatment unit in which the flow of water is vertical.

sonic meter A meter that sends sound pulses alternately in opposite diagonal directions across the pipe. The difference between the frequency of the sound signal traveling with the flow of water and the signal against the flow of water is an accurate indication of the water's velocity.

SPADNS method A colorimetric procedure used to determine the concentration of fluoride ion in water.

specific capacity A measurement of the well yield per unit (usually per foot) of drawdown. Mathematically, it is the well yield divided by the drawdown.

specific gravity   The ratio of the density of a substance to a standard density. For solids and liquids, the density is compared to the density of water (62.4 lb/ft³). For gases, the density is compared to the density of air (0.075 lb/ft³).

specific ion meter   A sensitive voltmeter used to measure the concentration of specific ions, such as fluoride in the water.

specific yield   A measure of well yield per unit of drawdown.

specific-capacity method   A testing method for determining the adequacy of an aquifer or well.

spectrophotometer   A photometer that uses a diffraction grating or a prism to control the light wavelengths used for specific analysis.

split-tee fitting   A special sleeve that is bolted around a main to allow a wet tap to be made. Also called tapping sleeve.

spray tower   A tower built around a spray aerator to keep the wind from blowing the spray and to prevent the water from freezing during cold temperatures.

spring line   The horizontal centerline of a pipe.

squirrel-cage induction motor   The most common type of induction electric motor. Rotor consists of a series of aluminum or copper bars parallel to the shaft, resembling a squirrel cage. Also known as split-phase motor.

stability   A measure of a water's tendency to corrode pipes or deposit scale in pipes.

stabilization   The water treatment process intended to reduce the corrosive or scale-forming tendencies of water.

standard compression hydrant   A type of dry barrel hydrant in which the main valve closes upward with the water pressure, creating a positive seal. Compare to *eddy hydrant*.

standpipe   A ground-level water storage tank for which the height is greater than the diameter.

static discharge head   The difference in height between the pump centerline and the level of the discharge-free water surface.

static mixer   A device designed to produce turbulence and mixing of chemicals with water, by means of fixed sloping vanes within the unit, without the need for any application of power.

static suction head (lift)   The difference in elevation between the pump centerline and the free water surface of the reservoir feeding the pump.

static water level   The water level in a well measured when no water is being taken from the aquifer, either by pumping or by free flow.

sterilization   The removal of all life from water, as contrasted with *disinfection*.

strain gauge   A type of pressure sensor that is commonly used in modern instrumentation systems, consisting of a thin, flexible sheet with imbedded electrical conducting elements.

streaming current detector   An online instrument used to adjust coagulant dosage at a water treatment plant. Within the instrument a streaming current is generated by charged particles in the water; this current indicates the degree of destabilization of the particles.

streaming current monitor   An instrument that passes a continuous sample of coagulated water past a streaming current detector.

stringer   The horizontal member of a shoring system, running parallel to the trench, to which the trench braces are attached.

stringing (hydrants)   The practice of dropping a weighted string down the barrel of a hydrant to check if the barrel has fully drained.

stuffing box   A portion of the pump casing through which the shaft extends and in which packing or a mechanical seal is placed to prevent leakage.

submersible pump   A vertical turbine pump with the motor placed below the impellers, designed to be submerged in water.

suction lift   The condition existing when the source of water supply is below the centerline of the pump.

supervisory control and data acquisition   A methodology involving equipment that both acquires data on an operation and provides limited to total control of equipment in response to the data.

surface overflow rate   A measurement of the amount of water leaving a sedimentation tank per unit of tank surface area. Mathematically, it is the flow rate from the tank divided by the tank surface area.

surface runoff   (1) That portion of the runoff of a drainage basin that has not percolated beneath the surface after precipitation. (2) The water that reaches a stream by traveling over the soil surface or by falling directly into the stream channels, including not only the large permanent streams but also the tiny rills and rivulets.

surface water   All water on the surface, as distinguished from *groundwater*.

surface water system   A water system using water from a lake or stream for its supply.

Surface Water Treatment Rule   A federal regulation established by the US Environmental Protection Agency under the Safe Drinking Water Act that imposes specific monitoring and treatment requirements on all public drinking water systems that draw water from a surface water source.

surge pressure   A momentary increase of water pressure in a pipeline resulting from a sudden change in water velocity or direction of flow.

surge tank   In cross-connection control, the receiving, nonpressurized storage vessel immediately downstream of an air gap.

surging and bailing   A method used to develop a well by alternately increasing and decreasing water pressure against the walls of the well to dislodge small particles.

suspended solids   Solid organic and inorganic particles that are held in suspension by the action of flowing water and are not dissolved. See also *total suspended solids*.

swab   A polyurethane foam plug, similar to a pig but more flexible and less durable.

synthetic organic chemical   A carbon-containing chemical that has been manufactured, as opposed to occurring in nature.

tablet method A method of disinfecting new or repaired water mains in which calcium hypochlorite tablets are placed in a section of pipe. As the water fills the pipe, the tablets dissolve, producing a chlorine concentration in the water.

tachometer generator A sensor for measuring the rotational speed of a shaft.

tapping The process of connecting laterals and service lines to mains and/or other laterals.

tapping sleeve A split fitting designed to fit over an existing water main to allow a connection to be made with a tapping machine while the main is under pressure.

tapping valve A special shutoff valve used with a tapping sleeve.

temporary hardness Another term for carbonate hardness, derived from the fact that the hardness-causing carbonate compounds precipitate when water is heated.

tensile strength A measure of the ability of pipe or other material to resist breakage when it is pulled lengthwise.

terminal head loss The head loss in a filter at which water can no longer be filtered at the desired rate because the suspended matter fills the voids in the filter and greatly increases the resistance to flow (head loss).

thermal stratification The layering of water as a result of temperature differences.

thermistor A semiconductor type of sensor that measures temperature.

thermocline The temperature transition zone in a stratified lake, located between the epilimnion and the hypolimnion.

thermocouple A sensor, made of two wires of dissimilar metals, that measures temperature.

three-phase power Alternating current power in which the current flow reaches three peaks in each direction during each cycle.

threshold odor The minimum odor of a water sample that can just be detected after successive dilutions of odorless water.

threshold odor number A numerical designation of the intensity of odor in water.

throttling The act of opening or closing a valve to control the rate of flow. Usually used to describe closing the valve.

till A type of soil consisting of a mix of clay, sand, and gravel.

time composite A composite sample consisting of several equal-volume samples taken at specified times.

titration A method of analyzing the composition of a solution by adding known amounts of a standardized solution until a given reaction or end point (color change, precipitation, or conductivity change) is produced.

total alkalinity The combined effect of hydroxyl alkalinity, carbonate alkalinity, and bicarbonate alkalinity.

Total Coliform Rule A regulation that became effective Dec. 31, 1990, doing away with the previous maximum contaminant level relating to the density of organisms and relating only to the presence or absence of the organisms in water.

total coliform test   Refers to either the multiple-tube fermentation test or the membrane filter test. Both tests indicate the presence of the entire coliform group or total coliforms.

total dissolved solids   The weight per unit volume of solids remaining after a sample has been filtered to remove suspended and colloidal solids. The solids passing the filter are evaporated to dryness. The filter pore diameter and evaporation temperature are frequently specified.

total dynamic head   The difference in height between the hydraulic grade line on the discharge side of the pump and the hydraulic grade line on the suction side of the pump; a measure of the total energy that a pump must impart to the water to move it from one point to another.

total organic carbon   The results of a general analysis performed on a water sample to determine the total organic content of the water.

total static head   The total height that the pump must lift the water when moving it from one point to another.

total suspended solids   A measure of all suspended solids in a liquid. A well-mixed sample is filtered through a standard glass fiber filter, and the residue retained on the filter is dried to a constant weight at 217°F to 221°F (103°C to 105°C). The increase in the weight of the filter represents the total suspended solids.

total trihalomethanes   The total of the concentration of all the trihalomethane compounds found in the analysis of a water sample.

totalizer   A device for indicating the total quantity of flow through a flowmeter. Also called integrator.

tracer study   A study using a substance that can readily be identified in water (such as a dye) to determine the distribution and rate of flow in a basin, pipe, or channel.

traffic load   The load placed on a buried pipe by the traffic traveling over it.

transect   An imaginary line along which samples are taken at specified intervals. Transect sampling is usually done on large bodies of water such as rivers and lakes.

transition coupling   A pipe coupling for joining pipes of the same nominal size but with different outside diameters.

transmission channel   In a telemetry system, the wire, radio wave, fiber-optic line, or microwave beam that carries the data from the transmitter to the receiver.

transmission line   The pipeline or aqueduct used for water transmission (i.e., movement of water from the source to the treatment plant and from the plant to the distribution system).

transpiration   The process by which water vapor is lost to the atmosphere from living plants.

travel-stop nut   A nut, used in dry barrel hydrants, that is screwed on to the threaded section of the main rod. It bottoms at the base of the packing plate, or revolving nut, and terminates downward travel (opening) of the hydrant valve.

tree system   A distribution system layout that centers on a single arterial main, which decreases in size with length. Branches are taken off at right angles, with sub-branches from each branch.

trench brace   The horizontal member of a shoring system that runs across a trench, attached to the stringers.

trihalomethane   A compound formed when natural organic substances from decaying vegetation and soil (such as humic and fulvic acids) react with chlorine.

trivalent ion   An ion having three valence charges, which can be either positive or negative.

tube settlers   A series of plastic tubes about 2 in. (50 mm) square, used for shallow-depth sedimentation.

tuberculation   The growth of nodules (tubercules) on the pipe interior, which reduces the inside diameter and increases the pipe roughness.

tubercules   The knobs of rust formed on the interior of cast-iron pipes by the corrosion process.

turbidimeter   An instrument that measures the amount of light impeded or scattered by suspended particles in a water sample, using a standard suspension as a reference.

turbidity   A physical characteristic of water, caused by the presence of suspended matter, that makes the water appear cloudy.

turbine meter   A meter that measures flow rates by measuring the speed at which a turbine spins in water, indicating the velocity at which the water is moving through a conduit of known cross-sectional area.

turbine pump   (1) A centrifugal pump in which fixed guide vanes (diffusers) partially convert the velocity energy of the water into pressure head as the water leaves the impeller. (2) A regenerative turbine pump.

turnover   The vertical circulation of water in large water bodies caused by the mixing effects of temperature changes and wind.

ultratfiltration   A pressure-driven membrane process that separates sub-micron particles (down to a 0.01-micrometer size or less) and dissolved solutes (down to a molecular weight cutoff of approximately 1,000 daltons) from a feed stream by using a sieving mechanism that is dependent on the pore size rating of the membrane.

ultrasonic flowmeter   A water meter that measures flow rate by measuring the difference in the velocity of sound beams directed through the water.

ultraviolet disinfection   Disinfection using an ultraviolet light.

underdrain   The bottom part of a filter that collects the filtered water and uniformly distributes the backwash water.

uniform corrosion   A form of corrosion that attacks a material at the same rate over the entire area of its surface.

unreasonable risk to health   A determination that must be made before a water system can be granted a variance or an exemption. To determine that there is no unreasonable risk to health, factors including the degree to which the maximum contaminant level is exceeded, the adverse health effects involved, the duration of the problem and the expected exceedance, and the type of population exposed are considered.

unstable   Corrosive or scale forming.

upper valve plate   A portion of the main valve assembly of a standard compression-type hydrant that closes against the seat.

vacuum   Any absolute pressure that is less than atmospheric (i.e., less than 14.7 psi [101 kPa] at sea level).

vacuum breaker   A mechanical device that prevents backflow by using a siphoning action created by a partial vacuum to allow air into the piping system to break the vacuum.

vacuum pump   A pump used to provide a partial vacuum, needed for filtering operations such as the membrane filter test.

valve box   A metal, concrete, or composite box or vault set over a valve stem at ground surface to allow access to the stem so that the valve can be opened and closed.

valve key   A metal wrench with a socket to fit a valve-operating nut, which is inserted into the valve box to operate the valve.

valve stem   The rod used to open or close a valve.

valve-and-hydrant map   A mapped record that pinpoints the location of valves throughout the distribution system. Generally of plat-and-list or intersection type.

Van der Waals's force   The attractive force existing between colloidal particles that allows the coagulation process to take place.

vault   An underground structure, normally made of concrete, that houses valves and other appurtenances.

velocity head   A measurement of the amount of energy in water that results from its velocity, or motion.

velocity meter   A meter that measures water velocity by using a rotor with vanes (such as a propeller). Operates on the principle that the vanes move at about the same velocity as the flowing water.

velocity pump   The general class of pumps that use a rapidly turning impeller to impart kinetic energy or velocity to fluids. The pump casing then converts this velocity head, in part, to pressure head. Also known as kinetic pump.

Venturi   An hourglass-shaped device based on the hydraulic principle that states that as the velocity of fluid flow increases, the pressure decreases. Used in a multitude of ways to measure flow, to feed chemicals, and to pump water.

Venturi meter   A pressure-differential meter used for measuring flow of water or other fluids through closed conduits or pipes, consisting of a Venturi tube and a flow-registering device.

Venturi tube   A type of primary element used in a pressure-differential meter that measures flow velocity based on the amount of pressure drop through the tube. Also used in a filter rate-of-flow controller.

vertical turbine pump   A centrifugal pump, commonly of the multistage diffuser type, in which the pump shaft is mounted vertically.

viscosity   The resistance of a fluid to flowing that results from internal molecular forces.

viscous   Having a sticky quality.

volatile   Capable of turning to vapor (evaporating) easily.

volatile organic compound   A class of manufactured, synthetic chemicals, generally used as industrial solvents. Classified as known or suspected carcinogens or as causing other adverse health effects.

voltmeter   An instrument for measuring electromotive force (electrical pressure), which is expressed in volts.

volumetric feeder   A chemical feeder that adds specific volumes of dry chemical.

volute   The expanding section of a pump casing (in a volute centrifugal pump) that converts velocity head to pressure head.

wash-water trough   A trough placed above the filter media to collect the backwash water and carry it to the drainage system.

water audit   A procedure that combines flow measurements and listening surveys in an attempt to give a reasonably accurate accounting of all water entering and leaving a system.

water hammer   The potentially damaging slam, bang, or shudder that occurs in a pipe when a sudden change in water velocity (usually as a result of too rapidly starting a pump or operating a valve) creates a great increase in water pressure.

water horsepower   The portion of the power delivered to a pump that is actually used to lift water. Compare to *brake horsepower* and *motor horsepower*.

water meter   A device installed in a pipe under pressure for measuring and registering the quantity of water passing through.

water table   The upper surface of the zone of saturation closest to the ground surface.

watercourse   (1) A running stream of water. (2) A natural or artificial channel for the passage of water.

water-table aquifer   An aquifer confined only by a lower impermeable layer.

water-table well   A well constructed in a water-table aquifer.

wear rings   The rings made of brass or bronze placed on the impeller and/or casing of a centrifugal pump to control the amount of water that is allowed to leak from the discharge to the suction side of the pump.

weighting agent   A material, such as bentonite, added to low-turbidity waters to provide additional particles for good floc formation.

weir   An obstruction to the flow of water, placed in an open channel. Measures flow rate by measuring the depth of the water flowing through a precisely sized and shaped notch in the weir.

weir overflow rate   A measurement of the number of gallons per day of water flowing over each foot of weir in a sedimentation tank or circular clarifier. Mathematically, the gallons-per-day flow over the weir divided by the total length of the weir in feet.

well development   The process of removing particles from an aquifer to produce potable water.

well point   A perforated metal tube or screen connected to the bottom of a suction pipe. The device is jetted or driven into the earth, and groundwater is withdrawn through it.

well screen   A sleeve with slots, holes, gauze, or wire wrap placed at the end of a well casing to allow water to enter the well. The screen prevents sand from entering the water supply.

well yield   The volume of water that is discharged from a well during a specified time period. Mathematically, the total volume discharged, divided by the time during which the discharge was monitored.

wellhead protection   A process of controlling potential groundwater contamination for a specific groundwater source.

wet barrel hydrant   A fire hydrant with no main valve. Under normal, nonemergency conditions, the barrel is full and pressurized (as long as the lateral piping to the hydrant is under pressure and the gate valve ahead of the hydrant is open). Each outlet has an independent valve that controls discharge from that outlet and has no drain mechanism. Used mainly in areas where temperatures do not drop below freezing.

wet scrubber   A device installed to remove dust from a dry chemical feeder by means of a continuous water spray.

wet tap   A connection made to a main that is full or under pressure. Compare to *dry tap*.

wet top hydrant   A dry barrel hydrant in which the threaded end of the main rod and the revolving or operating nut are not sealed from water in the barrel when the main valve of the hydrant is open and the hydrant is in use.

wheeler bottom   A patented filter underdrain system using small porcelain spheres of various sizes in conical depressions.

Winkler titration   An iodometric titration method for volumetrically determining dissolved oxygen in water. It is used both to determine dissolved oxygen and to calibrate other methods of determining dissolved oxygen. Several modifications of the procedure are available to account for certain interferences.

wire-mesh screen   A screen made of a wire fabric attached to a metal frame. Usually equipped with a motor so that it can move continuously through the water and be automatically cleaned with a water spray. Used to remove finer debris from the water than the bar screen is able to remove.

wire-to-water efficiency   The ratio of the total power input (electric current expressed as motor horsepower) to a motor and pump assembly to the total power output (water horsepower), expressed as a percent.

wound-rotor induction motor   A type of electric motor, similar to a squirrel-cage induction motor but easier to start and capable of variable-speed operation.

yoke   A fitting designed to assist in easy meter installation and to maintain electrical continuity between the incoming and outgoing lines even if the meter is removed for service.

zeta potential   The resistance between suspended particles in water.
zone of saturation   The part of an aquifer that has all the water it can contain.

Additional Resources

Supporting the almost 55,000 public and private water systems in the United States are a number of professional organizations, with activities ranging from basic and applied research, through technological advances in water distribution and treatment, to operator certification and public outreach. The leading US association for water is the American Water Works Association (AWWA), and this section lists all of AWWA's standards and manuals. Other water-related organizations are also listed, accompanied by their URLs and brief descriptions of each organization's mission and focus.

LIST OF ANSI/AWWA STANDARDS _____

Source

Groundwater and Wells

    A100   Water Wells

Treatment

Filtration

    B100   Granular Filter Media
    B101   Precoat Filter Media
    B102   Manganese Greensand for Filters
    B110   Membrane Systems

Softening

    B200   Sodium Chloride
    B201   Soda Ash
    B202   Quicklime and Hydrated Lime

Disinfection Chemicals

    B300   Hypochlorites
    B301   Liquid Chlorine
    B302   Ammonium Sulfate
    B303   Sodium Chlorite
    B304   Liquid Oxygen for Ozone Generation for Water, Wastewater and Reclaimed Water Systems
    B305   Anhydrous Ammonia
    B306   Aqua Ammonia (Liquid Ammonium Hydroxide)

Coagulation

    B402   Ferrous Sulfate
    B403   Aluminum Sulfate—Liquid, Ground, or Lump
    B404   Liquid Sodium Silicate
    B405   Sodium Aluminate
    B406   Ferric Sulfate
    B407   Liquid Ferric Chloride
    B408   Liquid Polyaluminum Chloride

B451   Poly (Diallyldimethylammonium Chloride)
B452   EPI-DMA Polyamines
B453   Polyacrylamide

Scale and Corrosion Control

B501   Sodium Hydroxide (Caustic Soda)
B502   Sodium Polyphosphate, Glassy (Sodium Hexametaphosphate)
B503   Sodium Tripolyphosphate
B504   Monosodium Phosphate, Anhydrous
B505   Disodium Phosphate, Anhydrous
B506   Zinc Orthophosphate
B510   Carbon Dioxide
B511   Potassium Hydroxide
B512   Sulfur Dioxide
B550   Calcium Chloride

Taste and Odor Control

B600   Powdered Activated Carbon
B601   Sodium Metabisulfite
B602   Copper Sulfate
B603   Permanganates
B604   Granular Activated Carbon
B605   Reactivation of Granular Activated Carbon

Fluorides

B701   Sodium Fluoride
B702   Sodium Fluorosilicate
B703   Fluorosilicic Acid

Pipe and Accessories

Ductile-Iron Pipe and Fittings

C104/A21.4   Cement–Mortar Lining for Ductile-Iron Pipe and Fittings
C105/A21.10   Polyethylene Encasement of Ductile-Iron Pipe Systems
C110/A21.10   Ductile-Iron and Gray-Iron Fittings

C111/A21.11   Rubber-Gasket Joints for Ductile-Iron
              Pressure Pipe and Fittings
C115/A21.15   Flanged Ductile-Iron Pipe with Ductile-Iron or
              Gray-Iron Threaded Flanges
C116/A21.16   Protective Fusion-Bonded Epoxy Coatings for
              the Interior and Exterior Surfaces of Ductile-
              Iron and Gray-Iron Fittings
C150/A21.50   Thickness Design of Ductile-Iron Pipe
C151/A21.51   Ductile-Iron Pipe, Centrifugally Cast
C153/A21.53   Ductile-Iron Compact Fittings

Steel Pipe

C200   Steel Water Pipe—6 In. (150 mm) and Larger
C203   Coal-Tar Protective Coatings and Linings for Steel
       Water Pipelines—Enamel and Tape—Hot-Applied
C205   Cement–Mortar Protective Lining and Coating for
       Steel Water Pipe—4 In. (100 mm) and Larger—Shop
       Applied
C206   Field Welding of Steel Water Pipe
C207   Steel Pipe Flanges for Waterworks Service—Sizes 4 In.
       Through 144 In. (100 mm Through 3,600 mm)
C208   Dimensions for Fabricated Steel Water Pipe Fittings
C209   Cold-Applied Tape Coatings for the Exterior of
       Special Sections, Connections, and Fittings for Steel
       Water Pipelines
C210   Liquid-Epoxy Coating Systems for the Interior and
       Exterior of Steel Water Pipelines
C213   Fusion-Bonded Epoxy Coating for the Interior and
       Exterior of Steel Water Pipelines
C214   Tape Coating Systems for the Exterior of Steel Water
       Pipelines
C215   Extruded Polyolefin Coatings for the Exterior of Steel
       Water Pipelines
C216   Heat-Shrinkable Cross-Linked Polyolefin Coatings for
       the Exterior of Special Sections, Connections, and
       Fittings for Steel Water Pipelines

Additional Resources

Valves and Hydrants

C500   Metal-Seated Gate Valves for Water Supply Service

C502   Dry-Barrel Fire Hydrants

C503   Wet-Barrel Fire Hydrants

C504   Rubber-Seated Butterfly Valves, 3 In. (75 mm)
       Through 72 In. (1,800 mm)

C507   Ball Valves, 6 In. Through 60 In. (150 mm Through
       1,500 mm)

C508   Swing-Check Valves for Waterworks Service, 2-In.
       Through 24 In. (50 mm Through 600 mm) NPS

C509   Resilient-Seated Gate Valves for Water Supply Service

C510   Double Check Valve Backflow Prevention Assembly

C511   Reduced-Pressure Principle Backflow Prevention
       Assembly

C512   Air Release, Air/Vacuum, and Combination Air Valves
       for Waterworks Service

C515   Reduced-Wall, Resilient-Seated Gate Valves for Water
       Supply Service

C516   Large-Diameter Rubber-Seated Butterfly Valves, Sizes
       78 In. (2,000 mm) and Larger

C517   Resilient-Seated Cast-Iron Eccentric Plug Valves

C518   Dual-Disc Swing-Check Valves for Waterworks Service

C520   Knife Gate Valves, Sizes 2 In. (50 mm) Through 96 In.
       (2,400 mm)

C530   Pilot-Operated Control Valves

C541   Hydraulic and Pneumatic Cylinder and Vane-Type
       Actuators for Valves and Slide Gates

C542   Electric Motor Actuators for Valves and Slide Gates

C550   Protective Interior Coatings for Valves and Hydrants

C560   Cast-Iron Slide Gates

C561   Fabricated Stainless-Steel Slide Gates

C562   Fabricated Aluminum Slide Gates

C563   Fabricated Composite Slide Gates

Pipe Installation

C600   Installation of Ductile-Iron Water Mains and Their
       Appurtenances

Service Lines

C800   Underground Service Line Valves and Fittings

Plastic Pipe

C900   Polyvinyl Chloride (PVC) Pressure Pipe and Fabricated Fittings, 4 In. Through 12 In. (100 mm Through 300 mm), for Water Transmission and Distribution

C901   Polyethylene (PE) Pressure Pipe and Tubing, ½ In. (13 mm) Through 3 In. (76 mm), for Water Service

C903   Polyethylene–Aluminum–Polyethylene & Cross-Linked Polyethylene–Aluminum–Cross-Linked Polyethylene Composite Pressure Pipes, ½ In. (12 mm) Through 2 In. (50 mm), for Water Service

C904   Cross-Linked Polyethylene (PEX) Pressure Pipe, ½ In. (12 mm) Through 3 In. (76 mm), for Water Service

C905   Polyvinyl Chloride (PVC) Pressure Pipe and Fabricated Fittings, 14 In. Through 48 In. (350 mm Through 1,200 mm)

C906   Polyethylene (PE) Pressure Pipe and Fittings, 4 In. (100 mm) Through 63 In. (1,600 mm), for Water Distribution and Transmission

C907   Injected-Molded Polyvinyl Chloride (PVC) Pressure Fittings, 4 In. Through 12 In. (100 mm Through 300 mm), for Water, Wastewater, and Reclaimed Water Service

C909   Molecularly Oriented Polyvinyl Chloride (PVC) Pressure Pipe, 4 In. Through 24 In. (100 mm Through 600 mm), for Water Distribution

C950   Fiberglass Pressure Pipe

Storage

D100   Welded Carbon Steel Tanks for Water Storage

D102   Coating Steel Water-Storage Tanks

D103   Factory-Coated Bolted Carbon Steel Tanks for Water Storage

Pumps

Plant Equipment

Utility Management

Additional Resources

LIST OF AWWA MANUALS

Additional Resources

ONLINE RESOURCES

American Backflow Prevention Association (ABPA)

<http://www.abpa.org/>

> A nonprofit organization that works to protect drinking water from contamination through cross-connections.

American Ground Water Trust

<http://www.agwt.org/index.htm>

> Works to promote public awareness of the environmental and economic importance of protecting America's groundwater through public information articles, workshops, and conferences.

American Indian Environmental Office (AIEO)

<http://www.epa.gov/indian/>

> Coordinates the USEPA-wide effort to strengthen public health and environmental protection in Indian Country, with a special emphasis on building Tribal capacity to administer their own environmental programs.

American Water Resources Association (AWRA)

<http://www.awra.org/>

> A member association dedicated to water resources research, planning, management, development, and education.

American Water Works Association (AWWA)

<http://www.awwa.org/>

> An international nonprofit scientific and educational society dedicated to the improvement of drinking water quality and supply. AWWA is the authoritative resource for knowledge, information, and advocacy to improve the quality and supply of drinking water in North America and beyond.

Association of Boards of Certification

<http://www.abccert.org/>

> A membership organization dedicated to advancing environmental certification programs.

Association of Metropolitan Water Agencies (AMWA)

<http://www.amwa.net/>

> Formed in 1981 by the general managers of the nation's largest water suppliers to represent them before Congress and federal agencies.

Association of State Drinking Water Administrators (ASDWA)

<http://www.asdwa.org/>

A nonprofit membership association serving state drinking water programs, working to protect public health by assuring high quality drinking water.

Centers for Disease Control and Prevention (CDC)

<http://www.cdc.gov/>

Information on current health issues as they relate to water and wastewater treatment along with contacts for individual state and local health departments. Large listing of links to other health and laboratory sites.

Chlorine Chemistry Council

<http://c3.org/>

Strives to achieve policies that promote the continuing responsible use of chlorine and chlorine-based products. Provides a variety of online information on the key roles that chlorine products play in our lives.

Green Communities

<http://www.epa.gov/greenkit/q5_facil.htm#drink>

An EPA Web site for communities with populations of 10,000 and under—policy and planning tools, regulatory tools, technical tools, financial tools, and other tools.

Green Mountain Water Environment Association

<http://www.gmwea.org/>

A nonprofit group of water supply and wastewater treatment personnel formed by the merger of Vermont's two long-active but separate water and wastewater organizations.

Groundwater Foundation

<http://www.groundwater.org/>

A nonprofit membership organization that informs and motivates people to care about and for groundwater by providing information about wise water management, protection, and policies.

Institute for Tribal Environmental Professionals (ITEP)
<http://www4.nau.edu/itep/>

Supports Native American student environmental research projects, nationwide training for tribal environmental staffs, and communication and outreach with tribes on resource protection issues.

International Bottled Water Association (IBWA)
<http://www.bottledwater.org/>

A trade organization representing the bottled water industry.

International Private Water Association, Inc.
<http://www.ipwa.org/>

A nonprofit corporation that promotes opportunities for private water project development and investment.

International Water Resources Association (IWRA)
<http://www.iwra.siu.edu/>

Strives to improve water management worldwide through dialogue, education, and research.

Local Government Environmental Assistance Network
<http://www.lgean.org/>

Provides environmental management, planning, and regulatory information for government officials, managers, and staff. Operates a free research service, publishes a quarterly newsletter, and provides a toll-free service.

Mountain Empire Community College
<http://www.wateredu.com/>

Water/wastewater courses are now on the Web and can be applied to an environmental associate degree at Mountain Empire Community College. Courses include applied information on physics, chemistry, microbiology, electricity, hydraulics, and hydrology, along with other science and engineering information.

National Association of Water Companies (NAWC)
<http://www.nawc.org/>

A nonprofit trade association serving the private and investor-owned water utility industry.

National Congress of American Indians (NCAI)
<http://www.ncai.org/>
>A private, nonprofit membership organization dedicated to the protection and furtherance of the rights of Native Americans.

National Council for the Public-Private Partnerships
<http://www.ncppp.org/>
>A nonprofit association dedicated to helping public and private sectors work cooperatively to provide services and develop financing and construction assistance.

National Drinking Water Clearinghouse (NDWC)
<http://www.nesc.wvu.edu/ndwc/>
>Intended for communities with fewer than 10,000 people to assist those who work with them by collecting, developing, and providing timely information relevant to drinking water issues.

National Environmental Training Association (NETA)
<http://www.ehs-training.org/>
>An international association of specialists for engineering, health, environmental, and safety training professionals.

National Ground Water Association (NGWA)
<http://www.ngwa.org/>
>A nonprofit professional society and trade association representing the groundwater industry.

National Rural Water Association (NRWA)
<http://www.nrwa.org/>
>A nonprofit membership organization that provides technical assistance, training, and materials through a network of 45 affiliated state associations representing over 18,000 small water and wastewater utilities. Individual state contacts can be found on this site.

National Small Flows Clearinghouse (NSFC)
<http://www.nesc.wvu.edu/nsfc/>
>Funded by the USEPA, provides information about innovative, low-cost wastewater treatments for small communities with populations less than 10,000.

National Tribal Environmental Council

<http://www.ntec.org/>

A tribal membership organization dedicated to protecting, promoting, and preserving tribal lands and effective resource management.

North American Membrane Society

<http://www.che.utexas.edu/nams/NAMSHP.html>

Through the University of Texas chemistry department. Dedicated to fostering the development and dissemination of knowledge in membrane science and technology.

NSF International

<http://www.nsf.org/>

Develops standards, product testing criteria, and certification services in public health, safety, and environmental protection.

Rural Community Assistance Program (RCAP)

<http://www.rcap.org/>

Works to improve rural community health, increase rural development, and enhance the quality of rural life. Listings of RCAP contacts can be found on this site.

United States Environmental Protection Agency (USEPA)

<http://www.epa.gov/>

Search for information regarding a variety of topics, including regulations, funding programs, policy statements, state contacts, and other environmental issues.

Water Environment Federation (WEF)

<http://www.wef.org/>

An international not-for-profit technical and educational organization of water professionals and specialists from around the world.

Water Quality Association

<http://www.wqa.org/>

An international trade association representing the household, commercial, and industrial water quality improvement products industry. Member companies manufacture and sell POU/POE equipment, packaged water treatment plants, and custom treatment systems.

Water Quality Information Center (WQIC)
<http://www.nal.usda.gov/wqic/>

Electronic access to information about water and agriculture, environmental news, databases, and open forum discussions.

Water Research Foundation
<http://www.waterresearchfoundation.org/>

A member-supported, international, nonprofit organization that sponsors research to enable water utilities, public health agencies, and other professionals to provide safe and affordable drinking water to consumers.

Water Surplus
<http://www.watersurplus.com/>

Provides the global water treatment community with a venue to buy, sell, or trade new and surplus water treatment assets. In addition to providing recycling and investment recovery services, they offer industry leaders a mechanism to provide third-world children with access to potable drinking water through an innovative tax-deductible inventory donation program.

Water Technology Online
<http://www.waternet.com/>

Disseminates current water treatment information as well as offers a purchasing directory and listing of new products of interest to water treatment professionals.

World Health Organization (WHO)
<http://www.who.int/en/>

Access to public health information and health-related issues around the world.

Index

NOTE: *f.* indicates figure. Abbreviations and acronyms are listed on pages 342–355, glossary terms on pages 358–399, and units of measure on pages 12–30, and accordingly are not cited individually in this index.

Code of Ethics for Water System Operators

1. The water system operator shall, at all times, recognize his or her primary obligation is to protect the safety, health, and welfare of the public in the performance of his or her duties. If his or her judgment is overruled under circumstances where the safety, health, and welfare of the public are endangered, he or she shall inform his or her employer of the possible consequences and notify such other proper authority about the situation, as may be appropriate.
2. The water system operator shall accept and perform water operations assignments only when qualified by education or experience in the specific technical area and levels of water operations involved. The water system operator may accept an assignment requiring education or experience outside of his or her own field of competence, but only under the direct supervision of licensed, qualified co-workers, consultants, or employees.
3. The water system operator shall be completely objective and truthful in all professional reports, statements, and testimony. He or she shall include all relevant and pertinent information in such reports, statements, and testimony.
4. The water system operator shall avoid conflicts of interest with his or her employer or customer, but, when unavoidable, the water system operator shall promptly disclose the circumstances to his or her employer or customer of any business association, interest, or circumstances which could influence his or her judgment or the quality of his or her work. The water system operator shall not review or influence the decision of his or her employees' work for any public body on which he or she may serve.
5. The water system operator shall not solicit or accept financial or other valuable items from material or equipment suppliers for specifying their product.
6. The water system operator shall not solicit or accept gratuities from contractors or other parties dealing with his or her customers or employer in connection with work for which he or she is responsible.
7. The water system operator shall not falsify his or her academic or professional qualifications. He or she shall not misrepresent or exaggerate his or her degree of responsibility in prior assignments, duties, or accomplishments to enhance his or her qualifications and work.
8. The water system operator shall not knowingly associate with or permit the use of his or her name or employer's name in a business venture by any person or company which he or she knows, or has reason to believe, is engaging in business or professional practices of fraudulent or dishonest nature.
9. If the water system operator has knowledge or reason to believe that another person or water company may be in violation of any of these rules, he or she shall present such information to the appropriate regulatory agency in writing and shall cooperate with the regulatory agency in furnishing information or assistance as may be required by the agency.